ADVANCED ELECTRONIC COMMUNICATIONS SYSTEMS

Wayne Tomasi
Mesa Community College

PRENTICE-HALL, INC., Englewood Cliffs, New Jersey 07632

Library of Congress Cataloging-in-Publication Data

TOMASI, WAYNE.
  Advanced electronic communications systems.

  Includes index.
  1. Digital communications. I. Title.
TK5103.7.T65 1987   621.38   86–16970
ISBN 0-13-011214-3

Editorial/production supervision and
interior design: **Kathryn Pavelec**
Cover design: **20/20 Services Inc.**
Manufacturing buyer: **Carol Bystrom**

In loving memory of my father, Harry Tomasi

Printed in the United States of America

10 9 8 7 6 5 4 3 2 1

ISBN 0-13-011214-3 025

PRENTICE-HALL INTERNATIONAL (UK) LIMITED, *London*
PRENTICE-HALL OF AUSTRALIA PTY. LIMITED, *Sydney*
PRENTICE-HALL CANADA INC., *Toronto*
PRENTICE-HALL HISPANOAMERICANA, S.A., *Mexico*
PRENTICE-HALL OF INDIA PRIVATE LIMITED, *New Delhi*
PRENTICE-HALL OF JAPAN, INC., *Tokyo*
PRENTICE-HALL OF SOUTHEAST ASIA PTE. LTD., *Singapore*
EDITORA PRENTICE-HALL DO BRASIL, LTDA., *Rio de Janeiro*

CONTENTS

Chapter 8   **SATELLITE COMMUNICATIONS**   266

Chapter 9   **SATELLITE MULTIPLE-ACCESS ARRANGEMENTS**   305

Chapter 10   **FIBER OPTICS COMMUNICATIONS**   324

PREFACE

During the past decade, the electronic communications industry has undergone some remarkable technological changes. The development of large-scale digital and linear integrated circuits has paved the way for many new and innovative approaches to communications. The need for these changes is attributed to the continuing increase in the number of digital and data communications systems. The purpose of this book is to expand the knowledge of the reader in the field of communications and to enhance his or her understanding of modern digital communications systems.

This book was written so that a reader with previous knowledge in basic electronic communications concepts (i.e., AM and FM transmission), basic digital theory, a basic knowledge of tuned circuits and filter concepts, and an understanding of mathematics through trigonometry will have little trouble grasping the concepts presented. Also, the order in which the topics are covered does not need to be followed strictly. Digital, data, and analog communications techniques are separated in such a way that almost any chapter sequence can be used. Within the text, there are numerous examples that emphasize the important concepts, and questions and problems are included at the end of each chapter. Also, answers to the odd-numbered problems are given at the end of the book.

Chapter 1 introduces the concepts of digital transmission and digital modulation. In this chapter the most common modulation schemes used in modern digital radio systems—FSK, PSK, and QAM—are described. The concepts of information capacity and bandwidth efficiency are explained. Chapter 2 introduces the field of data communications. Detailed explanations are given for numerous data communications concepts, including transmission methods, circuit configurations, topologies, character codes, error control mechanisms, data formats, and data modems. Chapter 3 describes data communications protocols. Synchronous and asynchronous data protocols are first defined, then explicit examples are given for each. The most popular character- and bit-oriented protocols are described. In Chapter 3 the basic concepts of a public data network and a local area network are outlined, and the international user-to-network packet switching protocol, X.25, is explained. Chapter 4 introduces digital transmission techniques. This includes a detailed explanation of pulse code modulation. The concepts of sampling, encoding, and companding (both analog and digital) are explained. Chapter 4 also includes descriptions of two lesser-known digital transmission techniques: adaptive delta modulation PCM and differential PCM. Chapter 5 explains the multiplexing of digital signals. Time-division multiplexing is discussed in detail and the operation of a modern LSI combo chip is explained. The North American Digital Hierarchy for digital transmission is outlined, including explanations of line encoding schemes, error detection/correction methods, and synchronization techniques. In Chapter 6 analog multiplexing is explained and AT&T's North American frequency-division-multiplexing hierarchy is described. Several methods are explained in which digital information can be transmitted with analog signals over the same communications medium. Chapter 7 introduces microwave radio communications and the concept of system gain. A block diagram approach to the operation of a microwave radio system is presented and numerous examples are included. In Chapter 8 satellite communications is introduced and the basic concepts of orbital patterns, radiation patterns, geosynchronous, and nonsynchronous systems are covered. System parameters and link equations are discussed and a detailed explanation of a satellite link budget is given. Chapter 9 extends the coverage of satellite systems to methods of multiple accessing. The three predominant methods for multiple accessing—frequency-division, time-division, and code-division multiple accessing—are explained. Chapter 10 covers the basic concepts of a fiber optic communications system. A detailed explanation is given for light-wave propagation through a guided fiber. Also, several light sources and light detectors are discussed, contrasting their advantages and disadvantages.

WAYNE F. TOMASI

Acknowledgments

I would like to acknowledge the following individuals for their contributions to this book: Gregory Burnell, Senior Managing Editor, Electronic Technology; Kathryn Pavelec, Production Editor; and the three reviewers of my manuscript—John Browne, SUNY-Farmingdale; Robert E. Greenwood, Ryerson Polytechnical Institute; and James W. Stewart, DeVry, Woodbridge.

Chapter 1

DIGITAL COMMUNICATIONS

INTRODUCTION

During the past several years, traditional analog communications systems that use conventional amplitude modulation (AM), frequency modulation (FM), or phase modulation (PM) have gradually been replaced with more modern digital communications systems. Digital communications systems offer several outstanding advantages over traditional analog systems: ease of processing, ease of multiplexing, and noise immunity.

The term *digital communications* covers a broad area of communications techniques, including digital transmission and digital radio. Digital transmission is the transmittal of digital pulses between two points in a communications system. Digital radio is the transmittal of digitally modulated analog carriers between two points in a communications system. Digital transmission systems require a physical facility between the transmitter and receiver, such as a metallic wire pair, a coaxial cable, or a fiber optic cable. In digital radio systems, the transmission medium is free space or the earth's atmosphere.

Figure 1-1 shows simplified block diagrams of both a digital transmission system and a digital radio system. In a digital transmission system, the original source information may be in digital or analog form. If it is in analog form, it must be converted to digital pulses prior to transmission and converted back to analog form at the receive end. In a digital radio system, the modulating input signal and the demodulated output signal are digital pulses. The digital pulses could originate from a digital transmission system, from a digital source such as a mainframe computer, or from the binary encoding of an analog signal.

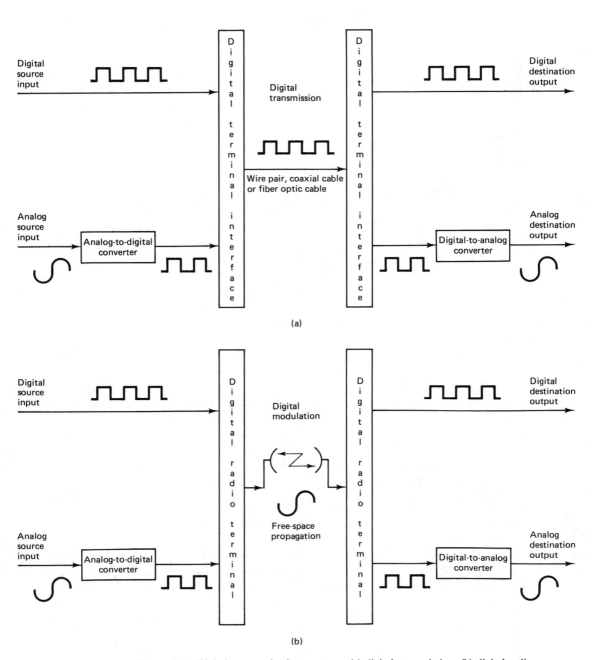

Figure 1-1   Digital communications systems: (a) digital transmission; (b) digital radio.

INFORMATION CAPACITY

The *information capacity* of a communications system represents the number of independent symbols that can be carried through the system in a given unit of time. The most basic symbol is the *binary digit* (bit). Therefore, it is often convenient to express the information capacity of a system in *bits per second* (bps). In 1928, R. Hartley of Bell Telephone Laboratories developed a useful relationship among bandwidth, transmission time, and information capacity. Simply stated, *Hartley's law* is

$$C \propto B \times T \qquad (1\text{-}1)$$

where

C = information capacity
B = bandwidth
T = transmission time

From Equation 1-1 it can be seen that the information capacity is a linear function and is directly proportional to both the system bandwidth and the transmission time. If either the bandwidth or the transmission time is changed, the information capacity will change proportionally.

In 1948, C. E. Shannon (also of Bell Telephone Laboratories) published a paper in the *Bell System Technical Journal* relating the information capacity of a communications channel to bandwidth, transmission time, and signal-to-noise ratio. Mathematically stated, the *Shannon limit for information capacity* is

$$C = B \log_2 \left(1 + \frac{S}{N}\right) \qquad (1\text{-}2)$$

where

C = information capacity (bps)
B = bandwidth
$\frac{S}{N}$ = signal power-to-noise power ratio

For a standard voice band communications channel with a signal-to-noise ratio of 1000 (30 dB) and a bandwidth of 2.7 kHz, the Shannon limit for information capacity is

$$C = 2700 \log_2 (1 + 1000)$$

$$= 26.9 \text{ kbps}$$

Shannon's formula is often misunderstood. The results of the preceding example indicate that 26.9 kbps can be transferred through a 2.7-kHz channel. This may be true, but it cannot be done with a binary system. To achieve an information transmission rate of 26.9 kbps through a 2.7-kHz channel, each symbol transmitted must contain more than one bit of information. Therefore, to achieve the Shannon limit for information capacity, digital transmission systems that have more than two output conditions must be used. Several such systems are described in the following chapters. These systems include both analog and digital modulation techniques and the transmission of both digital and analog signals.

DIGITAL RADIO

The property that distinguishes a digital radio system from a conventional AM, FM, or PM radio system is that in a digital radio system the modulating and demodulated signals are digital pulses rather than analog waveforms. Digital radio uses analog carriers just as conventional systems do. Essentially, there are three digital modulation techniques that are commonly used in digital radio systems: *frequency shift keying* (FSK), *phase shift keying* (PSK), and *quadrature amplitude modulation* (QAM).

FREQUENCY SHIFT KEYING

Frequency shift keying (FSK) is a relatively simple, low-performance form of digital modulation. FSK is a constant-envelope form of angle modulation similar to conventional frequency modulation except that the modulating signal is a binary pulse stream that varies between two discrete voltage levels rather than a continuously changing waveform.

FSK Transmitter

With binary FSK, the center or carrier frequency is shifted (deviated) by the binary input data. Consequently, the output of an FSK modulator is a step function in the frequency domain. As the binary input signal changes from a logic 0 to a logic 1, and vice versa, the FSK output shifts between two frequencies: a *mark* or *logic 1 frequency* and a *space* or *logic 0 frequency*. With FSK, there is a change in the output frequency each time the logic condition of the binary input signal changes. Consequently, the output rate of change is equal to the input rate of change. In digital modulation, the rate of change at the input to the modulator is called the *bit rate* and has the units of bits per second (bps). The rate of change at the output of the modulator is called *baud* or *baud rate* and is equal to the reciprocal of the time of one output signaling element. In FSK, the input and output rates of change are equal; therefore, the bit rate and baud rate are equal. A simple binary FSK transmitter is shown in Figure 1-2.

Bandwidth Considerations of FSK

As with all electronic communications systems, bandwidth is one of the primary considerations when designing an FSK transmitter. FSK is similar to conventional frequency modulation and so can be described in a similar manner.

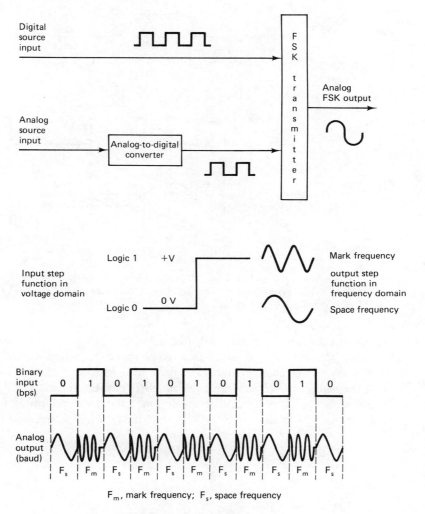

Figure 1-2 Binary FSK transmitter.

Figure 1-3 shows an FSK modulator. An FSK modulator is a type of FM transmitter and is very often a *voltage-controlled oscillator* (VCO). It can be seen that the fastest input rate of change occurs when the binary input is a series of alternating 1's and 0's: namely, a square wave. The *fundamental frequency* of a binary square wave is equal to one-half of the bit rate. Consequently, if only the fundamental frequency of the input is considered, the highest modulating frequency to the FSK modulator is equal to one-half of the input bit rate.

The rest frequency of the VCO is chosen such that it falls halfway between the mark and space frequencies. A logic 1 condition at the input shifts the VCO

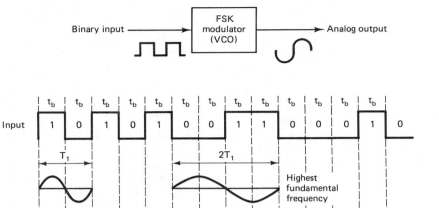

Figure 1-3 FSK modulator. t_b, Time of one bit $= 1/\mathrm{bps}$; F_m, mark frequency; F_s, space frequency; T1, period of shortest cycle; $1/\mathrm{T1}$, fundamental frequency of binary square wave; F_b, input bit rate (bps).

from its rest frequency to the mark frequency, and a logic 0 condition at the input shifts the VCO from its rest frequency to the space frequency. Consequently, as the input binary signal changes from a logic 1 to a logic 0, and vice versa, the VCO output frequency *shifts* or *deviates* back and forth between the mark and space frequencies. Because FSK is a form of frequency modulation, the formula for *modulation index* used in FM is also valid for FSK. Modulation index is given as

$$\mathrm{MI} = \frac{\Delta F}{F_a}$$

where

$\mathrm{MI} =$ modulation index
$\Delta F =$ frequency deviation (Hz)
$F_a =$ modulating frequency (Hz)

The worst-case modulation index is the modulation index that yields the widest output bandwidth, called the *deviation ratio*. The worst-case or widest bandwidth occurs when both the frequency deviation and the modulating frequencies are at their maximum values.

In an FSK modulator, ΔF is the peak frequency deviation of the carrier and is equal to the difference between the rest frequency and either the mark or space frequency (or half the difference between the mark and space frequencies). The peak frequency deviation depends on the amplitude of the modulating signal. In a binary digital signal, all logic 1's have the same voltage and all logic 0's have the same voltage; consequently, the frequency deviation is constant and always at its maximum value. F_a is equal to the fundamental frequency of the binary input which under the worst-case condition (alternating 1's and 0's) is equal to one-half of the bit rate. Consequently, for FSK,

$$MI = \frac{\left|\dfrac{F_m - F_s}{2}\right|}{\dfrac{F_b}{2}} = \frac{|F_m - F_s|}{F_b} \qquad (1\text{-}3)$$

where

$$\frac{|F_m - F_s|}{2} = \text{peak frequency deviation}$$

$$\frac{F_b}{2} = \text{fundamental frequency of the binary input signal}$$

With conventional FM, the bandwidth is directly proportional to the modulation index. Consequently, in FSK the modulation index is generally kept below 1.0, thus producing a relatively narrowband FM output spectrum. The minimum bandwidth required to propagate a signal is called the *minimum Nyquist bandwidth* (F_n). When modulation is used and a double-sided output spectrum is generated, the minimum bandwidth is called the *minimum double-sided Nyquist bandwidth* or the *minimum IF bandwidth*.

EXAMPLE 1-1

For an FSK modulator with space, rest, and mark frequencies of 60, 70, and 80 MHz, respectively and an input bit rate of 20 Mbps, determine the output baud and the minimum required bandwidth.

Solution Substituting into Equation 1-3, we have

$$MI = \frac{|F_m - F_s|}{F_b} = \frac{|80 \text{ MHz} - 60 \text{ MHz}|}{20 \text{ Mbps}}$$

$$= \frac{20 \text{ MHz}}{20 \text{ Mbps}} = 1.0$$

From the Bessel chart (Table 1-1), a modulation index of 1.0 yields three sets of significant side frequencies. Each side frequency is separated from the center frequency or an adjacent side frequency by a value equal to the modulating frequency, which in this example is 10 MHz ($F_b/2$). The output spectrum for this modulator is shown in Figure 1-4. It can be seen that the minimum double-sided Nyquist bandwidth is 60 MHz and the baud rate is 20 megabaud, the same as the bit rate.

Because FSK is a form of narrowband frequency modulation, the minimum bandwidth is dependent on the modulation index. For a modulation index between

TABLE 1-1   BESSEL FUNCTION CHART

MI	J_0	J_1	J_2	J_3	J_4
0.0	1.00				
0.25	0.98	0.12			
0.5	0.94	0.24	0.03		
1.0	0.77	0.44	0.11	0.02	
1.5	0.51	0.56	0.23	0.06	0.01
2.0	0.22	0.58	0.35	0.13	0.03

0.5 and 1, either two or three sets of significant side frequencies are generated. Thus the minimum bandwidth is two to three times the input bit rate.

FSK Receiver

The most common circuit used for demodulating FSK signals is the *phase-locked loop* (PLL), which is shown in Figure 1-5. A PLL-FSK demodulator works very much like a PLL-FM demodulator. As the input to the PLL shifts between the mark and space frequencies, the *dc error voltage* at the output of the phase comparator follows the frequency shift. Because there are only two input frequencies (mark and space), there are also only two output error voltages. One represents a logic 1 and the other a logic 0. Therefore, the output is a two-level (binary) representation of the FSK input. Generally, the natural frequency of the PLL is made equal to the center frequency of the FSK modulator. As a result, the changes in the dc error voltage follow the changes in the analog input frequency and are symmetrical around 0 V dc.

FSK has a poorer error performance than PSK or QAM and, consequently, is seldom used for high-performance digital radio systems. Its use is restricted to low-performance, low-cost, asynchronous data modems that are used for data communications over analog, voice band telephone lines (see Chapter 2).

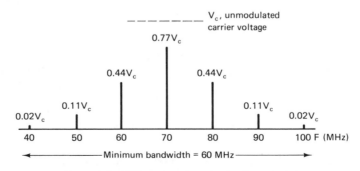

Figure 1-4   FSK output spectrum for Example 1-1.

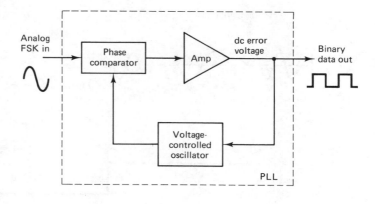

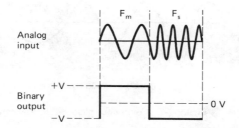

Figure 1-5 PLL-FSK demodulator.

Minimum Shift-Keying FSK

Minimum shift-keying FSK (MSK) is a form of *continuous-phase* frequency shift keying (CPFSK). Essentially, MSK is binary FSK except that the mark and space frequencies are *synchronized* with the input binary bit rate. Synchronous simply means that there is a precise time relationship between the two; it does not mean they are equal. With MSK, the mark and space frequencies are selected such that they are separated from the center frequency by an exact odd multiple of one-half of the bit rate [F_m and $F_s = n(F_b/2)$, where $n =$ any odd whole integer]. This ensures that there is a smooth phase transition in the analog output signal when it changes from a mark to a space frequency, or vice versa. Figure 1-6 shows a *noncontinuous* FSK waveform. It can be seen that when the input changes from a logic 1 to a logic 0, and vice versa, there is an abrupt phase discontinuity in the analog output signal. When this occurs, the demodulator has trouble following the frequency shift; consequently, an error may occur.

Figure 1-7 shows a continuous-phase MSK waveform. Notice that when the output frequency changes, it is a smooth, continuous transition. Each transition occurs at a zero crossing; consequently, there are no phase discontinuities. MSK has a better bit-error performance than conventional FSK for a given signal-to-noise ratio. The

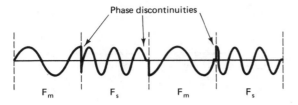

Phase discontinuities

F_m F_s F_m F_s

Figure 1-6 Noncontinuous FSK waveform.

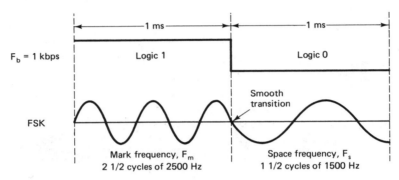

F_b = 1 kbps

Logic 1

Logic 0

1 ms   1 ms

FSK

Smooth transition

Mark frequency, F_m
2 1/2 cycles of 2500 Hz

Space frequency, F_s
1 1/2 cycles of 1500 Hz

$F_m = 5 F_b/2 = 5\ 1000/2 = 2500$ Hz $F_s = 3 F_b/2 = 3\ 1000/2 = 1500$ Hz

Figure 1-7 Continuous-phase MSK waveform.

disadvantage of MSK is that it requires synchronizing circuits and is therefore more expensive to implement.

PHASE SHIFT KEYING

Phase shift keying (PSK) is another form of angle-modulated, constant-envelope digital modulation. PSK is similar to conventional phase modulation except that with PSK the input signal is a binary digital signal and a limited number of output phases are possible.

BINARY PHASE SHIFT KEYING

With *binary phase shift keying* (BPSK), two output phases are possible for a single carrier frequency ("binary" meaning "2"). One output phase represents a logic 1 and the other a logic 0. As the input digital signal changes, the phase of the output carrier shifts between two angles that are 180° out of phase. Another name for BPSK is *phase reversal keying* (PRK).

BPSK Transmitter

Figure 1-8 shows a simplified block diagram of a BPSK modulator. The balanced modulator acts like a phase reversing switch. Depending on the logic condition of the digital input, the carrier is transferred to the output either in phase or 180° out of phase with the reference carrier oscillator.

Figure 1-9a shows the schematic diagram of a balanced ring modulator. The balanced modulator has two inputs: a carrier that is in phase with the reference oscillator and the binary digital data. For the balanced modulator to operate properly, the digital input voltage must be much greater than the peak carrier voltage. This ensures that the digital input controls the on/off state of diodes D1–D4. If the binary input is a logic 1 (positive voltage), diodes D1 and D2 are forward biased and "on," while diodes D3 and D4 are reverse biased and "off" (Figure 1-9b). With the polarities shown, the carrier voltage is developed across transformer T2 in phase with the carrier voltage across T1. Consequently, the output signal is in phase with the reference oscillator.

If the binary input is a logic 0 (negative voltage), diodes D1 and D2 are reverse biased and "off," while diodes D3 and D4 are forward biased and "on" (Figure 1-9c). As a result, the carrier voltage is developed across transformer T2 180° out of phase with the carrier voltage across T1. Consequently, the output signal is 180° out of phase with the reference oscillator. Figure 1-10 shows the truth table, phasor diagram, and constellation diagram for a BPSK modulator. A *constellation diagram*, which is sometimes called a *signal state-space diagram*, is similar to a phasor diagram except that the entire phasor is not drawn. In a constellation diagram, only the relative positions of the peaks of the phasors are shown.

Bandwidth Considerations of BPSK

A balanced modulator is a *product modulator*; the output signal is the product of the two input signals. In a BPSK modulator, the carrier input signal is multiplied by the binary data. If +1 V is assigned to a logic 1 and −1 V is assigned to a logic 0, the input carrier ($\sin \omega_c t$) is multiplied by either a + or − 1. Consequently, the output signal is either $+1 \sin \omega_c t$ or $-1 \sin \omega_c t$; the first represents a signal that is

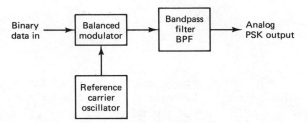

Figure 1-8   BPSK modulator.

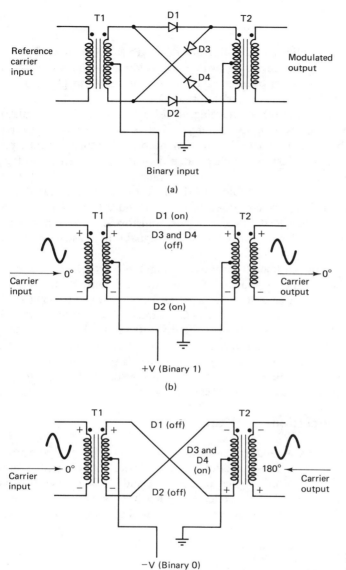

Figure 1-9 (a) Balanced ring modulator; (b) logic 1 input; (c) logic 0 input.

in phase with the reference oscillator, the latter a signal that is 180° out of phase with the reference oscillator. Each time the input logic condition changes, the output phase changes. Consequently, for BPSK, the output rate of change (baud) is equal to the input rate of change (bps), and the widest output bandwidth occurs when the input binary data are an alternating 1/0 sequence. The fundamental frequency

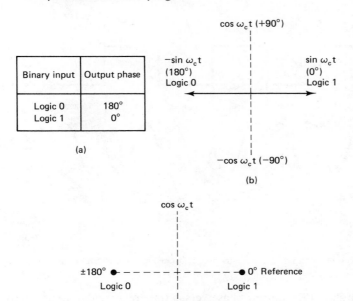

Binary input	Output phase
Logic 0	180°
Logic 1	0°

(a)

(b)

(c)

Figure 1-10 BPSK modulator: (a) truth table; (b) phasor diagram; (c) constellation diagram.

(F_a) of an alternating 1/0 bit sequence is equal to one-half of the bit rate ($F_b/2$). Mathematically, the output of a BPSK modulator is

$$\text{output} \quad = \quad \underbrace{(\sin \omega_a t)}_{\substack{\text{fundamental frequency} \\ \text{of the binary} \\ \text{modulating signal}}} \quad \times \quad \underbrace{(\sin \omega_c t)}_{\text{carrier}}$$

or

$$\tfrac{1}{2} \cos (\omega_c t - \omega_a t) - \tfrac{1}{2} \cos (\omega_c t + \omega_a t)$$

Consequently, the minimum double-sided Nyquist bandwidth (F_n) is

$$\begin{array}{c} \omega_c t + \omega_a t \\ \underline{- (\omega_c t - \omega_a t)} \end{array} \quad \text{or} \quad \begin{array}{c} \omega_c t + \omega_a t \\ \underline{- \omega_c t + \omega_a t} \\ 2\omega_a t \end{array}$$

and because $\omega_a t = F_b/2$,

$$F_n = 2 \left(\frac{F_b}{2} \right) = F_b$$

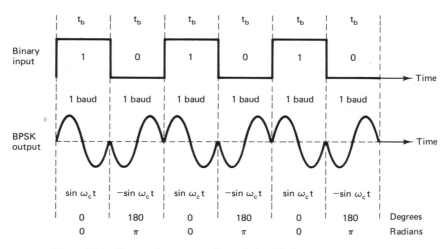

Figure 1-11   Output phase versus time relationship for a BPSK modulator.

Figure 1-11 shows the output phase versus time relationship for a BPSK waveform. It can be seen that the output spectrum from a BPSK modulator is simply a double-sideband suppresssed carrier signal where the upper and lower side frequencies are separated from the carrier frequency by a value equal to one-half of the bit rate. Consequently, the minimum bandwidth (F_n) required to pass the worst-case BPSK output signal is equal to the input bit rate.

EXAMPLE 1-2

For a BPSK modulator with a carrier frequency of 70 MHz and an input bit rate of 10 Mbps, determine the maximum and minimum upper and lower side frequencies, draw the output spectrum, determine the minimum Nyquist bandwidth, and calculate the baud.

Solution   Substituting into the equation for the output of a balanced modulator yields

$$\text{output} = (\sin \omega_a t)(\sin \omega_c t)$$

$$= [\sin 2\pi(5\text{ MHz})t)][\sin 2\pi(70\text{ MHz})t]$$

$$= \underbrace{\tfrac{1}{2}\cos 2\pi(70\text{ MHz} - 5\text{ MHz})t}_{\text{lower side frequency}} - \underbrace{\tfrac{1}{2}\cos 2\pi(70\text{ MHz} + 5\text{ MHz})t}_{\text{upper side frequency}}$$

Minimum lower side frequency (LSF):

$$\text{LSF} = 70\text{ MHz} - 5\text{ MHz} = 65\text{ MHz}$$

Maximum upper side frequency (USF):

$$\text{USF} = 70\text{ MHz} + 5\text{ MHz} = 75\text{ MHz}$$

Therefore, the output spectrum for the worst-case binary input conditions is as follows:

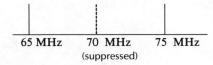

65 MHz   70 MHz   75 MHz
(suppressed)

The minimum Nyquist bandwidth (F_n) is

$$F_n = 75 \text{ MHz} - 65 \text{ MHz} = 10 \text{ MHz}$$

and the baud $= F_b$ or 10 megabaud.

BPSK Receiver

Figure 1–12 shows the block diagram of a BPSK receiver. The input signal may be $+\sin \omega_c t$ or $-\sin \omega_c t$. The coherent carrier recovery circuit detects and regenerates a carrier signal that is both frequency and phase coherent with the original transmit carrier. The balanced modulator is a product detector; the output is the product of the two inputs (the BPSK signal and the recovered carrier). The low-pass filter (LPF) separates the recovered binary data from the complex demodulated spectrum. Mathematically, the demodulation process is as follows.

For a BPSK input signal of $+\sin \omega_c t$ (logic 1), the output of the balanced modulator is

$$\text{output} = (\sin \omega_c t)(\sin \omega_c t) = \sin^2 \omega_c t$$

(filtered out)

or

$$\sin^2 \omega_c t = \tfrac{1}{2}(1 - \cos 2\omega_c t) = \tfrac{1}{2} - \tfrac{1}{2} \cos 2\omega ct$$

$$\text{output} = +\tfrac{1}{2} \text{ V dc} = \text{logic 1}$$

It can be seen that the output of the balanced modulator contains a positive dc voltage ($+\tfrac{1}{2}$ V) and a cosine wave at twice the carrier frequency ($2\omega_c t$). The LPF has a cutoff frequency much lower than $2\omega_c t$ and thus blocks the second harmonic of the carrier and passes only the positive dc component. A positive dc voltage represents a demodulated logic 1.

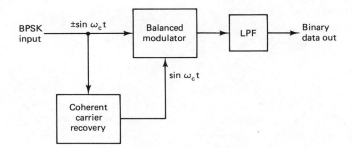

Figure 1-12 BPSK receiver.

For a BPSK input signal of $-\sin \omega_c t$ (logic 0), the output of the balanced modulator is

$$\text{output} = (-\sin \omega_c t)(\sin \omega_c t) = -\sin^2 \omega_c t$$

$$\text{(filtered out)}$$

or

$$-\sin^2 \omega_c t = -\tfrac{1}{2}(1 - \cos 2\omega_c t) = -\tfrac{1}{2} + \tfrac{1}{2}\cos 2\omega_c t$$

$$\text{output} = -\tfrac{1}{2} \text{ V dc} = \text{logic 0}$$

The output of the balanced modulator contains a negative dc voltage $(-\tfrac{1}{2} \text{ V})$ and a cosine wave at twice the carrier frequency $(2\omega_c t)$. Again, the LPF blocks the second harmonic of the carrier and passes only the negative dc component. A negative dc voltage represents a demodulated logic 0.

M-ary Encoding

M-ary is a term derived from the word "binary." *M* is simply a digit that represents the number of conditions possible. The two digital modulation techniques discussed thus far (binary FSK and BPSK) are binary systems; there are only two possible output conditions. One represents a logic 1 and the other a logic 0; thus they are *M*-ary systems where $M = 2$. With digital modulation, very often it is advantageous to encode at a level higher than binary. For example, a PSK system with four possible output phases is an *M*-ary system where $M = 4$. If there were eight possible output phases, $M = 8$, and so on. Mathematically,

$$N = \log_2 M$$

where

N = number of bits
M = number of output conditions possible with n bits

For example, if 2 bits were allowed to enter a modulator before the output were allowed to change,

$$2 = \log_2 M \quad \text{and} \quad 2^2 = M \quad \text{thus } M = 4$$

An $M = 4$ indicates that with 2 bits, four different output conditions are possible. For $N = 3$, $M = 2^3$ or 8, and so on.

QUATERNARY PHASE SHIFT KEYING

Quaternary phase shift keying (QPSK), or *quadrature PSK* as it is sometimes called, is another form of angle-modulated, constant-envelope digital modulation. QPSK is an *M*-ary encoding technique where $M = 4$ (hence the name "quaternary," meaning

"4"). With QPSK four output phases are possible for a single carrier frequency. Because there are four different output phases, there must be four different input conditions. Because the digital input to a QPSK modulator is a binary (base 2) signal, to produce four different input conditions it takes more than a single input bit. With 2 bits, there are four possible conditions: 00, 01, 10, and 11. Therefore, with QPSK, the binary input data are combined into groups of 2 bits called *dibits*. Each dibit code generates one of the four possible output phases. Therefore, for each 2-bit dibit clocked into the modulator, a single output change occurs. Therefore, the rate of change at the output (baud rate) is one-half of the input bit rate.

QPSK Transmitter

A block diagram of a QPSK modulator is shown in Figure 1-13. Two bits (a dibit) are clocked into the bit spliter. After both bits have been serially inputted, they are simultaneously parallel outputted. One bit is directed to the I channel and the other to the Q channel. The I bit modulates a carrier that is in phase with the reference oscillator (hence the name "I" for "in phase" channel), and the Q bit modulates a carrier that is 90° out of phase or in quadrature with the reference carrier (hence the name "Q" for "quadrature" channel).

It can be seen that once a dibit has been split into the I and Q channels, the operation is the same as in a BPSK modulator. Essentially, a QPSK modulator is

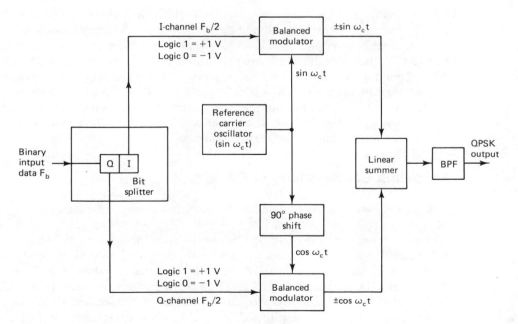

Figure 1-13 QPSK modulator.

two BPSK modulators combined in parallel. Again, for a logic 1 = +1 V and a logic 0 = −1 V, two phases are possible at the output of the I balanced modulator (+sin $\omega_c t$ and −sin $\omega_c t$), and two phases are possible at the output of the Q balanced modulator (+cos $\omega_c t$ and −cos $\omega_c t$). When the linear summer combines the two quadrature (90° out of phase) signals, there are four possible resultant phases: +sin $\omega_c t$ + cos $\omega_c t$, +sin $\omega_c t$ − cos $\omega_c t$, −sin $\omega_c t$ + cos $\omega_c t$, and −sin $\omega_c t$ − cos $\omega_c t$.

EXAMPLE 1-3

For the QPSK modulator shown in Figure 1-13, construct the truth table, phasor diagram, and constellation diagram.

Solution   For a binary data input of Q = 0 and I = 0, the two inputs to the I balanced modulator are −1 and sin $\omega_c t$, and the two inputs to the Q balanced modulator are −1 and cos $\omega_c t$. Consequently, the outputs are

$$\text{I balanced modulator} = (-1)(\sin \omega_c t) = -1 \sin \omega_c t$$

$$\text{Q balanced modulator} = (-1)(\cos \omega_c t) = -1 \cos \omega_c t$$

and the output of the linear summer is

$$-1 \cos \omega_c t - 1 \sin \omega_c t = 1.414 \sin \omega_c t - 135°$$

For the remaining dibit codes (01, 10, and 11), the procedure is the same. The results are shown in Figure 1-14.

In Figure 1-14b it can be seen that with QPSK each of the four possible output phasors has exactly the same amplitude. Therefore, the binary information must be encoded entirely in the phase of the output signal. This is the most important characteristic of PSK that distinguishes it from QAM, which is explained later in this chapter. Also, from Figure 1-14b it can be seen that the angular separation between any two adjacent phasors in QPSK is 90°. Therefore, a QPSK signal can undergo almost a +45° or −45° shift in phase during transmission and still retain the correct encoded information when demodulated at the receiver. Figure 1-15 shows the output phase versus time relationship for a QPSK modulator.

Bandwidth Considerations of QPSK

With QPSK, since the input data are divided into two channels, the bit rate in either the I or the Q channel is equal to one-half of the input data ($F_b/2$). (Essentially, the bit splitter stretches the I and Q bits to twice their input bit length.) Consequently, the highest fundamental frequency present at the data input to the I or the Q balanced modulator is equal to one-fourth of the input data rate (one-half of $F_b/2 = F_b/4$). As a result, the output of the I and Q balanced modulators requires a minimum double-sided Nyquist bandwidth equal to one-half of the incoming bit rate (F_n = twice $F_b/4 = F_b/2$). Thus with QPSK, a bandwidth compression is realized (the

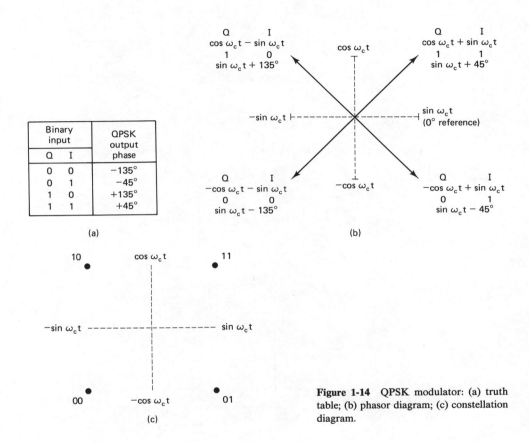

Binary input		QPSK output phase
Q	I	
0	0	−135°
0	1	−45°
1	0	+135°
1	1	+45°

(a)

(c)

Figure 1-14 QPSK modulator: (a) truth table; (b) phasor diagram; (c) constellation diagram.

minimum bandwidth is less than the incoming bit rate). Also, since the QPSK output signal does not change phase until 2 bits (a dibit) have been clocked into the bit splitter, the fastest output rate of change (baud) is also equal to one-half of the input bit rate. As with BPSK, the minimum bandwidth and the baud are equal. This relationship is shown in Figure 1-16.

In Figure 1-16 it can be seen that the worst-case input condition to the I or Q balanced modulator is an alternating 1/0 pattern, which occurs when the binary input data has a 1100 repetitive pattern. One cycle of the fastest binary transition (a 1/0 sequence) in the I or Q channel takes the same time as 4 input data bits. Consequently, the highest fundamental frequency at the input and fastest rate of

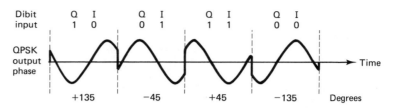

Figure 1-15 Output phase versus time relationship for a QPSK modulator.

change at the output of the balanced modulators is equal to one-fourth of the binary input bit rate.

The output of the balanced modulators can be expressed mathematically as

$$\theta = (\sin \omega_a t)(\sin \omega_c t)$$

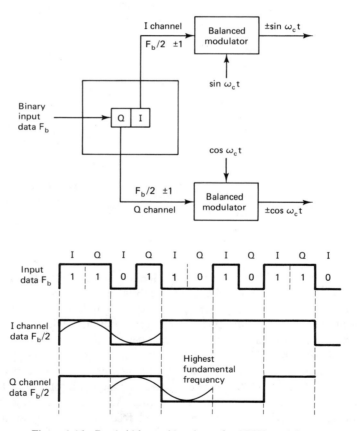

Figure 1-16 Bandwidth considerations of a QPSK modulator.

where

$$\underbrace{\omega_a t = 2\pi \frac{F_b t}{4}}_{\substack{\text{modulating} \\ \text{signal}}} \quad \text{and} \quad \underbrace{\omega_c t = 2\pi F_c t}_{\text{carrier}}$$

Thus

$$\theta = \left(\sin 2\pi \frac{F_b t}{4}\right)(\sin 2\pi F_c t)$$

$$= \frac{1}{2}\cos 2\pi \left(F_c - \frac{F_b}{4}\right)t - \frac{1}{2}\cos 2\pi \left(F_c + \frac{F_b}{4}\right)t$$

The output frequency spectrum extends from $F_c + F_b/4$ to $F_c - F_b/4$ and the minimum bandwidth (F_N) is

$$\left(F_c + \frac{F_b}{4}\right) - \left(F_c - \frac{F_b}{4}\right) = \frac{2F_b}{4} = \frac{F_b}{2}$$

EXAMPLE 1-4

For a QPSK modulator with an input data rate (F_b) equal to 10 Mbps and a carrier frequency of 70 MHz, determine the minimum double-sided Nyquist bandwidth (F_N) and the baud. Also, compare the results with those achieved with the BPSK modulator in Example 1-2. Use the QPSK block diagram shown in Figure 1-13 as the modulator model.

Solution The bit rate in both the I and Q channels is equal to one-half of the transmission bit rate or

$$F_{bQ} = F_{bI} = \frac{F_b}{2} = \frac{10 \text{ Mbps}}{2} = 5 \text{ Mbps}$$

The highest fundamental frequency presented to either balanced modulator is

$$F_a = \frac{F_{bQ}}{2} \quad \text{or} \quad \frac{F_{bI}}{2} = \frac{5 \text{ Mbps}}{2} = 2.5 \text{ MHz}$$

The output wave from each balanced modulator is

$$(\sin 2\pi F_a t)(\sin 2\pi F_c t)$$

$$\tfrac{1}{2}\cos 2\pi(F_c - F_a)t - \tfrac{1}{2}\cos 2\pi(F_c + F_a)t$$

$$\tfrac{1}{2}\cos 2\pi[(70 - 2.5) \text{ MHz}]t - \tfrac{1}{2}\cos 2\pi[(70 + 2.5) \text{ MHz}]t$$

$$\tfrac{1}{2}\cos 2\pi(67.5 \text{ MHz})t - \tfrac{1}{2}\cos 2\pi(72.5 \text{ MHz})t$$

The minimum Nyquist bandwidth is

$$F_N = (72.5 - 67.5) \text{ MHz} = 5 \text{ MHz}$$

The baud equals the bandwidth; thus

$$\text{baud} = 5 \text{ megabaud}$$

The output spectrum is as follows:

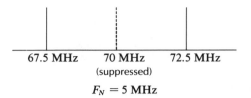

67.5 MHz   70 MHz   72.5 MHz
                (suppressed)

$$F_N = 5 \text{ MHz}$$

It can be seen that for the same input bit rate the minimum bandwidth required to pass the output of the QPSK modulator is equal to one-half of that required for the BPSK modulator in Example 1-2. Also, the baud rate for the QPSK modulator is one-half that of the BPSK modulator.

QPSK Receiver

The block diagram of a QPSK receiver is shown in Figure 1-17. The power splitter directs the input QPSK signal to the I and Q product detectors and the carrier recovery circuit. The carrier recovery circuit reproduces the original transmit carrier oscillator signal. The recovered carrier must be frequency and phase coherent with the transmit reference carrier. The QPSK signal is demodulated in the I and Q product detectors, which generate the original I and Q data bits. The outputs of the product detectors are fed to the bit combining circuit, where they are converted from parallel I and Q data channels to a single binary output data stream.

The incoming QPSK signal may be any one of the four possible output phases shown in Figure 1-14. To illustrate the demodulation process, let the incoming QPSK signal be $-\sin \omega_c t + \cos \omega_c t$. Mathematically, the demodulation process is as follows.

The receive QPSK signal ($-\sin \omega_c t + \cos \omega_c t$) is one of the inputs to the I product detector. The other input is the recovered carrier ($\sin \omega_c t$). The output of the I product detector is

$$I = \underbrace{(-\sin \omega_c t + \cos \omega_c t)}_{\text{QPSK input signal}} \underbrace{(\sin \omega_c t)}_{\text{carrier}}$$

$$= (-\sin \omega_c t)(\sin \omega_c t) + (\cos \omega_c t)(\sin \omega_c t)$$

$$= -\sin^2 \omega_c t + (\cos \omega_c t)(\sin \omega_c t)$$

$$= -\tfrac{1}{2}(1 - \cos 2\omega_c t) + \tfrac{1}{2} \sin (\omega_c t + \omega_c t) + \tfrac{1}{2} \sin (\omega_c t - \omega_c t)$$

$$\overset{\text{(filtered out)}}{} \qquad \overset{\text{(equals 0)}}{}$$

$$I = -\tfrac{1}{2} + \tfrac{1}{2} \cos 2\omega_c t + \tfrac{1}{2} \sin 2\omega_c t + \tfrac{1}{2} \sin 0$$

$$= -\tfrac{1}{2} \text{ V dc (logic 0)}$$

Again, the receive QPSK signal ($-\sin \omega_c t + \cos \omega_c t$) is one of the inputs to the Q product detector. The other input is the recovered carrier shifted 90° in phase ($\cos \omega_c t$). The output of the Q product detector is

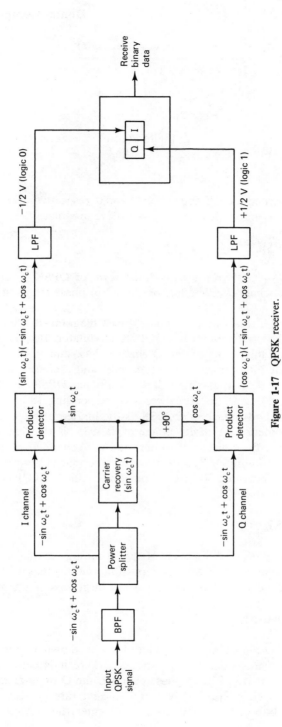

Figure 1-17 QPSK receiver.

$$Q = \underbrace{(-\sin \omega_c t + \cos \omega_c t)}_{\text{QPSK input signal}} \underbrace{(\cos \omega_c t)}_{\text{carrier}}$$

$$= \cos^2 \omega_c t - (\sin \omega_c t)(\cos \omega_c t)$$

$$= \tfrac{1}{2}(1 + \cos 2\omega_c t) - \tfrac{1}{2}\sin(\omega_c t + \omega_c t) - \tfrac{1}{2}\sin(\omega_c t - \omega_c t)$$

$$Q = \tfrac{1}{2} + \tfrac{1}{2}\cos 2\omega_c t \overset{\text{(filtered out)}}{-} \tfrac{1}{2}\sin 2\omega_c t - \overset{\text{(equals 0)}}{\tfrac{1}{2}\sin 0}$$

$$= \tfrac{1}{2} \text{ V dc (logic 1)}$$

The demodulated I and Q bits (1 and 0, respectively) correspond to the constellation diagram and truth table for the QPSK modulator shown in Fig. 1-14.

Offset QPSK

Offset QPSK (OQPSK) is a modified form of QPSK where the bit waveforms on the I and Q channels are offset or shifted in phase from each other by one-half of a bit time.

Figure 1-18 shows a simplified block diagram, the bit sequence alignment, and the constellation diagram for a OQPSK modulator. Because changes in the I channel occur at the midpoints of the Q-channel bits, and vice versa, there is never more than a single bit change in the dibit code, and therefore there is never more than a 90° shift in the output phase. In conventional QPSK, a change in the input dibit from 00 to 11 or 01 to 10 causes a corresponding 180° shift in the output phase. Therefore, an advantage of OQPSK is the limited phase shift that must be imparted during modulation. A disadvantage of OQPSK is that changes in the output phase occur at twice the data rate in either the I or Q channels. Consequently, with OQPSK the baud and minimum bandwidth are twice that of conventional QPSK for a given transmission bit rate. OQPSK is sometimes called OKQPSK (*offset-keyed PSK*).

EIGHT-PHASE PSK

Eight-phase PSK (8PSK) is an *M*-ary encoding technique where $M = 8$. With an 8PSK modulator, there are eight possible output phases. To encode eight different phases, the incoming bits are considered in groups of 3 bits, called *tribits* ($2^3 = 8$).

8PSK Transmitter

A block diagram of an 8PSK modulator is shown in Figure 1-19. The incoming serial bit stream enters the bit splitter, where it is converted to a parallel, three-channel output (the I or in-phase channel, the Q or in-quadrature channel, and the C or control channel). Consequently, the bit rate in each of the three channels is $F_b/3$. The bits in the I and C channels enter the I-channel 2-to-4-level converter,

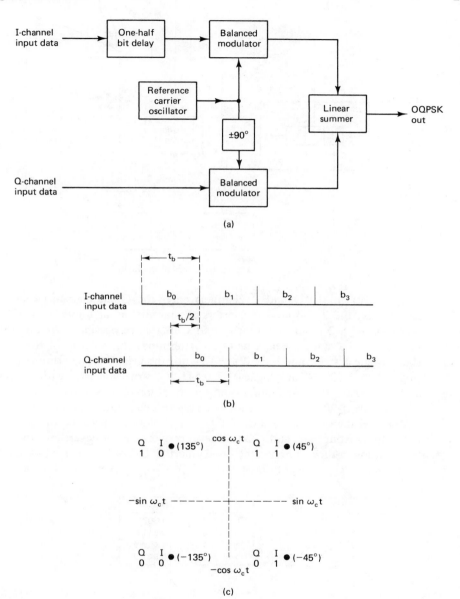

Figure 1-18 Offset keyed PSK (OQPSK): (a) block diagram; (b) bit alignment; (c) constellation diagram.

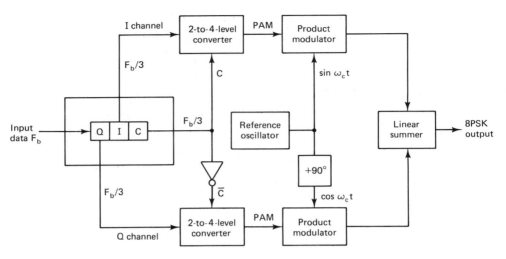

Figure 1-19 8PSK modulator.

and the bits in the Q and $\overline{C}$ channels enter the Q-channel 2-to-4-level converter. Essentially, the 2-to-4-level converters are parallel-input *digital-to-analog converters* (DACs). With 2 input bits, four output voltages are possible. The algorithm for the DACs is quite simple. The I or Q bit determines the polarity of the output analog signal (logic 1 = +V and logic 0 = −V) while the C or $\overline{C}$ bit determines the magnitude (logic 1 = 1.307 V and logic 0 = 0.541 V). Consequently, with two magnitudes and two polarities, four different output conditions are possible.

Figure 1-20 shows the truth table and corresponding output conditions for the 2-to-4-level converters. Because the C and $\overline{C}$ bits can never be the same logic state, the outputs from the I and Q 2-to-4-level converters can never have the same magnitude, although they can have the same polarity. The output of a 2-to-4-level converter is an *M*-ary, *pulse-amplitude-modulated* (PAM) signal where $M = 4$.

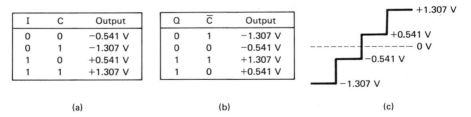

Figure 1-20 I- and Q-channel 2-to-4-level converters: (a) I-channel truth table; (b) Q-channel truth table; (c) PAM levels.

EXAMPLE 1-5

For a tribit input of Q = 0, I = 0, and C = 0 (000), determine the output phase for the 8PSK modulator shown in Figure 1-19.

Solution The inputs to the I-channel 2-to-4-level converter are I = 0 and C = 0. From Figure 1-20 the output is −0.541 V. The inputs to the Q-channel 2-to-4-level converter are Q = 0 and $\overline{C}$ = 1. Again from Figure 1-20, the output is −1.307 V.

Thus the two inputs to the I-channel product modulator are −0.541 and sin $\omega_c t$. The output is

$$I = (-0.541)(\sin \omega_c t) = -0.541 \sin \omega_c t$$

The two inputs to the Q-channel product modulator are −1.307 V and cos $\omega_c t$. The output is

$$Q = (-1.307)(\cos \omega_c t) = -1.307 \cos \omega_c t$$

The outputs of the I- and Q-channel product modulators are combined in the linear summer and produce a modulated output of

$$\text{summer output} = -0.541 \sin \omega_c t - 1.307 \cos \omega_c t$$

$$= 1.41 \sin \omega_c t - 112.5°$$

For the remaining tribit codes (001, 010, 011, 100, 101, 110, and 111), the procedure is the same. The results are shown in Figure 1-21.

From Figure 1-21 it can be seen that the angular separation between any two adjacent phasors is 45°, half what it is with QPSK. Therefore, an 8PSK signal can undergo almost a ±22.5° phase shift during transmission and still retain its integrity. Also, each phasor is of equal magnitude; the tribit condition (actual information) is again contained only in the phase of the signal. The PAM levels of 1.307 and 0.541 are relative values. Any levels may be used as long as their ratio is 0.541/1.307 and their arc tangent is equal to 22.5°. For example, if their values were doubled to 2.614 and 1.082, the resulting phase angles would not change, although the magnitude of the phasor would increase proportionally.

It should also be noted that the tribit code between any two adjacent phases changes by only one bit. This type of code is called the *Gray code* or, sometimes, the *maximum distance code*. This code is used to reduce the number of transmission errors. If a signal were to undergo a phase shift during transmission, it would most likely be shifted to an adjacent phasor. Using the Gray code results in only a single bit being received in error.

Figure 1-22 shows the output phase versus time relationship of an 8PSK modulator.

Bandwidth Considerations of 8PSK

With 8PSK, since the data are divided into three channels, the bit rate in the I, Q, or C channel is equal to one-third of the binary input data rate ($F_b/3$). (The bit splitter stretches the I, Q, and C bits to three times their input bit length.) Because the I, Q, and C bits are outputted simultaneously and in parallel, the 2-to-4-level converters also see a change in their inputs (and consequently their outputs) at a rate equal to $F_b/3$.

Figure 1-23 shows the bit timing relationship between the binary input data;

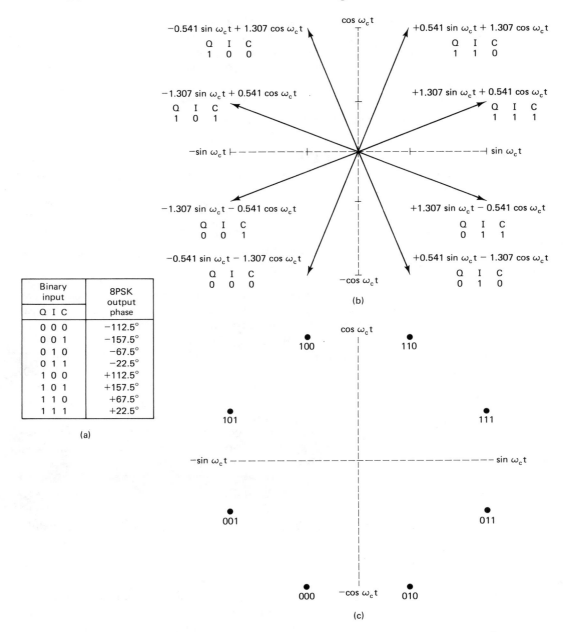

Figure 1-21   8PSK modulator: (a) truth table; (b) phasor diagram; (c) constellation diagram.

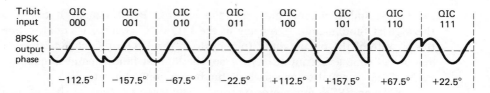

Tribit input	QIC 000	QIC 001	QIC 010	QIC 011	QIC 100	QIC 101	QIC 110	QIC 111
8PSK output phase	$-112.5°$	$-157.5°$	$-67.5°$	$-22.5°$	$+112.5°$	$+157.5°$	$+67.5°$	$+22.5°$

Figure 1-22 Output phase versus time relationship for an 8PSK modulator.

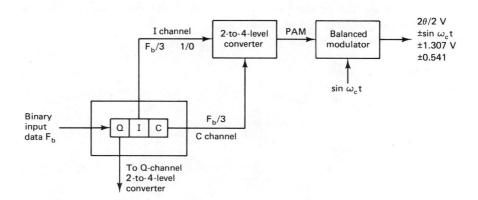

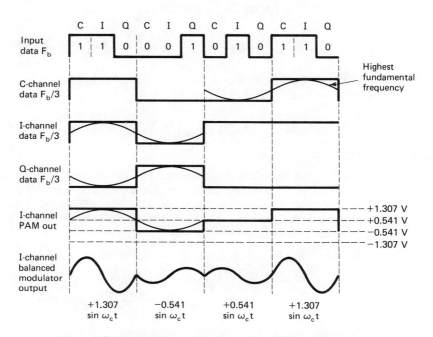

Figure 1-23 Bandwidth considerations of an 8PSK modulator.

the I-, Q-, and C-channel data; and the I and Q PAM signals. It can be seen that the highest fundamental frequency in the I, Q or C channel is equal to one-sixth of the bit rate of the binary input (one cycle in the I, Q, or C channel takes the same amount of time as six input bits). Also, the highest fundamental frequency in either PAM signal is equal to one-sixth of the binary input bit rate.

With an 8PSK modulator, there is one change in phase at the output for every 3 data input bits. Consequently, the baud for 8PSK equals $F_b/3$, the same as the minimum bandwidth. Again, the balanced modulators are product modulators; their outputs are the product of the carrier and the PAM signal. Mathematically, the output of the balanced modulators is

$$\theta = (X \sin \omega_a t)(\sin \omega_c t)$$

where

$$\underbrace{\omega_a t = 2\pi \frac{F_b t}{6}}_{\text{modulating signal}} \quad \text{and} \quad \underbrace{\omega_c t = 2\pi F_c t}_{\text{carrier}}$$

and

$$X = \pm 1.307 \quad \text{or} \quad \pm 0.541$$

Thus

$$\theta = \left(X \sin 2\pi \frac{F_b t}{6} \right)(\sin 2\pi F_c t)$$

$$= \frac{X}{2} \cos 2\pi \left(F_c - \frac{F_b}{6} \right) t - \frac{X}{2} \cos 2\pi \left(F_c + \frac{F_b}{6} \right) t$$

The output frequency spectrum extends from $F_c + F_b/6$ to $F_c - F_b/6$ and the minimum bandwidth (F_N) is

$$\left(F_c + \frac{F_b}{6} \right) - \left(F_c - \frac{F_b}{6} \right) = \frac{2F_b}{6} = \frac{F_b}{3}$$

EXAMPLE 1-6

For an 8PSK modulator with an input data rate (F_b) equal to 10 Mbps and a carrier frequency of 70 MHz, determine the minimum double-sided Nyquist bandwidth (F_N) and the baud. Also, compare the results with those achieved with the BPSK and QPSK modulators in Examples 1-2 and 1-4. Use the 8PSK block diagram shown in Figure 1-19 as the modulator model.

Solution The bit rate in the I, Q, and C channels is equal to one-third of the input bit rate, or

$$F_{bC} = F_{bQ} = F_{bI} = \frac{10 \text{ Mbps}}{3} = 3.33 \text{ Mbps}$$

Therefore, the fastest rate of change and highest fundamental frequency presented to either balanced modulator is

$$F_a = \frac{F_{bC}}{2} \quad \text{or} \quad \frac{F_{bQ}}{2} \quad \text{or} \quad \frac{F_{bI}}{2} = \frac{3.33 \text{ Mbps}}{2} = 1.667 \text{ Mbps}$$

The output wave from the balance modulators is

$$(\sin 2\pi F_a t)(\sin 2\pi F_c t)$$

$$\tfrac{1}{2}\cos 2\pi(F_c - F_a)t - \tfrac{1}{2}\cos 2\pi(F_c + F_a)t$$

$$\tfrac{1}{2}\cos 2\pi[(70 - 1.667)\text{ MHz}]t - \tfrac{1}{2}\cos 2\pi[(70 + 1.667)\text{ MHz}]t$$

$$\tfrac{1}{2}\cos 2\pi(68.333\text{ MHz})t - \tfrac{1}{2}\cos 2\pi(71.667\text{ MHz})t$$

The minimum Nyquist bandwidth is

$$F_N = (71.667 - 68.333)\text{ MHz} = 3.333\text{ MHz}$$

Again, the baud equals the bandwidth; thus

$$\text{baud} = 3.333 \text{ megabaud}$$

The output spectrum is as follows:

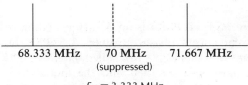

68.333 MHz   70 MHz   71.667 MHz
(suppressed)

$$F_N = 3.333 \text{ MHz}$$

It can be seen that for the same input bit rate the minimum bandwidth required to pass the output of an 8PSK modulator is equal to one-third that of the BPSK modulator in Example 1-2 and 50% less than that required for the QPSK modulator in Example 1-4. Also, in each case the baud has been reduced by the same proportions.

8PSK Receiver

Figure 1-24 shows a block diagram of an 8PSK receiver. The power splitter directs the input 8PSK signal to the I and Q product detectors and the carrier recovery circuit. The carrier recovery circuit reproduces the original reference oscillator signal. The incoming 8PSK signal is mixed with the recovered carrier in the I product detector and with a quadrature carrier in the Q product detector. The outputs of the product detectors are 4-level PAM signals that are fed to the 4-to-2-level *analog-to-digital converters* (ADCs). The outputs from the I-channel 4-to-2-level converter are the I and C bits, while the outputs from the Q-channel 4-to-2-level converter are the Q and $\overline{\text{C}}$ bits. The parallel-to-serial logic circuit converts the I/C and Q/$\overline{\text{C}}$ bit pairs to serial I, Q, and C output data streams.

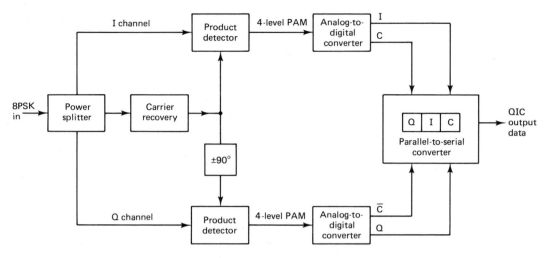

Figure 1-24 8PSK receiver.

SIXTEEN-PHASE PSK

Sixteen-phase PSK (16PSK) is an *M*-ary encoding technique where $M = 16$; there are 16 different output phases possible. A 16PSK modulator acts on the incoming data in groups of 4 bits ($2^4 = 16$), called *quadbits*. The output phase does not change until 4 bits have been inputted into the modulator. Therefore, the output rate of change (baud) and the minimum bandwidth are equal to one-fourth of the incoming bit rate ($F_b/4$). The truth table and constellation diagram for a 16PSK transmitter are shown in Figure 1-25.

With 16PSK, the angular separation between adjacent output phases is only 22.5°. Therefore, a 16PSK signal can undergo almost a ±11.25° phase shift during transmission and still retain its integrity. Because of this, 16PSK is highly susceptible to phase impairments introduced in the transmission medium and is therefore seldom used.

QUADRATURE AMPLITUDE MODULATION

Quadrature amplitude modulation (QAM) is a form of digital modulation where the digital information is contained in both the amplitude and phase of the transmitted carrier.

EIGHT QAM

Eight QAM (8QAM) is an *M*-ary encoding technique where $M = 8$. Unlike 8PSK, the output signal from an 8QAM modulator is not a constant-amplitude signal.

Bit code	Phase	Bit code	Phase
0000	11.25°	1000	191.25°
0001	33.75°	1001	213.75°
0010	56.25°	1010	236.25°
0011	78.75°	1011	258.75°
0100	101.25°	1100	281.25°
0101	123.75°	1101	303.75°
0110	146.25°	1110	326.25°
0111	168.75°	1111	348.75°

(a)

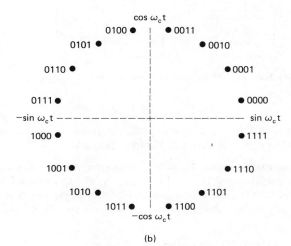

(b)

Figure 1-25 16PSK: (a) truth table; (b) constellation diagram.

8QAM Transmitter

Figure 1-26 shows the block diagram of an 8QAM transmitter. As you can see, the only difference between the 8QAM transmitter and the 8PSK transmitter shown in Figure 1-19 is the omission of the inverter between the C channel and the Q product modulator. As with 8PSK, the incoming data are divided into groups of three (tribits): the I, Q, and C channels, each with a bit rate equal to one-third of the incoming data rate. Again, the I and Q bits determine the polarity of the PAM signal at the output of the 2-to-4-level converters, and the C channel determines the magnitude. Because the C bit is fed uninverted to both the I- and Q-channel 2-to-4-level converters, the magnitudes of the I and Q PAM signals are always equal. Their polarities depend on the logic condition of the I and Q bits and therefore may be different. Figure 1-27 shows the truth table for the I- and Q-channel 2-to-4-level converters; they are the same.

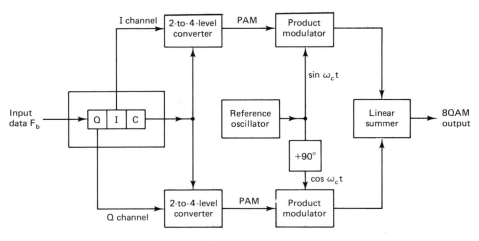

Figure 1-26 8QAM transmitter block diagram.

EXAMPLE 1-7

For a tribit input of Q = 0, I = 0, and C = 0 (000), determine the output amplitude and phase for the 8QAM modulator shown in Figure 1-26.

Solution The inputs to the I-channel 2-to-4-level converter are I = 0 and C = 0. From Figure 1-27 the output is −0.541 V. The inputs to the Q-channel 2-to-4-level converter are Q = 0 and C = 0. Again from Figure 1-27, the output is −0.541 V.

Thus the two inputs to the I-channel product modulator are −0.541 and sin $\omega_c t$. The output is

$$I = (-0.541)(\sin \omega_c t) = -0.541 \sin \omega_c t$$

The two inputs to the Q-channel product modulator are −0.541 and cos $\omega_c t$. The output is

$$Q = (-0.541)(\cos \omega_c t) = -0.541 \cos \omega_c t$$

The outputs from the I- and Q-channel product modulators are combined in the linear summer and produce a modulated output of

$$\text{summer output} = -0.541 \sin \omega_c t - 0.541 \cos \omega_c t$$

$$= 0.765 \sin \omega_c t - 135°$$

For the remaining tribit codes (001, 010, 011, 100, 101, 110, and 111), the procedure is the same. The results are shown in Figure 1-28.

I/Q	C	Output
0	0	−0.541
0	1	−1.307 V
1	0	+0.541
1	1	+1.307 V

Figure 1-27 Truth table for the I- and Q-channel 2-to-4-level converters.

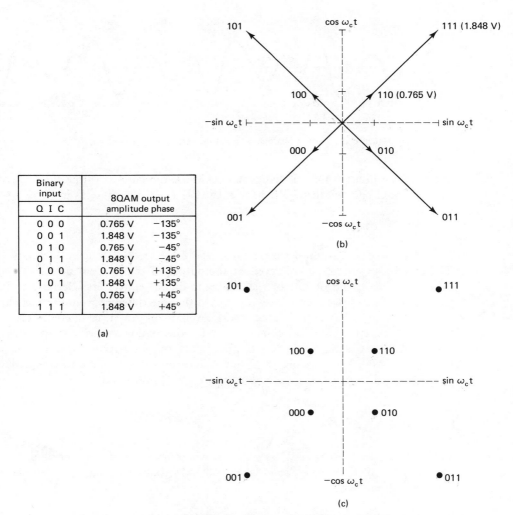

Binary input		8QAM output	
Q I C		amplitude	phase
0 0 0		0.765 V	−135°
0 0 1		1.848 V	−135°
0 1 0		0.765 V	−45°
0 1 1		1.848 V	−45°
1 0 0		0.765 V	+135°
1 0 1		1.848 V	+135°
1 1 0		0.765 V	+45°
1 1 1		1.848 V	+45°

(a)

Figure 1-28 8QAM modulator: (a) truth table; (b) phasor diagram; (c) constellation diagram.

Figure 1-29 shows the output phase versus time relationship for an 8QAM modulator. Note that there are two output amplitudes and only four phases are possible.

Bandwidth Considerations of 8QAM

In 8QAM, the bit rate in the I and Q channels is one-third of the input binary rate, the same as in 8PSK. As a result, the highest fundamental modulating frequency

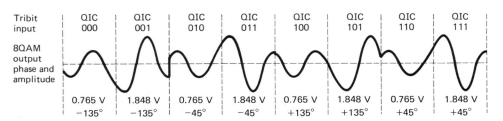

Figure 1-29   Output phase and amplitude versus time relationship for 8QAM.

and the fastest output rate of change in 8QAM are the same as with 8PSK. Therefore, the minimum bandwidth required for 8QAM is $F_b/3$, the same as in 8PSK.

8QAM Receiver

An 8QAM receiver is almost identical to the 8PSK receiver shown in Figure 1-24. The differences are the PAM levels at the output of the product detectors and the binary signals at the output of the analog-to-digital converters. Because there are two transmit amplitudes possible with 8QAM that are different from those achievable with 8PSK, the four demodulated PAM levels in 8QAM are different from those in 8PSK. Therefore, the conversion factor for the analog-to-digital converters must also be different. Also, with 8QAM the binary output signals from the I-channel analog-to-digital converter are the I and C bits, and the binary output signals from the Q-channel analog-to-digital converter are the Q and C bits.

SIXTEEN QAM

Like 16PSK, 16QAM is an M-ary system where $M = 16$. The input data are acted on in groups of four ($2^4 = 16$). As with 8QAM, both the phase and amplitude of the transmit carrier are varied.

16QAM Transmitter

The block diagram for a 16QAM transmitter is shown in Figure 1-30. The input binary data are divided into four channels: the I, $\bar{\text{I}}$, Q, and $\bar{\text{Q}}$. The bit rate in each channel is equal to one-fourth of the input bit rate ($f_b/4$). Four bits are serially clocked into the bit splitter; then they are outputted simultaneously and in parallel with the I, $\bar{\text{I}}$, Q, and $\bar{\text{Q}}$ channels. The I and Q bits determine the polarity at the output of the 2-to-4-level converters (a logic 1 = positive and a logic 0 = negative). The $\bar{\text{I}}$ and $\bar{\text{Q}}$ bits determine the magnitude (a logic 1 = 0.821 V and a logic 0 = 0.22 V). Consequently, the 2-to-4-level converters generate a 4-level PAM signal. Two polarities and two magnitudes are possible at the output of each 2-to-4-level converter. They are ±0.22 V and ±0.821 V. The PAM signals modulate the in-

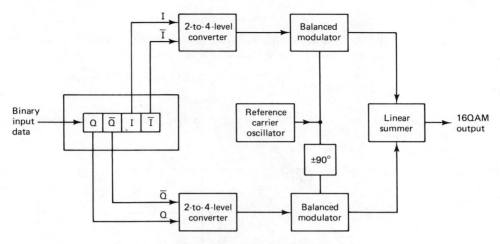

Figure 1-30 16QAM transmitter block diagram.

I	Ī	Output
0	0	−0.22 V
0	1	−0.821 V
1	0	+0.22 V
1	1	+0.821 V

(a)

Q	Q̄	Output
0	0	−0.22 V
0	1	−0.821 V
1	0	+0.22 V
1	1	+0.821 V

(b)

Figure 1-31 Truth tables for the I- and Q-channel 2-to-4-level converters: (a) I channel; (b) Q channel.

phase and quadrature carriers in the product modulators. Four outputs are possible for each product modulator. For the I product modulator they are $+0.821 \sin \omega_c t$, $-0.821 \sin \omega_c t$, $+0.22 \sin \omega_c t$, and $-0.22 \sin \omega_c t$. For the Q product modulator they are $+0.821 \cos \omega_c t$, $+0.22 \cos \omega_c t$, $-0.821 \cos \omega_c t$, and $-0.22 \cos \omega_c t$. The linear summer combines the outputs from the I- and Q-channel product modulators and produces the 16 output conditions necessary for 16 QAM. Figure 1-31 shows the truth table for the I- and Q-channel 2-to-4-level converters.

EXAMPLE 1-8

For a quadbit input of $I = 0$, $\bar{I} = 0$, $Q = 0$, and $\bar{Q} = 0$ (0000), determine the output amplitude and phase for the 16QAM modulator shown in Figure 1-30.

Solution The inputs to the I-channel 2-to-4-level converter are $I = 0$ and $\bar{I} = 0$. From Figure 1-31 the output is −0.22 V. The inputs to the Q-channel 2-to-4-level converter are $Q = 0$ and $\bar{Q} = 0$. Again from Figure 1-31, the output is −0.22 V.

Thus the two inputs to the I-channel product modulator are −0.22 V and $\sin \omega_c t$. The output is

$$I = (-0.22)(\sin \omega_c t) = -0.22 \sin \omega_c t$$

The two inputs to the Q-channel product modulator are −0.22 V and $\cos \omega_c t$. The output is

$$Q = (-0.22)(\cos \omega_c t) = -0.22 \cos \omega_c t$$

The outputs from the I- and Q-channel product modulators are combined in the linear summer and produce a modulated output of

$$\text{summer output} = -0.22 \sin \omega_c t - 0.22 \cos \omega_c t$$

$$= 0.31 \sin \omega_c t - 135°$$

For the remaining quadbit codes the procedure is the same. The results are shown in Figure 1-32.

Binary input				16QAM output	
Q	Q̄	I	Ī		
0	0	0	0	0.311 V	−135°
0	0	0	1	0.850 V	−165°
0	0	1	0	0.311 V	−45°
0	0	1	1	0.850 V	−15°
0	1	0	0	0.850 V	−105°
0	1	0	1	1.161 V	−135°
0	1	1	0	0.850 V	−75°
0	1	1	1	1.161 V	−45°
1	0	0	0	0.311 V	135°
1	0	0	1	0.850 V	175°
1	0	1	0	0.850 V	45°
1	0	1	1	0.850 V	15°
1	1	0	0	0.850 V	105°
1	1	0	1	1.161 V	135°
1	1	1	0	0.850 V	75°
1	1	1	1	1.161 V	45°

(a)

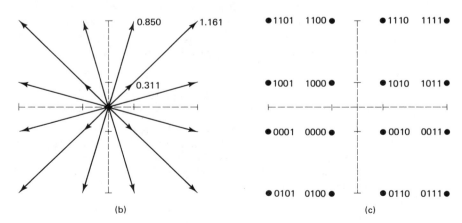

(b)   (c)

Figure 1-32   16QAM modulator: (a) truth table; (b) phasor diagram; (c) constellation diagram.

Bandwidth Considerations of 16QAM

With 16QAM, since the input data are divided into four channels, the bit rate in the I, $\bar{\text{I}}$, Q, or $\bar{\text{Q}}$ channel is equal to one-fourth of the binary input data rate ($F_b/4$). (The bit splitter stretches the I, $\bar{\text{I}}$, Q, and $\bar{\text{Q}}$ bits to four times their input bit length.) Also, because the I, $\bar{\text{I}}$, Q, and $\bar{\text{Q}}$ bits are outputted simultaneously and in parallel, the 2-to-4-level converters see a change in their inputs and outputs at a rate equal to one-fourth of the input data rate.

Figure 1-33 shows the bit timing relationship between the binary input data;

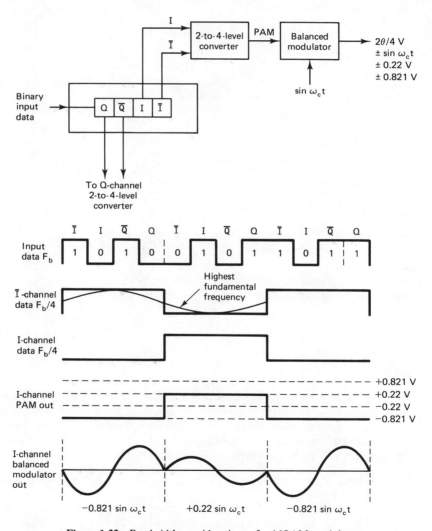

Figure 1-33 Bandwidth considerations of a 16QAM modulator.

the I, Ī, Q, and Q̄ channel data; and the I and Q PAM signals. It can be seen that the highest fundamental frequency in the I, Ī, Q, or Q̄ channel is equal to one-eighth of the bit rate of the binary input data (one cycle in the I, Ī, Q, or Q̄ channel takes the same amount of time as 8 input bits). Also, the highest fundamental frequency of either PAM signal is equal to one-eighth of the binary input bit rate.

With a 16QAM modulator, there is one change in the output signal (either its phase, amplitude, or both) for every 4 input data bits. Consequently, the baud equals $F_b/4$, the same as the minimum bandwidth.

Again, the balanced modulators are product modulators and their outputs can be represented mathematically as

$$\theta = (X \sin \omega_a t)(\sin \omega_c t)$$

where

$$\omega_a t = 2\pi \frac{F_b t}{8} \quad \text{and} \quad \omega_c t = 2\pi F_c t$$

$$\underbrace{\qquad\qquad}_{\text{modulating signal}} \qquad \underbrace{\qquad}_{\text{carrier}}$$

and

$$X = \pm 0.22 \quad \text{or} \quad \pm 0.821$$

Thus

$$\theta = \left(X \sin 2\pi \frac{F_b t}{8}\right)(\sin 2\pi F_c t)$$

$$= \frac{X}{2} \cos 2\pi \left(F_c - \frac{F_b}{8}\right) t - \frac{X}{2} \cos 2\pi \left(F_c + \frac{F_b}{8}\right) t$$

The output frequency spectrum extends from $F_c + F_b/8$ to $F_c - F_b/8$ and the minimum bandwidth (F_N) is

$$\left(F_c + \frac{F_b}{8}\right) - \left(F_c - \frac{F_b}{8}\right) = \frac{2F_b}{8} = \frac{F_b}{4}$$

EXAMPLE 1-9

For a 16QAM modulator with an input data rate (F_b) equal to 10 Mbps and a carrier frequency of 70 MHz, determine the minimum double-sided Nyquist frequency (F_N) and the baud. Also, compare the results with those achieved with the BPSK, QPSK, and 8PSK modulators in Examples 1-2, 1-4, and 1-6. Use the 16QAM block diagram shown in Figure 1-26 as the modulator model.

Solution The bit rate in the I, Ī, Q, and Q̄ channels is equal to one-fourth of the input bit rate or

$$F_{bI} = F_{b\bar{I}} = F_{bQ} = F_{b\bar{Q}} = \frac{F_b}{4} = \frac{10 \text{ Mbps}}{4} = 2.5 \text{ Mbps}$$

Therefore, the fastest rate of change and highest fundamental frequency presented to either balanced modulator is

$$F_a = \frac{F_{bI}}{2} \text{ or } \frac{F_{b\bar{I}}}{2} \text{ or } \frac{F_{bQ}}{2} \text{ or } \frac{F_{b\bar{Q}}}{2} = \frac{2.5 \text{ Mbps}}{2} = 1.25 \text{ Mhz}$$

The output wave from the balanced modulator is

$$(\sin 2\pi F_a t)(\sin 2\pi F_c t)$$

$$\tfrac{1}{2} \cos 2\pi (F_c - F_a)t - \tfrac{1}{2} \cos 2\pi (F_c + F_a)t$$

$$\tfrac{1}{2} \cos 2\pi [(70 - 1.25) \text{ MHz}]t - \tfrac{1}{2} \cos 2\pi [(70 + 1.25) \text{ MHz}]t$$

$$\tfrac{1}{2} \cos 2\pi (68.75 \text{ MHz})t - \tfrac{1}{2} \cos 2\pi (71.25 \text{ MHz})t$$

The minimum Nyquist bandwidth is

$$F_N = (71.25 - 68.75) \text{ MHz} = 2.5 \text{ MHz}$$

The baud equals the bandwidth; thus

$$\text{baud} = 2.5 \text{ megabaud}$$

The output spectrum is as follows:

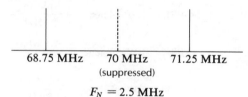

68.75 MHz   70 MHz   71.25 MHz
(suppressed)

$$F_N = 2.5 \text{ MHz}$$

For the same input bit rate, the minimum bandwidth required to pass the output of a 16QAM modulator is equal to one-fourth that of the BPSK modualtor, one-half that of QPSK, and 25% less than with 8PSK. For each modulation technique, the baud is also reduced by the same proportions.

BANDWIDTH EFFICIENCY

Bandwidth efficiency (or *information density* as it is sometimes called) is often used to compare the performance of one digital modulation technique to another. In essence, it is the ratio of the transmission bit rate to the minimum bandwidth required for a particular modulation scheme. Bandwidth efficiency is generally normalized to a 1-Hz bandwidth and thus indicates the number of bits that can be propagated through a medium for each hertz of bandwidth. Mathematically, bandwidth efficiency is

$$\text{BW efficiency} = \frac{\text{transmission rate (bps)}}{\text{minimum bandwidth (Hz)}} \tag{1-4}$$

$$= \frac{\text{bits/second}}{\text{hertz}} = \frac{\text{bits/second}}{\text{cycles/second}} = \frac{\text{bits}}{\text{cycle}}$$

EXAMPLE 1-10

Determine the bandwidth efficiencies for the following modulation schemes: BPSK, QPSK, 8PSK, and 16QAM.

Solution   Recall from Examples 1-2, 1-4, 1-6, and 1-9 the minimum bandwidths required to propagate a 10-Mbps transmission rate with the following modulation schemes:

Modulation scheme	Minimum bandwidth (MHz)
BPSK	10
QPSK	5
8PSK	3.33
16QAM	2.5

Substituting into Equation 1-4, the bandwidth efficiencies are determined as follows:

$$\text{BPSK:}\quad \text{BW efficiency} = \frac{10\ \text{Mbps}}{10\ \text{MHz}} = \frac{1\ \text{bps}}{\text{Hz}} = \frac{1\ \text{bit}}{\text{cycle}}$$

$$\text{QPSK:}\quad \text{BW efficiency} = \frac{10\ \text{Mbps}}{5\ \text{MHz}} = \frac{2\ \text{bps}}{\text{Hz}} = \frac{2\ \text{bits}}{\text{cycle}}$$

$$\text{8PSK:}\quad \text{BW efficiency} = \frac{10\ \text{Mbps}}{3.33\ \text{MHz}} = \frac{3\ \text{bps}}{\text{Hz}} = \frac{3\ \text{bits}}{\text{cycle}}$$

$$\text{16QAM:}\quad \text{BW efficiency} = \frac{10\ \text{Mbps}}{2.5\ \text{MHz}} = \frac{4\ \text{bps}}{\text{Hz}} = \frac{4\ \text{bits}}{\text{cycle}}$$

The results indicate that BPSK is the least efficient and 16QAM is the most efficient. 16QAM requires one-fourth as much bandwidth as BPSK for the same bit rate.

PSK AND QAM SUMMARY

The various forms of FSK, PSK, and QAM are summarized in Table 1-2.

TABLE 1-2   DIGITAL MODULATION SUMMARY

Modulation	Encoding	Bandwidth (Hz)	Baud	Bandwidth efficiency (bps/Hz)
FSK	Single bit	$\geq F_b$	F_b	≤ 1
BPSK	Single bit	F_b	F_b	1
QPSK	Dibit	$F_b/2$	$F_b/2$	2
8PSK	Tribit	$F_b/3$	$F_b/3$	3
8QAM	Tribit	$F_b/3$	$F_b/3$	3
16PSK	Quadbit	$F_b/4$	$F_b/4$	4
16QAM	Quadbit	$F_b/4$	$F_b/4$	4

CARRIER RECOVERY

Carrier recovery is the process of extracting a phase-coherent reference carrier from a received carrier waveform. This is sometimes called *phase referencing*. In the phase modulation techniques described thus far, the binary data were encoded as a precise phase of the transmitted carrier. (This is referred to as *absolute phase encoding*.) Depending on the encoding method, the angular separation between adjacent phasors varied between 15 and 180°. To correctly demodulate the data, a phase-coherent carrier was recovered and compared with the received signal in a product detector. To determine the absolute phase of the received signal, it is necessary to reproduce a carrier at the receiver that is phase coherent with the transmit reference oscillator. This is the function of the carrier recovery circuit.

With PSK and QAM, the carrier is suppressed in the balanced modulators and is therefore not transmitted. Consequently, at the receiver the carrier cannot simply be tracked with a standard phase-locked loop. A more sophisticated method is used.

One common method of achieving carrier recovery for BPSK is the *squaring loop*. Figure 1-34 shows the block diagram of a squaring loop. The received BPSK waveform is filtered and then squared. The filtering reduces the spectral width of the received noise. The squaring circuit removes the modulation and generates the second harmonic of the carrier frequency. This harmonic is phase tracked by the PLL. The VCO output frequency from the PLL is then divided by 2 and used as the phase reference for the product detectors.

With BPSK, only two output phases are possible: $+\sin \omega_c t$ and $-\sin \omega_c t$. Mathematically, the operation of the squaring circuit can be described as follows.

For a receive signal of $+\sin \omega_c t$ the output of the squaring circuit is

$$\text{output} = (+\sin \omega_c t)(+\sin \omega_c t) = +\sin^2 \omega_c t$$

$$\text{(filtered out)}$$

$$= \tfrac{1}{2}(1 - \cos 2\omega_c t) = \tfrac{1}{2} - \tfrac{1}{2}\cos 2\omega_c t$$

For a received signal of $-\sin \omega_c t$ the output of the squaring circuit is

$$\text{output} = (-\sin \omega_c t)(-\sin \omega_c t) = +\sin^2 \omega_c t$$

$$\text{(filtered out)}$$

$$= \tfrac{1}{2}(1 - \cos 2\omega_c t) = \tfrac{1}{2} - \tfrac{1}{2}\cos 2\omega_c t$$

It can be seen that in both cases the output from the squaring circuit contained a dc voltage ($-\tfrac{1}{2}$ V) and a signal at twice the carrier frequency ($\cos 2\omega_c t$). The dc voltage is removed by filtering, leaving only $\cos 2\omega_c t$.

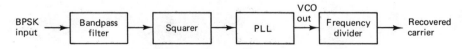

Figure 1-34 Squaring loop carrier recovery circuit for a BPSK receiver.

A more elaborate carrier recovery circuit is the *Costas* or *quadrature loop*, which combines carrier recovery with noise suppression and can therefore accurately recover a carrier from a poorer-quality received signal than can a conventional squaring loop.

Carrier recovery circuits for higher-than-binary encoding techniques are similar to BPSK except that circuits which raise the receive signal to the fourth, eighth, and higher powers are used.

DIFFERENTIAL PHASE SHIFT KEYING

Differential phase shift keying (DPSK) is an alternative form of digital modulation where the binary input information is contained in the difference between two successive signaling elements rather than the absolute phase. With DPSK it is not necessary to recover a phase-coherent carrier. Instead, a received signaling element is delayed by one signaling element time slot and then compared to the next received signaling element. The difference in the phase of the two signaling elements determines the logic condition of the data.

DIFFERENTIAL BPSK

DBPSK Transmitter

Figure 1-35a shows a simplified block diagram of a *differential binary phase shift keying* (DBPSK) transmitter. An incoming information bit is XNORed with the preceding bit prior to entering the BPSK modulator (balanced modulator). For the

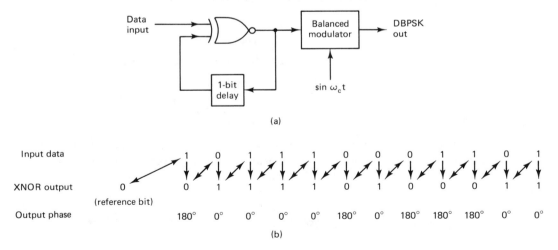

Figure 1-35   DBPSK modulator: (a) block diagram; (b) timing diagram.

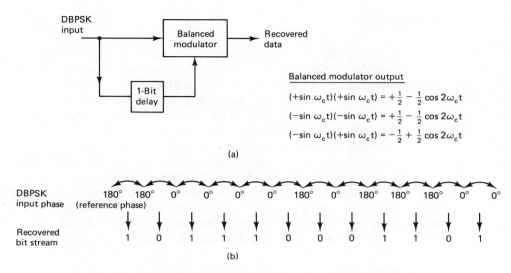

Figure 1-36 DBPSK demodulator: (a) block diagram; (b) timing sequence.

first data bit, there is no preceding bit with which to compare it. Therefore, an initial reference bit is assumed. Figure 1-35b shows the relationship between the input data, the XNOR output data, and the phase at the output of the balanced modulator. If the initial reference bit is assumed a logic 1, the output from the XNOR circuit is simply the complement of that shown.

In Figure 1-35b the first data bit is XNORed with the reference bit. If they are the same, the XNOR output is a logic 1; if they are different, the XNOR output is a logic 0. The balanced modulator operates the same as a conventional BPSK modulator; a logic 1 produces $+\sin \omega_c t$ at the output and a logic 0 produces $-\sin \omega_c t$ at the output.

DBPSK Receiver

Figure 1-36 shows the block diagram and timing sequence for a DBPSK receiver. The received signal is delayed by one bit time, then compared with the next signaling element in the balanced modulator. If they are the same, a logic 1 (+ voltage) is generated. If they are different, a logic 0 (− voltage) is generated. If the reference phase is incorrectly assumed, only the first demodulated bit is in error. Differential encoding can be implemented with higher-than-binary digital modulation schemes, although the differential algorithms are much more complicated than for DBPSK.

The primary advantage of DPSK is the simplicity with which it can be implemented. With DPSK, no carrier recovery circuit is needed. A disadvantage of DPSK is that it requires between 1 and 3 dB more signal-to-noise ratio to achieve the same bit error rate as that of absolute PSK.

CLOCK RECOVERY

As with any digital system, digital radio requires precise timing or clock synchronization between the transmit and the receive circuitry. Because of this, it is necessary to regenerate clocks at the receiver that are synchronous with those at the transmitter.

Figure 1-37a shows a simple circuit that is commonly used to recover clocking information from the received data. The recovered data are delayed by one-half a bit time and then compared with the original data in an XOR circuit. The frequency of the clock that is recovered with this method is equal to the received data rate (F_b). Figure 1-37b shows the relationship between the data and the recovered clock timing. From Figure 1-37b it can be seen that as long as the receive data contain a substantial number of transitions (1/0 sequences), the recovered clock is maintained. If the receive data were to undergo an extended period of successive 1's or 0's, the recovered clock would be lost. To prevent this from occurring, the data are scrambled at the transmit end and descrambled at the receive end.

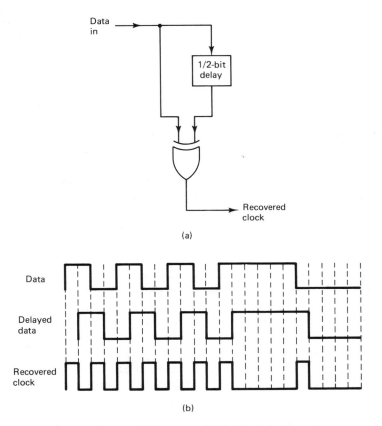

Figure 1-37   (a) Clock recovery circuit; (b) timing diagram.

PROBABILITY OF ERROR AND BIT ERROR RATE

Probability of error $P(e)$ and *bit error rate* (BER) are often used interchangeably, although they do have slightly different meanings. $P(e)$ is a theoretical (mathematical) expectation of the error rate for a given system. BER is an empirical (historical) record of a system's actual error performance. For example, if a system has a $P(e)$ of 10^{-5}, this means that mathematically, you can expect one bit error in every 100,000 bits transmitted $(1/10^{-5} = 1/100,000)$. If a system has a BER of 10^{-5}, this means that in the past there was one bit error for every 100,000 bits transmitted.

Probability of error is a function of the receiver *carrier-to-noise ratio*. Depending on the M-ary used and the desired $P(e)$, the minimum carrier-to-noise ratio varies. In general, the minimum carrier-to-noise ratio required for a QAM system is less than that required for a comparable PSK system (see Table 1-3). Also, the higher the level of encoding used, the higher the minimum carrier-to-noise ratio. Another parameter often used for comparing digital system performances is the energy of the bit-to-noise ratio (E_b/N_0). E_b/N_0 is explained in Chapter 8, where several examples are shown for determining the minimum carrier-to-noise ratio for a given M-ary system and desired $P(e)$.

TABLE 1-3   PERFORMANCE COMPARISON OF VARIOUS DIGITAL MODULATION SCHEMES (BER $= 10^{-6}$)

Modulation technique	C/N ratio (dB)	E_b/N_0 ratio (dB)
BPSK	13.6	10.6
QPSK	13.6	10.6
8QAM	13.6	10.6
8PSK	18.8	14
16PSK	24.3	18.3
16QAM	20.5	14.5
32QAM	24.4	17.4
64QAM	26.6	18.8

APPLICATIONS FOR DIGITAL MODULATION

A digitally modulated transceiver (*trans*mitter-re*ceiver*) that uses FSK, PSK, or QAM has many applications. They are used in digitally modulated microwave radio and satellite systems (Chapter 8) with carrier frequencies from tens of megahertz to several gigahertz, and they are also used for voice band data modems (Chapter 2) with carrier frequencies between 300 and 3000 Hz.

QUESTIONS

1-1. Explain *digital transmission* and *digital radio*.

1-2. Define *information capacity*.

1-3. What are the three most predominant modulation schemes used in digital radio systems?

1-4. Explain the relationship between bits per second and baud for an FSK system.

1-5. Define the following terms for FSK modulation: frequency deviation, modulation index, and deviation ratio.

1-6. Explain the relationship between (a) the minimum bandwidth required for an FSK system and the bit rate, and (b) the mark and space frequencies.

1-7. What is the difference between standard FSK and MSK? What is the advantage of MSK?

1-8. Define *PSK*.

1-9. Explain the relationship between bits per second and baud for a BPSK system.

1-10. What is a constellation diagram, and how is it used with PSK?

1-11. Explain the relationship between the minimum bandwidth required for a BPSK system and the bit rate.

1-12. Explain *M-ary*.

1-13. Explain the relationship between bits per second and baud for a QPSK system.

1-14. Explain the significance of the I and Q channels in a QPSK modulator.

1-15. Define *dibit*.

1-16. Explain the relationship between the minimum bandwidth required for a QPSK system and the bit rate.

1-17. What is a coherent demodulator?

1-18. What advantage does OQPSK have over conventional QPSK? What is a disadvantage of OQPSK?

1-19. Explain the relationship between bits per second and baud for an 8PSK system.

1-20. Define *tribit*.

1-21. Explain the relationship between the minimum bandwidth required for an 8PSK system and the bit rate.

1-22. Explain the relationship between bits per second and baud for a 16PSK system.

1-23. Define *quadbit*.

1-24. Define *QAM*.

1-25. Explain the relationship between the minimum bandwidth required for a 16QAM system and the bit rate.

1-26. What is the difference between PSK and QAM?

1-27. Define *bandwidth efficiency*.

1-28. Define *carrier recovery*.

1-29. Explain the differences between absolute PSK and differential PSK.

1-30. What is the purpose of a clock recovery circuit? When is it used?

1.31. What is the difference between probability of error and bit error rate?

PROBLEMS

1-1. For an FSK modulator with space, rest, and mark frequencies of 40, 50, and 60 MHz, respectively, and an input bit rate of 10 Mbps, determine the output baud and minimum bandwidth. Sketch the output spectrum.

1-2. Determine the minimum bandwidth and baud for a BPSK modulator with a carrier frequency of 40 MHz and an input bit rate of 500 kbps. Sketch the output spectrum.

1-3. For the QPSK modulator shown in Figure 1-13, change the $+90°$ phase-shift network to $-90°$ and sketch the new constellation diagram.

1-4. For the QPSK demodulator shown in Figure 1-17, determine the I and Q bits for an input signal of $\sin \omega_c t - \cos \omega_c t$.

1-5. For an 8PSK modulator with an input data rate (F_b) equal to 20 Mbps and a carrier frequency of 100 MHz, determine the minimum double-sided Nyquist bandwidth (F_N) and the baud. Sketch the output spectrum.

1-6. For the 8PSK modulator shown in Figure 1-19, change the reference oscillator to cos $\omega_c t$ and sketch the new constellation diagram.

1-7. For a 16QAM modulator with an input bit rate (F_b) equal to 20 Mbps and a carrier frequency of 100 MHz, determine the minimum double-sided Nyquist bandwidth (F_N) and the baud. Sketch the output spectrum.

1-8. For the 16QAM modulator shown in Figure 1-26, change the reference oscillator to cos $\omega_c t$ and determine the output expressions for the following I, Ī, Q, and Q̄ input conditions: 0000, 1111, 1010, and 0101.

1-9. Determine the bandwidth efficiency for the following modulators.
  (a) QPSK, $F_b = 10$ Mbps
  (b) 8PSK, $F_b = 21$ Mbps
  (c) 16QAM, $F_b = 20$ Mbps

1-10. For the DBPSK modulator shown in Figure 1-35, determine the output phase sequence for the following input bit sequence: 00110011010101 (assume that the reference bit = 1).

Chapter 2

DATA COMMUNICATIONS

INTRODUCTION

Data communications can be defined as the transmission of digital information (usually in binary form) from a source to a destination. The original source data are in digital form and the received data are in digital form, although the data can be transmitted in analog or digital form. The source information can be binary-coded alpha/numeric characters such as ASCII or EBCDIC, microprocessor op-codes, control words, user addresses, program data, or data base information.

A data communications network can be as simple as two personal computers connected together through the public telephone network, or it can comprise a complex network of one or more mainframe computers and hundreds of remote terminals. Data communications networks are used to connect automatic teller machines (ATMs) to bank computers or they can be used to interface computer terminals (CTs) or keyboard displays (KDs) directly to application programs in mainframe computers. Data communications networks are used for airline and hotel reservation systems and for mass media and news networks such as the Associated Press (AP) or United Press International (UPI). The list of applications for data communications networks goes on almost indefinitely.

HISTORY OF DATA COMMUNICATIONS

It is highly likely that data communications began long before recorded time in the form of smoke signals or tom-tom drums, although it is improbable that these signals were binary coded. If we limit the scope of data communications to methods that use electrical signals to transmit binary-coded information, then data communications began in 1837 with the invention of the *telegraph* and the development of the *Morse code* by Samuel F. B. Morse. With telegraph, dots and dashes (analogous to binary 1's and 0's) are transmitted across a wire using electromechanical induction. Various combinations of these dots and dashes were used to represent binary codes for letters, numbers, and punctuation. Actually, the first telegraph was invented in England by Sir Charles Wheatstone and Sir Willaim Cooke, but their contraption required six different wires for a single telegraph line. In 1840, Morse secured an American patent for the telegraph and in 1844 the first telegraph line was established between Baltimore and Washington, D.C. In 1849, the first slow-speed telegraph printer was invented, but it was not until 1860 that high-speed (15 bps) printers were available. In 1850, the Western Union Telegraph Company was formed in Rochester, New York, for the purpose of carrying coded messages from one person to another.

In 1874, Emile Baudot invented a telegraph *multiplexer*, which allowed signals from up to six different telegraph machines to be transmitted simultaneously over a single wire. The telephone was invented in 1876 by Alexander Graham Bell and, consequently, very little new evolved in telegraph until 1899, when Marconi succeeded in sending radio telegraph messages. Telegraph was the only means of sending information across large spans of water until 1920, when the first commercial radio stations were installed.

Bell Laboratories developed the first special-purpose computer in 1940 using electromechanical relays. The first general-purpose computer was an automatic sequence-controlled calculator developed jointly by Harvard University and International Business Machines Corporation (IBM). The UNIVAC computer, built in 1951 by Remington Rand Corporation (now Sperry Rand), was the first mass-produced electronic computer. Since 1951, the number of mainframe computers, small business computers, personal computers, and computer terminals has increased exponentially, creating a situation where more and more people have the need to exchange digital information with each other. Consequently, the need for data communications has also increased exponentially.

Until 1968, the AT&T operating tariff allowed only equipment furnished by AT&T to be connected to AT&T lines. In 1968, a landmark Supreme Court decision, the Carterfone decision, allowed non-Bell companies to interconnect to the vast AT&T communications network. This decision started the *interconnect industry*, which has led to competitive data communications offerings by a large number of independent companies.

DATA COMMUNICATIONS CIRCUITS

Figure 2-1 shows a simplified block diagram of a data communications circuit. There is a source of digital information, a transmission medium, and a destination. Both the source and destination equipment are digital; they process information in the

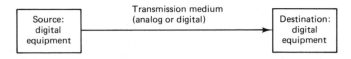

Figure 2-1 Data communications circuit: simplified block diagram.

form of binary pulses. The transmission medium may be a digital or an analog facility and could comprise one or more of the following: metallic wire pair, coaxial cable, microwave radio, satellite radio, or an optical fiber.

Data Communications Circuit Configurations and Topologies

Configurations.   Data communications circuits can be generally categorized as either two-point or multipoint. A *two-point* configuration involves only two locations or stations, whereas a *multipoint* configuration involves three or more stations. A two-point circuit can involve the transfer of information between a mainframe computer and a remote computer terminal, two mainframe computers, or two remote computer terminals. A multipoint circuit is generally used to interconnect a single mainframe computer (*host*) to many remote computer terminals, although any combination of three or more computers or computer terminals constitutes a multipoint circuit.

Topologies.   The topology or architecture of a data communications circuit identifies how the various locations within the network are interconnected. The most common topologies used are the *point to point*, the *star*, the *bus* or *multidrop*, the *ring* or *loop*, and the *mesh*. These are all multipoint configurations except the point to point. Figure 2-2 shows the various circuit configurations and topologies used for data communications networks.

Transmission Modes

Essentially, there are four modes of transmission for data communications circuits: *simplex*, *half duplex*, *full duplex*, and *full/full duplex*.

Simplex.   With simplex operation, data transmission is unidirectional; information can be sent only in one direction. Simplex lines are also called *receive-only* or *one-way-only* lines.

Half duplex (HDX).   In the half-duplex mode, data transmission is possible in both directions, but not at the same time. Half-duplex lines are also called two-way alternate lines.

Full duplex (FDX).   In the full-duplex mode, transmissions are possible in both directions simultaneously, but they must be between the same two stations. Full-duplex lines are also called two-way-simultaneous or simply *duplex* lines.

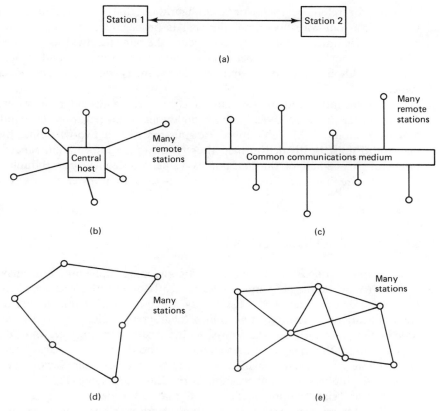

Figure 2-2   Data network topologies: (a) point to point; (b) star; (c) bus or multidrop; (d) ring or loop; (e) mesh.

Full/full duplex (F/FDX).   In the F/FDX mode, transmission is possible in both directions at the same time but not between the same two stations (i.e., one station is transmitting to a second station and receiving from a third station at the same time). F/FDX is possible only on multipoint circuits.

Two-Wire versus Four-Wire Operation

Two-wire, as the name implies, involves a transmission medium that either uses two wires (a signal and a reference lead) or a configuration that is equivalent to having only two wires. With two-wire operation, full- or half-duplex transmission is possible. For full-duplex operation, the signals propagating in opposite directions must occupy different bandwidths; otherwise, they will mix linearly and interfere with each other.

Four-wire, as the name implies, involves a transmission medium that uses four wires (two are used for signals that are propagating in opposite directions and two are used for reference leads) or a configuration that is equivalent to having four wires. With four-wire operation, the signals propagating in opposite directions are physically separated and therefore can occupy the same bandwidths without interfering with each other. Four-wire operation provides more isolation and is preferred over two-wire, although four-wire requires twice as many wires and, consequently, twice the cost.

A transmitter and its associated receiver are equivalent to a two-wire circuit. A transmitter and a receiver for both directions of propagation are equivalent to a four-wire circuit. With full-duplex transmission over a two-wire line, the available bandwidth must be divided in half, thus reducing the information capacity in either direction to one-half of the half-duplex value. Consequently, full-duplex operation over two-wire lines requires twice as much time to transfer the same amount of information.

DATA COMMUNICATIONS CODES

Data communications codes are used for encoding alpha/numeric characters and symbols (punctuation, etc.) and are consequently often called *character sets*, *character languages*, or *character codes*. Essentially, three types of characters are used in data communications codes: *data link control* characters, which are used to facilitate the orderly flow of data from the source to the destination; *graphic control* characters, which involve the syntax or presentation of the data at the receive terminal; and *alpha/numeric* characters, which are used to represent the various symbols used for letters, numbers, and punctuation in the English language.

The first data communications code that saw widespread usage was the Morse code. The Morse code used three unequal-length symbols (dot, dash, and space) to encode alpha/numeric characters, punctuation marks, and an interrogation word.

The Morse code is inadequate for use in modern digital computer equipment because all characters do not have the same number of symbols or take the same length of time to send, and each Morse code operator transmits code at a different rate. Also, with Morse code, there is an insufficient selection of graphic and data link control characters to facilitate the transmission and presentation of the data typically used in contemporary computer applications.

The three most common character sets presently used for character encoding are the Baudot code, the American Standard Code for Information Interchange (ASCII), and the Extended Binary-Coded Decimal Interchange Code (EBCDIC).

Baudot Code

The *Baudot code* (sometimes called the *Telex code*) was the first fixed-length character code. The Baudot code was developed by a French postal engineer, Thomas Murray, in 1875 and named after Emile Baudot, an early pioneer in telegraph printing. The

Baudot code is a 5-bit character code that is used primarily for low-speed teletype equipment such as the TWX/Telex system. With a 5-bit code there are only 2^5 or 32 codes possible, which is insufficient to represent the 26 letters of the alphabet, the 10 digits, and the various punctuation marks and control characters. Therefore, the Baudot code uses *figure* shift and *letter* shift characters to expand its capabilities to 58 characters. The latest version of the Baudot code is recommended by the CCITT as the International Alphabet No. 2. The Baudot code is still used by Western Union Company for their TWX and Telex teletype systems. The AP and UPI news services also use the Baudot code for sending news information around the world. The most recent version of the Baudot code is shown in Table 2-1.

TABLE 2-1 BAUDOT CODE

Character shift		Binary code				
Letter	*Figure*	*Bit:* 4	3	2	1	0
A	—	1	1	0	0	0
B	?	1	0	0	1	1
C	:	0	1	1	1	0
D	$	1	0	0	1	0
E	3	1	0	0	0	0
F	!	1	0	1	1	0
G	&	0	1	0	1	1
H	#	0	0	1	0	1
I	8	0	1	1	0	0
J	'	1	1	0	1	0
K	(	1	1	1	1	0
L	)	0	1	0	0	1
M	.	0	0	1	1	1
N	,	0	0	1	1	0
O	9	0	0	0	1	1
P	0	0	1	1	0	1
Q	1	1	1	1	0	1
R	4	0	1	0	1	0
S	bel	1	0	1	0	0
T	5	0	0	0	0	1
U	7	1	1	1	0	0
V	;	0	1	1	1	1
W	2	1	1	0	0	1
X	/	1	0	1	1	1
Y	6	1	0	1	0	1
Z	"	1	0	0	0	1
Figure shift		1	1	1	1	1
Letter shift		1	1	0	1	1
Space		0	0	1	0	0
Line feed (LF)		0	1	0	0	0
Blank (null)		0	0	0	0	0

TABLE 2-2   ASCII-77 CODE—ODD PARITY

	Binary code								Hex		Binary code								Hex
Bit:	7	6	5	4	3	2	1	0		Bit:	7	6	5	4	3	2	1	0	
NUL	1	0	0	0	0	0	0	0	00	@	0	1	0	0	0	0	0	0	40
SOH	0	0	0	0	0	0	0	1	01	A	1	1	0	0	0	0	0	1	41
STX	0	0	0	0	0	0	1	0	02	B	1	1	0	0	0	0	1	0	42
ETX	1	0	0	0	0	0	1	1	03	C	0	1	0	0	0	0	1	1	43
EOT	0	0	0	0	0	1	0	0	04	D	1	1	0	0	0	1	0	0	44
ENQ	1	0	0	0	0	1	0	1	05	E	0	1	0	0	0	1	0	1	45
ACK	1	0	0	0	0	1	1	0	06	F	0	1	0	0	0	1	1	0	46
BEL	0	0	0	0	0	1	1	1	07	G	1	1	0	0	0	1	1	1	47
BS	0	0	0	0	1	0	0	0	08	H	1	1	0	0	1	0	0	0	48
HT	1	0	0	0	1	0	0	1	09	I	0	1	0	0	1	0	0	1	49
NL	1	0	0	0	1	0	1	0	0A	J	0	1	0	0	1	0	1	0	4A
VT	0	0	0	0	1	0	1	1	0B	K	1	1	0	0	1	0	1	1	4B
FF	1	0	0	0	1	1	0	0	0C	L	0	1	0	0	1	1	0	0	4C
CR	0	0	0	0	1	1	0	1	0D	M	1	1	0	0	1	1	0	1	4D
SO	0	0	0	0	1	1	1	0	0E	N	1	1	0	0	1	1	1	0	4E
SI	1	0	0	0	1	1	1	1	0F	O	0	1	0	0	1	1	1	1	4F
DLE	0	0	0	1	0	0	0	0	10	P	1	1	0	1	0	0	0	0	50
DC1	0	0	0	1	0	0	0	1	11	Q	0	1	0	1	0	0	0	1	51
DC2	1	0	0	1	0	0	1	0	12	R	0	1	0	1	0	0	1	0	52
DC3	0	0	0	1	0	0	1	1	13	S	1	1	0	1	0	0	1	1	53
DC4	1	0	0	1	0	1	0	0	14	T	0	1	0	1	0	1	0	0	54
NAK	0	0	0	1	0	1	0	1	15	U	1	1	0	1	0	1	0	1	55
SYN	0	0	0	1	0	1	1	0	16	V	1	1	0	1	0	1	1	0	56
ETB	1	0	0	1	0	1	1	1	17	W	0	1	0	1	0	1	1	1	57
CAN	1	0	0	1	1	0	0	0	18	X	0	1	0	1	1	0	0	0	58
EM	0	0	0	1	1	0	0	1	19	Y	1	1	0	1	1	0	0	1	59
SUB	0	0	0	1	1	0	1	0	1A	Z	1	1	0	1	1	0	1	0	5A
ESC	1	0	0	1	1	0	1	1	1B	[	0	1	0	1	1	0	1	1	5B
FS	0	0	0	1	1	1	0	0	1C	\	1	1	0	1	1	1	0	0	5C
GS	1	0	0	1	1	1	0	1	1D	]	0	1	0	1	1	1	0	1	5D
RS	1	0	0	1	1	1	1	0	1E	∧	0	1	0	1	1	1	1	0	5E
US	0	0	0	1	1	1	1	1	1F	—	1	1	0	1	1	1	1	1	5F
SP	0	0	1	0	0	0	0	0	20	`	1	1	1	0	0	0	0	0	60
!	1	0	1	0	0	0	0	1	21	a	0	1	1	0	0	0	0	1	61
"	1	0	1	0	0	0	1	0	22	b	0	1	1	0	0	0	1	0	62
#	0	0	1	0	0	0	1	1	23	c	1	1	1	0	0	0	1	1	63
$	1	0	1	0	0	1	0	0	24	d	0	1	1	0	0	1	0	0	64
%	0	0	1	0	0	1	0	1	25	e	1	1	1	0	0	1	0	1	65
&	0	0	1	0	0	1	1	0	26	f	1	1	1	0	0	1	1	0	66
'	1	0	1	0	0	1	1	1	27	g	0	1	1	0	0	1	1	1	67
(	1	0	1	0	1	0	0	0	28	h	0	1	1	0	1	0	0	0	68
)	0	0	1	0	1	0	0	1	29	i	1	1	1	0	1	0	0	1	69
*	0	0	1	0	1	0	1	0	2A	j	1	1	1	0	1	0	1	0	6A
+	1	0	1	0	1	0	1	1	2B	k	0	1	1	0	1	0	1	1	6B
,	0	0	1	0	1	1	0	0	2C	l	1	1	1	0	1	1	0	0	6C
-	1	0	1	0	1	1	0	1	2D	m	0	1	1	0	1	1	0	1	6D

TABLE 2-2 (*continued*)

			Binary code						Hex				Binary code						Hex
Bit:	7	6	5	4	3	2	1	0			7	6	5	4	3	2	1	0	
										Bit:									
.	1	0	1	0	1	1	1	0	2E	n	0	1	1	0	1	1	1	0	6E
/	0	0	1	0	1	1	1	1	2F	o	1	1	1	0	1	1	1	1	6F
0	1	0	1	1	0	0	0	0	30	p	0	1	1	1	0	0	0	0	70
1	0	0	1	1	0	0	0	1	31	q	1	1	1	1	0	0	0	1	71
2	0	0	1	1	0	0	1	0	32	r	1	1	1	1	0	0	1	0	72
3	1	0	1	1	0	0	1	1	33	s	0	1	1	1	0	0	1	1	73
4	0	0	1	1	0	1	0	0	34	t	1	1	1	1	0	1	0	0	74
5	1	0	1	1	0	1	0	1	35	u	0	1	1	1	0	1	0	1	75
6	1	0	1	1	0	1	1	0	36	v	0	1	1	1	0	1	1	0	76
7	0	0	1	1	0	1	1	1	37	w	1	1	1	1	0	1	1	1	77
8	0	0	1	1	1	0	0	0	38	x	1	1	1	1	1	0	0	0	78
9	1	0	1	1	1	0	0	1	39	y	0	1	1	1	1	0	0	1	79
:	1	0	1	1	1	0	1	0	3A	z	0	1	1	1	1	0	1	0	7A
;	0	0	1	1	1	0	1	1	3B	{	1	1	1	1	1	0	1	1	7B
<	1	0	1	1	1	1	0	0	3C	¦	0	1	1	1	1	1	0	0	7C
=	0	0	1	1	1	1	0	1	3D	}	1	1	1	1	1	1	0	1	7D
>	0	0	1	1	1	1	1	0	3E	~	1	1	1	1	1	1	1	0	7E
?	1	0	1	1	1	1	1	1	3F	DEL	0	1	1	1	1	1	1	1	7F

NUL = null
SOH = start of heading
STX = start of text
ETX = end of text
EOT = end of transmission
ENQ = enquiry
ACK = acknowledge
BEL = bell
BS = back space
HT = horizontal tab
NL = new line
VT = vertical tab
FF = form feed
CR = carriage return
SO = shift-out
SI = shift-in
DLE = data link escape

DC1 = device control 1
DC2 = device control 2
DC3 = device control 3
DC4 = device control 4
NAK = negative acknowledge
SYN = synchronous
ETB = end of transmission block
CAN = cancel
SUB = substitute
ESC = escape
FS = field separator
GS = group separator
RS = record separator
US = unit separator
SP = space
DEL = delete

ASCII Code

In 1963, in an effort to standardize data communications codes, the United States adopted the Bell System model 33 teletype code as the United States of America Standard Code for Information Interchange (USASCII), better known simply as ASCII-63. Since its adoption, ASCII has generically progressed through the 1965, 1967, and 1977 versions, with the 1977 version being recommended by the CCITT as the International Alphabet No. 5. ASCII is a 7-bit character set which has 2^7 or 128 codes. With ASCII, the least significant bit (LSB) is designated b_0 and the most significant bit (MSB) is designated b_6. b_7 is not part of the ASCII code but is generally reserved for the parity bit, which is explained later in this chapter. Actually, with any character set, all bits are equally significant because the code does not represent a weighted binary number. It is common with character codes to refer to bits by their order; b_0 is the zero-order bit, b_1 is the first-order bit, b_7 is the seventh-order bit, and so on. With serial transmission, the bit transmitted first is called the LSB. With ASCII, the low-order bit (b_0) is the LSB and is transmitted first. ASCII is probably the code most often used today. The 1977 version of the ASCII code is shown in Table 2-2.

EBCDIC Code

EBCDIC is an 8-bit character code developed by IBM and used extensively in IBM and IBM-compatible equipment. With 8 bits, 2^8 or 256 codes are possible, making EBCDIC the most powerful character set. Note that with EBCDIC the LSB is designated b_7 and the MSB is designated b_0. Therefore, with EBCDIC, the high-order bit (b_7) is transmitted first and the low-order bit (b_0) is transmitted last. The EBCDIC code is shown in Table 2-3.

TABLE 2-3   EBCDIC CODE

Bit :	0	1	2	3	4	5	6	7	Hex	Bit :	0	1	2	3	4	5	6	7	Hex
NUL	0	0	0	0	0	0	0	0	00		1	0	0	0	0	0	0	0	80
SOH	0	0	0	0	0	0	0	1	01	a	1	0	0	0	0	0	0	1	81
STX	0	0	0	0	0	0	1	0	02	b	1	0	0	0	0	0	1	0	82
ETX	0	0	0	0	0	0	1	1	03	c	1	0	0	0	0	0	1	1	83
	0	0	0	0	0	1	0	0	04	d	1	0	0	0	0	1	0	0	84
PT	0	0	0	0	0	1	0	1	05	e	1	0	0	0	0	1	0	1	85
	0	0	0	0	0	1	1	0	06	f	1	0	0	0	0	1	1	0	86
	0	0	0	0	0	1	1	1	07	g	1	0	0	0	0	1	1	1	87
	0	0	0	0	1	0	0	0	08	h	1	0	0	0	1	0	0	0	88
	0	0	0	0	1	0	0	1	09	i	1	0	0	0	1	0	0	1	89
	0	0	0	0	1	0	1	0	0A		1	0	0	0	1	0	1	0	8A
	0	0	0	0	1	0	1	1	0B		1	0	0	0	1	0	1	1	8B
FF	0	0	0	0	1	1	0	0	0C		1	0	0	0	1	1	0	0	8C
	0	0	0	0	1	1	0	1	0D		1	0	0	0	1	1	0	1	8D
	0	0	0	0	1	1	1	0	0E		1	0	0	0	1	1	1	0	8E
	0	0	0	0	1	1	1	1	0F		1	0	0	0	1	1	1	1	8F
DLE	0	0	0	1	0	0	0	0	10		1	0	0	1	0	0	0	0	90
SBA	0	0	0	1	0	0	0	1	11	j	1	0	0	1	0	0	0	1	91
EUA	0	0	0	1	0	0	1	0	12	k	1	0	0	1	0	0	1	0	92
IC	0	0	0	1	0	0	1	1	13	l	1	0	0	1	0	0	1	1	93
	0	0	0	1	0	1	0	0	14	m	1	0	0	1	0	1	0	0	94
NL	0	0	0	1	0	1	0	1	15	n	1	0	0	1	0	1	0	1	95

TABLE 2-3 (*continued*)

	Bit:0	1	2	3	4	5	6	7	Hex	Bit:	0	1	2	3	4	5	6	7	Hex
	0	0	0	1	0	1	1	0	16	o	1	0	0	1	0	1	1	0	96
	0	0	0	1	0	1	1	1	17	p	1	0	0	1	0	1	1	1	97
	0	0	0	1	1	0	0	0	18	q	1	0	0	1	1	0	0	0	98
EM	0	0	0	1	1	0	0	1	19	r	1	0	0	1	1	0	0	1	99
	0	0	0	1	1	0	1	0	1A		1	0	0	1	1	0	1	0	9A
	0	0	0	1	1	0	1	1	1B		1	0	0	1	1	0	1	1	9B
DUP	0	0	0	1	1	1	0	0	1C		1	0	0	1	1	1	0	0	9C
SF	0	0	0	1	1	1	0	1	1D		1	0	0	1	1	1	0	1	9D
FM	0	0	0	1	1	1	1	0	1E		1	0	0	1	1	1	1	0	9E
ITB	0	0	0	1	1	1	1	1	1F		1	0	0	1	1	1	1	1	9F
	0	0	1	0	0	0	0	0	20		1	0	1	0	0	0	0	0	A0
	0	0	1	0	0	0	0	1	21	~	1	0	1	0	0	0	0	1	A1
	0	0	1	0	0	0	1	0	22	s	1	0	1	0	0	0	1	0	A2
	0	0	1	0	0	0	1	1	23	t	1	0	1	0	0	0	1	1	A3
	0	0	1	0	0	1	0	0	24	u	1	0	1	0	0	1	0	0	A4
	0	0	1	0	0	1	0	1	25	v	1	0	1	0	0	1	0	1	A5
ETB	0	0	1	0	0	1	1	0	26	w	1	0	1	0	0	1	1	0	A6
ESC	0	0	1	0	0	1	1	1	27	x	1	0	1	0	0	1	1	1	A7
	0	0	1	0	1	0	0	0	28	y	1	0	1	0	1	0	0	0	A8
	0	0	1	0	1	0	0	1	29	z	1	0	1	0	1	0	0	1	A9
	0	0	1	0	1	0	1	0	2A		1	0	1	0	1	0	1	0	AA
	0	0	1	0	1	0	1	1	2B		1	0	1	0	1	0	1	1	AB
	0	0	1	0	1	1	0	0	2C		1	0	1	0	1	1	0	0	AC
ENQ	0	0	1	0	1	1	0	1	2D		1	0	1	0	1	1	0	1	AD
	0	0	1	0	1	1	1	0	2E		1	0	1	0	1	1	1	0	AE
	0	0	1	0	1	1	1	1	2F		1	0	1	0	1	1	1	1	AF
	0	0	1	1	0	0	0	0	30		1	0	1	1	0	0	0	0	B0
	0	0	1	1	0	0	0	1	31		1	0	1	1	0	0	0	1	B1
SYN	0	0	1	1	0	0	1	0	32		1	0	1	1	0	0	1	0	B2
	0	0	1	1	0	0	1	1	33		1	0	1	1	0	0	1	1	B3
	0	0	1	1	0	1	0	0	34		1	0	1	1	0	1	0	0	B4
	0	0	1	1	0	1	0	1	35		1	0	1	1	0	1	0	1	B5
	0	0	1	1	0	1	1	0	36		1	0	1	1	0	1	1	0	B6
EOT	0	0	1	1	0	1	1	1	37		1	0	1	1	0	1	1	1	B7
	0	0	1	1	1	0	0	0	38		1	0	1	1	1	0	0	0	B8
	0	0	1	1	1	0	0	1	39		1	0	1	1	1	0	0	1	B9
	0	0	1	1	1	0	1	0	3A		1	0	1	1	1	0	1	0	BA
	0	0	1	1	1	0	1	1	3B		1	0	1	1	1	0	1	1	BB
RA	0	0	1	1	1	1	0	0	3C		1	0	1	1	1	1	0	0	BC
NAK	0	0	1	1	1	1	0	1	3D		1	0	1	1	1	1	0	1	BD
	0	0	1	1	1	1	1	0	3E		1	0	1	1	1	1	1	0	BE
SUB	0	0	1	1	1	1	1	1	3F		1	0	1	1	1	1	1	1	BF
SP	0	1	0	0	0	0	0	0	40	{	1	1	0	0	0	0	0	0	C0
	0	1	0	0	0	0	0	1	41	A	1	1	0	0	0	0	0	1	C1
	0	1	0	0	0	0	1	0	42	B	1	1	0	0	0	0	1	0	C2
	0	1	0	0	0	0	1	1	43	C	1	1	0	0	0	0	1	1	C3
	0	1	0	0	0	1	0	0	44	D	1	1	0	0	0	1	0	0	C4

TABLE 2-3 (*continued*)

Bit:	0	1	2	3	4	5	6	7	Hex	Bit:	0	1	2	3	4	5	6	7	Hex
	0	1	0	0	0	1	0	1	45	E	1	1	0	0	0	1	0	1	C5
	0	1	0	0	0	1	1	0	46	F	1	1	0	0	0	1	1	0	C6
	0	1	0	0	0	1	1	1	47	G	1	1	0	0	0	1	1	1	C7
	0	1	0	0	1	0	0	0	48	H	1	1	0	0	1	0	0	0	C8
	0	1	0	0	1	0	0	1	49	I	1	1	0	0	1	0	0	1	C9
¢	0	1	0	0	1	0	1	0	4A		1	1	0	0	1	0	1	0	CA
.	0	1	0	0	1	0	1	1	4B		1	1	0	0	1	0	1	1	CB
<	0	1	0	0	1	1	0	0	4C		1	1	0	0	1	1	0	0	CC
(	0	1	0	0	1	1	0	1	4D		1	1	0	0	1	1	0	1	CD
+	0	1	0	0	1	1	1	0	4E		1	1	0	0	1	1	1	0	CE
¦	0	1	0	0	1	1	1	1	4F		1	1	0	0	1	1	1	1	CF
&	0	1	0	1	0	0	0	0	50	}	1	1	0	1	0	0	0	0	D0
	0	1	0	1	0	0	0	1	51	J	1	1	0	1	0	0	0	1	D1
	0	1	0	1	0	0	1	0	52	K	1	1	0	1	0	0	1	0	D2
	0	1	0	1	0	0	1	1	53	L	1	1	0	1	0	0	1	1	D3
	0	1	0	1	0	1	0	0	54	M	1	1	0	1	0	1	0	0	D4
	0	1	0	1	0	1	0	1	55	N	1	1	0	1	0	1	0	1	D5
	0	1	0	1	0	1	1	0	56	O	1	1	0	1	0	1	1	0	D6
	0	1	0	1	0	1	1	1	57	P	1	1	0	1	0	1	1	1	D7
	0	1	0	1	1	0	0	0	58	Q	1	1	0	1	1	0	0	0	D8
	0	1	0	1	1	0	0	1	59	R	1	1	0	1	1	0	0	1	D9
!	0	1	0	1	1	0	1	0	5A		1	1	0	1	1	0	1	0	DA
$	0	1	0	1	1	0	1	1	5B		1	1	0	1	1	0	1	1	DB
*	0	1	0	1	1	1	0	0	5C		1	1	0	1	1	1	0	0	DC
)	0	1	0	1	1	1	0	1	5D		1	1	0	1	1	1	0	1	DD
;	0	1	0	1	1	1	1	0	5E		1	1	0	1	1	1	1	0	DE
¬	0	1	0	1	1	1	1	1	5F		1	1	0	1	1	1	1	1	DF
-	0	1	1	0	0	0	0	0	60	\	1	1	1	0	0	0	0	0	E0
/	0	1	1	0	0	0	0	1	61		1	1	1	0	0	0	0	1	E1
	0	1	1	0	0	0	1	0	62	S	1	1	1	0	0	0	1	0	E2
	0	1	1	0	0	0	1	1	63	T	1	1	1	0	0	0	1	1	E3
	0	1	1	0	0	1	0	0	64	U	1	1	1	0	0	1	0	0	E4
	0	1	1	0	0	1	0	1	65	V	1	1	1	0	0	1	0	1	E5
	0	1	1	0	0	1	1	0	66	W	1	1	1	0	0	1	1	0	E6
	0	1	1	0	0	1	1	1	67	X	1	1	1	0	0	1	1	1	E7
	0	1	1	0	1	0	0	0	68	Y	1	1	1	0	1	0	0	0	E8
	0	1	1	0	1	0	0	1	69	Z	1	1	1	0	1	0	0	1	E9
	0	1	1	0	1	0	1	0	6A		1	1	1	0	1	0	1	0	EA
,	0	1	1	0	1	0	1	1	6B		1	1	1	0	1	0	1	1	EB
%	0	1	1	0	1	1	0	0	6C		1	1	1	0	1	1	0	0	EC
	0	1	1	0	1	1	0	1	6D		1	1	1	0	1	1	0	1	ED
>	0	1	1	0	1	1	1	0	6E		1	1	1	0	1	1	1	0	EE
?	0	1	1	0	1	1	1	1	6F		1	1	1	0	1	1	1	1	EF
	0	1	1	1	0	0	0	0	70	0	1	1	1	1	0	0	0	0	F0
	0	1	1	1	0	0	0	1	71	1	1	1	1	1	0	0	0	1	F1
	0	1	1	1	0	0	1	0	72	2	1	1	1	1	0	0	1	0	F2
	0	1	1	1	0	0	1	1	73	3	1	1	1	1	0	0	1	1	F3

TABLE 2-3 (*continued*)

	Binary code								Hex	Bit:	Binary code								Hex
Bit:	0	1	2	3	4	5	6	7			0	1	2	3	4	5	6	7	
	0	1	1	1	0	1	0	0	74	4	1	1	1	1	0	1	0	0	F4
	0	1	1	1	0	1	0	1	75	5	1	1	1	1	0	1	0	1	F5
	0	1	1	1	0	1	1	0	76	6	1	1	1	1	0	1	1	0	F6
	0	1	1	1	0	1	1	1	77	7	1	1	1	1	0	1	1	1	F7
	0	1	1	1	1	0	0	0	78	8	1	1	1	1	1	0	0	0	F8
▲	0	1	1	1	1	0	0	1	79	9	1	1	1	1	1	0	0	1	F9
:	0	1	1	1	1	0	1	0	7A		1	1	1	1	1	0	1	0	FA
#	0	1	1	1	1	0	1	1	7B		1	1	1	1	1	0	1	1	FB
@	0	1	1	1	1	1	0	0	7C		1	1	1	1	1	1	0	0	FC
▲	0	1	1	1	1	1	0	1	7D		1	1	1	1	1	1	0	1	FD
=	0	1	1	1	1	1	1	0	7E		1	1	1	1	1	1	1	0	FE
"	0	1	1	1	1	1	1	1	7F		1	1	1	1	1	1	1	1	FF

DLE = data link escape

DUP = duplicate
EM = end of medium
ENQ = enquiry
EOT = end of transmission
ESC = escape
ETB = end of transmission block
ETX = end of text
EUA = erase unprotected to address
FF = form feed
FM = field mark
IC = insert cursor

ITB = end of intermediate transmission block
NUL = null
PT = program tab
RA = repeat to address
SBA = set buffer address
SF = start field
SOH = start of heading
SP = space
STX = start of text
SUB = substitute
SYN = synchronous
NAK = negative acknowledge

ERROR CONTROL

A data communications circuit can be as short as a few feet or as long as several thousand miles, and the transmission medium can be as simple as a piece of wire or as complex as a microwave, satellite, or fiber optic system. Therefore, due to the nonideal transmission characteristics that are associated with any communications system, it is inevitable that errors will occur and that it is necessary to develop and implement procedures for error control. Error control can be divided into two general categories: error detection and error correction.

Error Detection

Error detection is simply the process of monitoring the received data and determining when a transmission error has occurred. Error detection techniques do not identify which bit (or bits) is in error, only that an error has occurred. The purpose of error detection is not to prevent errors from occurring but to prevent undetected errors from occurring. How a system reacts to transmission errors is system dependent and varies considerably. The most common error detection techniques used for data communications circuits are: redundancy, exact-count encoding, parity, vertical and longitudinal redundancy checking, and cyclic redundancy checking.

Redundancy.   *Redundancy* involves transmitting each character twice. If the same character is not received twice in succession, a transmission error has occurred. The same concept can be used for messages. If the same sequence of characters is not received twice in succession, in exactly the same order, a transmission error has occurred.

Exact-count encoding.   With *exact-count encoding*, the number of 1's in each character is the same. An example of an exact-count encoding scheme is the ARQ code shown in Table 2-4. With the ARQ code, each character has three 1's in it, and therefore a simple count of the number of 1's received can determine if a transmission error has occurred.

Parity.   *Parity* is probably the simplest error detection scheme used for data communications systems and is used with both vertical and horizontal redundancy checking. With parity, a single bit (called a *parity bit*) is added to each character to force the total number of 1's in the character, including the parity bit, to be either an odd number (odd parity) or an even number (even parity). For example, the ASCII code for the letter "C" is 43 hex or P1000011 binary, with the P bit representing the parity bit. There are three 1's in the code, not counting the parity bit. If odd parity is used, the P bit is made a 0, keeping the total number of 1's at three, an odd number. If even parity is used, the P bit is made a 1 and the total number of 1's is four, an even number.

Taking a closer look at parity, it can be seen that the parity bit is independent of the number of 0's in the code and unaffected by pairs of 1's. For the letter "C," if all the 0 bits are dropped, the code is P1————11. For odd parity, the P bit is still a 0 and for even parity, the P bit is still a 1. If pairs of 1's are also excluded, the code is either P1————, P————1, or P————1—. Again, for odd parity the P bit is a 0, and for even parity the P bit is a 1.

The definition of parity is *equivalence* of *equality*. A logic gate that will determine when all its inputs are equal is the XOR gate. With an XOR gate, if all the inputs are equal (either all 0's or all 1's), the output is a 0. If all inputs are not equal, the output is a 1. Figure 2-3 shows two circuits that are commonly used to generate a parity bit. Essentially, both circuits go through a comparison process eliminating 0's and pairs of 1's. The circuit shown in Figure 2-3a uses *sequential (serial)* comparison, while the circuit shown in Figure 2-3b uses *combinational (parallel)* comparison. With the sequential parity generator b_0 is XORed with b_1, the result is XORed with b_2, and so on. The result of the last XOR operation is compared with a *bias* bit. If even parity is desired, the bias bit is made a logic 0. If odd parity is desired, the bias bit is made a logic 1. The output of the circuit is the parity bit, which is

TABLE 2-4   ARQ EXACT-COUNT CODE

		Binary code					Character	
Bit:	1 2 3 4 5 6 7						Letter	Figure
	0 0 0 1 1 1 0						Letter shift	
	0 1 0 0 1 1 0						Figure shift	
	0 0 1 1 0 1 0						A	—
	0 0 1 1 0 0 1						B	?
	1 0 0 1 1 0 0						C	:
	0 0 1 1 1 0 0						D	(WRU)
	0 1 1 1 0 0 0						E	3
	0 0 1 0 0 1 1						F	%
	1 1 0 0 0 0 1						G	@
	1 0 1 0 0 1 0						H	£
	1 1 1 0 0 0 0						I	8
	0 1 0 0 0 1 1						J	(bell)
	0 0 0 1 0 1 1						K	(
	1 1 0 0 0 1 0						L	)
	1 0 1 0 0 0 1						M	.
	1 0 1 0 1 0 0						N	,
	1 0 0 0 1 1 0						O	9
	1 0 0 1 0 1 0						P	0
	0 0 0 1 1 0 1						Q	1
	1 1 0 0 1 0 0						R	4
	0 1 0 1 0 1 0						S	'
	1 0 0 0 1 0 1						T	5
	0 1 1 0 0 1 0						U	7
	1 0 0 1 0 0 1						V	=
	0 1 0 0 1 0 1						W	2
	0 0 1 0 1 1 0						X	/
	0 0 1 0 1 0 1						Y	6
	0 1 1 0 0 0 1						Z	+
	0 0 0 0 1 1 1							(blank)
	1 1 0 1 0 0 0							(space)
	1 0 1 1 0 0 0							(line feed)
	1 0 0 0 0 1 1							(carriage return)

appended to the character code. With the parallel parity generator, comparisons are made in layers or levels. Pairs of bits (b_0 and b_1, b_2 and b_3, etc.) are XORed. The results of the first-level XOR gates are then XORed together. The process continues until only one bit is left, which is XORed with the bias bit. Again, if even parity is desired, the bias bit is made a logic 0 and if odd parity is desired, the bias bit is made a logic 1.

The circuits shown in Figure 2-3 can also be used for the parity checker in the receiver. A parity checker uses the same procedure as a parity generator except that the logic condition of the final comparison is used to determine if a parity violation has occurred (for odd parity a 1 indicates an error and a 0 indicates no error; for even parity, a 1 indicates an error and a 0 indicates no error).

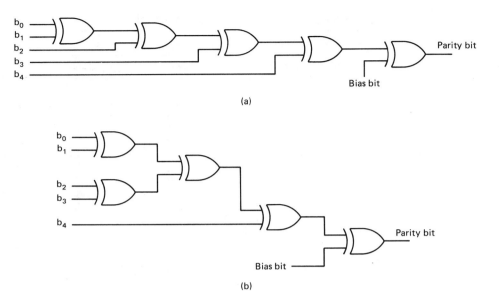

Figure 2-3 Parity generators: (a) serial; (b) parallel. 1, odd parity; 2, even parity.

The primary advantage of parity is its simplicity. The disadvantage is that when an even number of bits are received in error, the parity checker will not detect it (i.e., if the logic conditions of 2 bits are changed, the parity remains the same). Consequently, parity, over a long period of time, will detect only 50% of the transmission errors (this assumes an equal probability that an even or an odd number of bits could be in error).

Vertical and horizontal redundancy checking. *Vertical redundancy checking* (VRC) is an error detection scheme that uses parity to determine if a transmission error has occurred within a character. Therefore, VRC is sometimes called *character parity*. With VRC, each character has a parity bit added to it prior to transmission. It may use even or odd parity. The example shown under the topic "parity" involving the ASCII character "C" is an example of how VRC is used.

Horizontal or longitudinal redundancy checking (HRC or LRC) is an error detection scheme that uses parity to determine if a transmission error has occurred in a message and is therefore sometimes called *message parity*. With LRC, each bit position has a parity bit. In other words, b_0 from each character in the message is XORed with b_0 from all of the other characters in the message. Similarly, b_1, b_2, and so on, are XORed with their respective bits from all the other characters in the message. Essentially, LRC is the result of XORing the "characters" that make up a message, whereas VRC is the XORing of the bits within a single character. With LRC, only even parity is used.

The LRC bit sequence is computed in the transmitter prior to sending the data, then transmitted as though it were the last character of the message. At the receiver, the LRC is recomputed from the data and the recomputed LRC is compared with the LRC transmitted with the message. If they are the same, it is assumed that no transmission errors have occurred. If they are different, a transmission error must have occurred.

Example 2-1 shows how VRC and LRC are determined.

EXAMPLE 2-1

Determine the VRC and LRC for the following ASCII-encoded message: THE CAT. Use odd parity for VRC and even parity for LRC.

Character			T	H	E	sp	C	A	T	LRC
Hex			*54*	*48*	*45*	*20*	*43*	*41*	*54*	*2F*
LSB	b_0		0	0	1	0	1	1	0	1
	b_1		0	0	0	0	1	0	0	1
	b_2		1	0	1	0	0	0	1	1
ASCII code	b_3		0	1	0	0	0	0	0	1
	b_4		1	0	0	0	0	0	1	0
	b_5		0	0	0	1	0	0	0	1
MSB	b_6		1	1	1	0	1	1	1	0
VRC	b_7		0	1	0	0	0	1	0	0

The LRC is 2FH or 00101111 binary. In EBCDIC, this is the character BEL.

The VRC bit for each character is computed in the vertical direction, and the LRC bits are computed in the horizontal direction. This is the same scheme that was used with the early teletype paper tapes and keypunch cards and has subsequently been carried over to present-day data communications applications.

The group of characters that make up the message (i.e., THE CAT) is often called a *block* of data. Therefore, the bit sequence for the LRC is often called a *block check character* (BCC) or a *block check sequence* (BCS). BCS is more appropriate because the LRC has no function as a character (i.e., it is not an alpha/numeric, graphic, or data link control character); the LRC is simply a *sequence of bits* used for error detection.

Historically, LRC detects between 95 and 98% of all transmission errors. LRC will not detect transmission errors when an even number of characters have an error in the same bit position. For example, if b_4 in two different characters is in error, the LRC is still valid even though multiple transmission errors have occurred.

If VRC and LRC are used simultaneously, the only time an error would go undetected is when an even number of bits in an even number of characters were in error and the same bit positions in each character are in error, which is highly unlikely to happen. VRC does not identify which bit is in error in a character, and

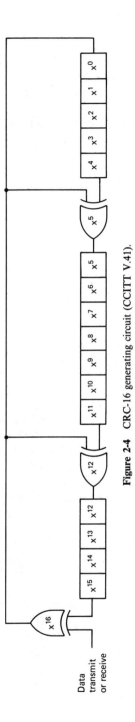

Figure 2-4 CRC-16 generating circuit (CCITT V.41).

LRC does not identify which character has an error in it. However, for single bit errors, VRC used together with LRC will identify which bit is in error. Otherwise, VRC and LRC only identify that an error has occurred.

Cyclic redundancy checking. Probably the most reliable scheme for error detection is *cyclic redundancy checking* (CRC). With CRC, approximately 99.95% of all transmission errors are detected. CRC is generally used with 8-bit codes such as EBCDIC or 7-bit codes when parity is not used.

In the United States, the most common CRC code is CRC-16, which is identical to the international standard, CCITT's V.41. With CRC-16, 16 bits are used for the BCS. Essentially, the CRC character is the remainder of a division process. A data message polynomial $G(x)$ is divided by a generator polynomial function $P(x)$, the quotient is discarded, and the remainder is truncated to 16 bits and added to the message as the BCS. With CRC generation, the division is not accomplished with a standard arithmetic division process. Instead of using straight subtraction, the remainder is derived from an XOR operation. At the receiver, the data stream and the BCS are divided by the same generating function $P(x)$. If no transmission errors have occurred, the remainder will be zero.

The generating polynomial for CRC-16 is

$$P(x) = x^{16} + x^{12} + x^5 + x^0$$

where

$$x^0 = 1.$$

The number of bits in the CRC code is equal to the highest exponent of the generating polynomial. The exponents identify the bit positions that contain a 1. Therefore, b_{16}, b_{12}, b_5, and b_0 are 1's and all of the other bit positions are 0's.

Figure 2-4 shows the block diagram for a circuit that will generate a CRC-16 BSC for the CCITT V.41 standard. Note that for each bit position of the generating polynomial that is a 1 there is an XOR gate.

EXAMPLE 2-2

Determine the BSC for the following data and CRC generating polynomials:

$$\text{data } G(x) = x^7 + x^5 + x^4 + x^2 + x^1 + x^0 \quad \text{or} \quad 10110111$$

$$\text{CRC } P(x) = x^5 + x^4 + x^1 + x^0 \quad \text{or} \quad 110011$$

Solution First $G(x)$ is multiplied by the number of bits in the CRC generating polynomial, 5.

$$x^5 (x^7 + x^5 + x^4 + x^2 + x^1 + x^0) = x^{12} + x^{10} + x^9 + x^7 + x^6 + x^5$$

$$= 1011011100000$$

```
                                 11010111
                       110011⌐1011011100000
                              110011
                                111101
                                110011
                                  111010
                                  110011
                                    100100
                                    110011
                                      101110
                                      110011
                                        111010
                                        110011
                                          01001 = CRC
```

The CRC is appended to the data to give the following transmitted data stream:

$$\underline{G(\underline{x})} \qquad CRC$$
$$10110111 \quad 01001$$

At the receiver, the transmitted data are again divided by $P(x)$.

```
                             11010111
                   110011⌐1011011101001
                          110011
                            111101
                            110011
                              111010
                              110011
                                100110
                                110011
                                  101010
                                  110011
                                    110011
                                    110011
                                    000000  remainder = 0
                                            no error occurred
```

Error Correction

Essentially, there are three methods of error correction: symbol substitution, retransmission, and forward error correction.

Symbol substitution. *Symbol substitution* was designed to be used in a human environment: when there is a human being at the receive terminal to analyze the received data and make decisions on its integrity. With symbol substitution, if a character is received in error, rather than revert to a higher level of error correction or display the incorrect character, a unique character that is undefined by the character code, such as a reverse question mark (ʕ), is substituted for the bad character. If the character in error cannot be discerned by the operator, retransmission is called for (i.e., symbol substitution is a form of selective retransmission). For example, if the message "Name" had an error in the first character, it would be displayed as "ʕame." An operator can discern the correct message by inspection and retransmission is unnecessary. However, if the message "$ʕ,000.00" were received, an operator could not determine the correct character, and retransmission is required.

Retransmission. *Retransmission*, as the name implies, is when a message is received in error and the receive terminal automatically calls for retransmission of the entire message. Retransmission is often called ARQ, which is an old radio communications term that means *automatic request for retransmission*. ARQ is probably the most reliable method of error correction, although it is not always the most efficient. Impairments on transmission media occur in bursts. If short messages are used, the likelihood that an impairment will occur during a transmission is small. However, short messages require more acknowledgments and line turnarounds than do long messages. Acknowledgments and line turnarounds for error control are forms of *overhead* (characters other than data that must be transmitted). With long messages, less turnaround time is needed, although the likelihood that a transmission error will occur is higher than for short messages. It can be shown statistically that message blocks between 256 and 512 characters are of optimum size when using ARQ for error correction.

Forward error correction. *Forward error correction* (FEC) is the only error correction scheme that actually detects and corrects transmission errors at the receive end without calling for retransmission.

With FEC, bits are added to the message prior to transmission. A popular error-correcting code is the *Hamming code*, developed by R. W. Hamming at Bell Laboratories. The number of bits in the Hamming code is dependent on the number of bits in the data character. The number of Hamming bits that must be added to a character is determined from the following expression:

$$2^n \geq m + n + 1 \tag{2-1}$$

where

$n =$ number of Hamming bits
$m =$ number of bits in the data character

EXAMPLE 2-3

For a 12-bit data string of 101100010010, determine the number of Hamming bits required, arbitrarily place the Hamming bits into the data string, determine the condition of each Hamming bit, assume an arbitrary single-bit transmission error, and prove that the Hamming code will detect the error.

Solution   Substituting into Equation 2-1, the number of Hamming bits is

$$2^n \geq m + n + 1$$

for $n = 4$:

$$2^4 = 16 \geq m + n + 1 = 12 + 4 + 1 = 17$$

$16 < 17$; therefore, 4 Hamming bits are insufficient.
For $n = 5$:

$$2^5 = 32 \geq m + n + 1 = 12 + 5 + 1 = 18$$

$32 > 18$; therefore, 5 Hamming bits are sufficient to meet the criterion of Equation 2-1. Therefore, a total of $12 + 5 = 17$ bits make up the data stream.

Arbitrarily place 5 Hamming bits into the data stream:

```
17 16 15 14 13 12 11 10 9 8 7 6 5 4 3 2 1
H  1  0  1  H  1  0  0  H H 0 1 0 H 0 1 0
```

To determine the logic condition of the Hamming bits, express all bit positions that contain a 1 as a 5-bit binary number and XOR them together.

Bit position	Binary number
2	00010
6	00110
XOR	00100
12	01100
XOR	01000
14	01110
XOR	00110
16	10000
XOR	10110 = Hamming code

$$b_{17} = 1, \quad b_{13} = 0, \quad b_9 = 1, \quad b_8 = 1, \quad b_4 = 0$$

The 17-bit encoded data stream becomes

```
H       H      H H        H
1 1 0 1 0 1 0 0 1 1 0 1 0 0 0 1 0
```

Assume that during transmission, an error occurs in bit position 14. The received data stream is

```
1 1 0 0 0 1 0 0 1 1 0 1 0 0 0 1 0
```

At the receiver to determine the bit position in error, extract the Hamming bits and XOR them with the binary code for each data bit position that contains a 1.

Bit position	Binary number
Hamming code	10110
2	10110
XOR	10100
6	00110
XOR	10010
12	01100
XOR	11110
16	10000
XOR	01110 = binary 14

Bit position 14 was received in error. To fix the error, simply complement bit 14.

The Hamming code described here will detect only single-bit errors. It cannot be used to identify multiple-bit errors or errors in the Hamming bits themselves. The Hamming code, like all FEC codes, requires the addition of bits to the data, consequently lengthening the transmitted message. The purpose of FEC codes is to reduce or eliminate the wasted time of retransmissions. However, the addition of the FEC bits to each message wastes transmission time in itself. Obviously, a trade-off is made between ARQ and FEC and system requirements determine which method is best suited to a particular system.

SYNCHRONIZATION

Synchronize means to coincide or agree in time. In data communications, there are four types of synchronization that must be achieved: bit or clock synchronization, modem or carrier synchronization, character synchronization, and message synchronization. The clock and carrier recovery circuits discussed in Chapter 1 accomplish bit and carrier synchronization, and message synchronization is discussed in Chapter 3.

Character Synchronization

Clock synchronization ensures that the transmitter and receiver agree on a precise time slot for the occurrence of a bit. When a continuous string of data is received, it is necessary to identify which bits belong to which characters and which bit is the least significant data bit, the parity bit, and the stop bit. In essence, this is character synchronization: identifying the beginning and the end of a character code. In data communications circuits, there are two formats used to achieve character synchronization: asynchronous and synchronous.

Asynchronous data format. With *asynchronous data*, each character is framed between a *start* and a *stop* bit. Figure 2-5 shows the format used to frame a character for asynchronous data transmission. The first bit transmitted is the start bit and is always a logic 0. The character code bits are transmitted next beginning with the LSB and continuing through the MSB. The parity bit (if used) is transmitted directly after the MSB of the character. The last bit transmitted is the stop bit, which is always a logic 1. There can be either 1, 1.5, or 2 stop bits.

A logic 0 is used for the start bit because an idle condition (no data transmission) on a data communications circuit is identified by the transmission of continuous 1's (these are often called *idle line 1's*). Therefore, the start bit of the first character is identified by a high-to-low transition in the received data, and the bit that immediately follows the start bit is the LSB of the character code. All stop bits are logic 1's, which guarantees a high-to-low transition at the beginning of each character. After the start bit is detected, the data and parity bits are clocked into the receiver. If data are transmitted in real time (i.e., as an operator types data into their computer terminal), the number of idle line 1's between each character will vary. During this *dead time*, the receiver will simply wait for the occurrence of another start bit before clocking in the next character.

Stop bit (1, 1.5, 2)	Parity bit	Data bits (5–7)							Start bit
1	1	1/0	b_6 MSB	b_5	b_4	b_3	b_2	b_1	b_0	0 LSB

Figure 2-5 Asynchronous data format.

EXAMPLE 2-4

For the following string of asynchronous ASCII-encoded data, identify each character (assume even parity and 2 stop bits).

```
  LSB     MSB
   |       |
1111000100001011010000001011111111101010001110100001011111   t →
     D         A             T         A
```

Synchronous data format. With *synchronous data*, rather than frame each character independently with start and stop bits, a unique synchronizing character called a SYN character is transmitted at the beginning of each message. For example, with ASCII code, the SYN character is 16H. The receiver disregards incoming data until it receives the SYN character, then it clocks in the next 7 bits and interprets them as a character. The character that is used to signify the end of a transmission varies with the type of protocol used and what kind of transmission it is. Message-terminating characters are discussed in Chapter 3.

With asynchronous data, it is not necessary that the transmit and receive clocks be continuously synchronized. It is only necessary that they operate at approximately the same rate and be synchronized at the beginning of each character. This was the purpose of the start bit, to establish a time reference for character synchronization. With synchronous data, the transmit and receive clocks must be synchronized because character synchronization occurs only once at the beginning of the message.

EXAMPLE 2-5

For the following string of synchronous ASCII-encoded data, identify each character (assume odd parity).

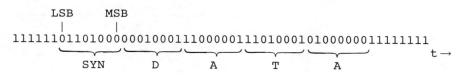

With asynchronous data, each character has 2 or 3 bits added to each character (1 start and 1 or 2 stop bits). These bits are additional overhead and thus reduce the efficiency of the transmission (i.e., the ratio of information bits to total transmitted bits). Synchronous data have two SYN characters (16 bits of overhead) added to each message. Therefore, asynchronous data are more efficient for short messages, and synchronous data are more efficient for long messages.

DATA COMMUNICATIONS HARDWARE

Figure 2-6 shows the block diagram of a multipoint data communications circuit that uses a bus topology. This arrangement is one of the most common configurations used for data communications circuits. At one station there is a mainframe computer and at each of the other two stations there is a *cluster* of computer terminals. The hardware and associated circuitry that connect the host computer to the remote computer terminals is called a *data communications link*. The station with the mainframe is called the *host* or *primary* and the other stations are called *secondaries* or simply *remotes*. An arrangement such as this is called a *centralized network*; there is one centrally located station (the host) with the responsiblity of ensuring an orderly flow of data between the remote stations and itself. Data flow is controlled by an applications program which is stored at the primary station.

At the primary station there is a mainframe computer, a *line control unit* (LCU), and a *data modem* (a data modem is commonly referred to simply as a *modem*). At each secondary station there is a modem, an LCU, and terminal equipment, such as computer terminals, printers, and so on. The mainframe is the host of the network and is where the applications program is stored for each circuit it serves. For simplicity, Figure 2-6 shows only one circuit served by the primary, although there can be many different circuits served by one mainframe computer. The primary station has the capability of storing, processing, or retransmitting the data it receives from the secondary stations. The primary also stores software for data base management.

The LCU at the primary station is more complicated than the LCUs at the

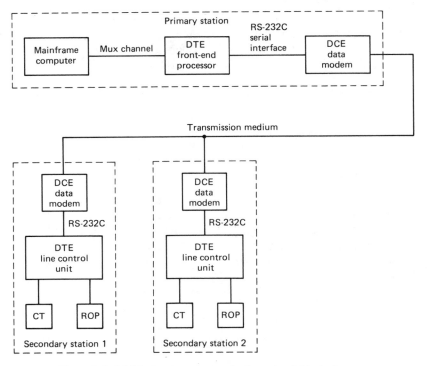

Figure 2-6 Multipoint data communications circuit block diagram.

secondary stations. The LCU at the primary station directs data traffic to and from many different circuits, which could all have different characteristics (i.e., different bit rates, character codes, data formats, etc.). The LCU at a secondary station directs data traffic between one data link and a few terminal devices which all operate at the same speed and use the same character code. Generally speaking, if the LCU has software associated with it, it is called a *front-end processor* (FEP). The LCU at the primary station is usually an FEP.

Line Control Unit

The LCU has several important functions. The LCU at the primary station serves as an interface between the host computer and the circuits that it serves. Each circuit served is connected to a different port on the LCU. The LCU directs the flow of input and output data between the different data communications links and their respective applications program. The LCU performs parallel-to-serial and serial-to-parallel conversion of data. The mux interface channel between the mainframe computer and the LCU transfers data in parallel. Data transfer between the modem and the LCU is done serially. The LCU also houses the circuitry that performs error detection and correction. Also, data link control (DLC) characters are inserted and deleted in the LCU. Data link control characters are explained in Chapter 3.

The LCU operates on the data when it is in digital form and is therefore called *data terminal equipment* (DTE). Essentially, any piece of equipment between the mainframe computer and the modem or the station equipment and its modem is classified as data terminal equipment. The modem is called *data communications equipment* (DCE) because it interfaces the digital DTE to the analog transmission line.

Within the LCU, there is a single integrated circuit that performs several of the LCU's functions. This circuit is called a UART when asynchronous transmission is used and a USRT when synchronous transmission is used.

Universal asynchronous receiver/transmitter (UART). The UART is used for asynchronous transmission of data between the DTE and the DCE. Asynchronous transmission means that an asynchronous data format is used and there is no clocking information transferred between the DTE and the DCE. The primary functions of the UART are:

1. To perform serial-to-parallel and parallel-to-serial conversion of data
2. To perform error detection by inserting and checking parity bits
3. To insert and detect start and stop bits

Functionally, the UART is divided into two sections: the transmitter and the receiver. Figure 2-7a shows a simplified block diagram of a UART transmitter.

Prior to transferring data in either direction, a *control* word must be programmed into the UART control register to indicate the nature of the data, such as the number of data bits; if parity is used, and if so, whether it is even or odd; and the number of stop bits. Essentially, the start bit is the only bit that is not optional; there is always only one start bit and it must be a logic 0. Figure 2-7b shows how to program the control word for the various functions. In the UART, the control word is used to set up the data-, parity-, and stop-bit steering logic circuit.

UART transmitter. The operation of the UART transmitter section is really quite simple. The UART sends a transmit buffer empty (TBMT) signal to the DTE to indicate that it is ready to receive data. When the DTE senses an active condition on TBMT, it sends a parallel data character to the transmit data lines (TD0–TD7) and strobes them into the transmit buffer register with the transmit data strobe signal (TDS). The contents of the transmit buffer register are transferred to the transit shift register when the transmit-end-of-character (TEOC) signal goes active (the TEOC signal simply tells the buffer register when the shift register is empty and available to receive data). The data pass through the steering logic circuit, where they pick up the appropriate start, stop, and parity bits. After data have been loaded into the transmit shift register, they are serially outputted on the transmit serial output (TSO) pin with a bit rate equal to the transmit clock (TCP) frequency. While the data in the transmit shift register are sequentially clocked out, the DTE loads the next charac-

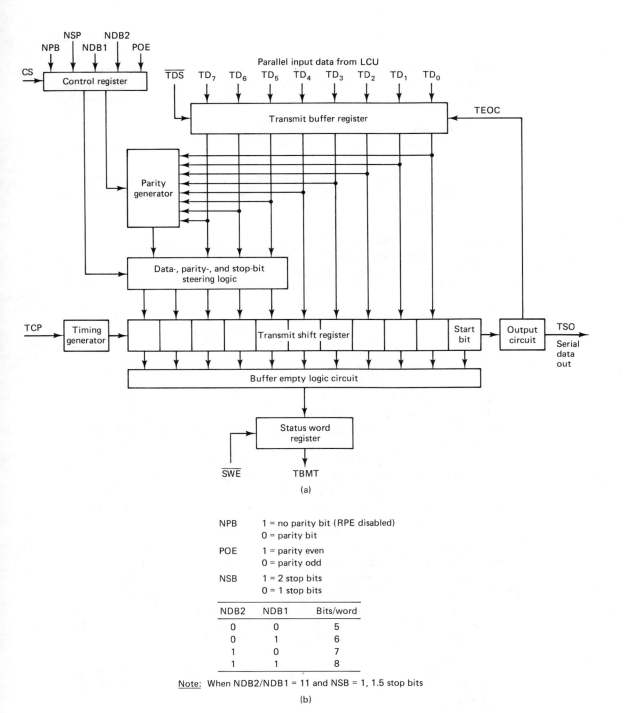

Figure 2-7 UART transmitter: (a) simplified block diagram; (b) control word.

ter into the buffer register. The process continues until the DTE has transferred all its data. The preceding sequence is shown in Figure 2-8.

UART receiver. A simplified block diagram of a UART receiver is shown in Figure 2-9. The number of stop bits, data bits, and the parity-bit information for the UART receiver are determined by the same control word that is used by the transmitter (i.e., the type of parity, the number of stop bits, and the number of data bits used for the UART receiver must be the same as that used for the UART transmitter).

The UART receiver ignores idle line 1's. When a valid start bit is detected by the start bit verification circuit, the data character is serially clocked into the receive shift register. If parity is used, the parity bit is checked in the parity check circuit. After one complete data character is loaded into the shift register, the character is transferred in parallel into the buffer register and the receive data available (RDA) flag is set in the status word register. To read the status register, the DTE activates status word enable ($\overline{\text{SWE}}$) and if it is active, reads the character from the buffer register by placing an active condition on the receive data enable (RDE) pin. After

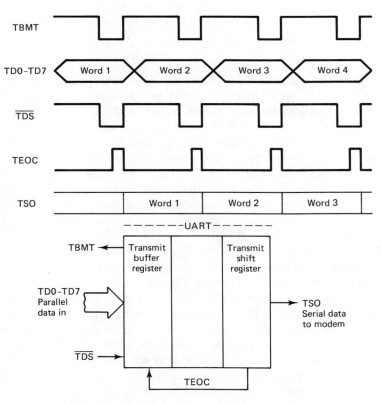

Figure 2-8 Timing diagram: UART transmitter.

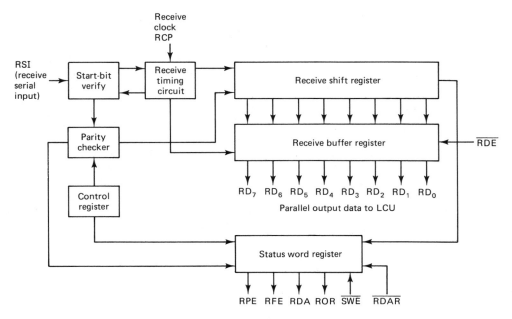

Figure 2-9 Simplified block diagram of a UART receiver.

reading the data, the DTE places an active signal on the receive data available reset ($\overline{\text{RDAR}}$) pin, which resets the RDA pin. Meanwhile, the next character is received and clocked into the receive shift register and the process repeats itself until all the data have been received. The preceding sequence is shown in Figure 2-10.

The status word register is also used for diagnostic information. The receive parity error (RPE) flag is set when a received character has a parity error in it. The receive framing error (RFE) flag is set when a character is received without any or an improper number of stop bits. The receive overrun (ROR) flag is set when a character in the buffer register is written over with another character (i.e., the DTE failed to service an active conditon on RDA before the next character was received by the shift register).

The receive clock for the UART (RCP) is 16 times higher than the receive data rate. This allows the start-bit verification circuit to determine if a high-to-low transition in the received data is actually a valid start bit and not simply a negative-going noise spike. Figure 2-11 shows how this is accomplished. The incoming idle line 1's (continuous high condition) are sampled at a rate 16 times the actual bit rate. This assures that a high-to-low transition is detected within $\frac{1}{16}$ of a bit time after it occurs. Once a low is detected, the verification circuit counts off seven clock pulses, then resamples the data. If it is still low, it is assumed that a valid start bit has been detected. If it has reverted to the high condition, it is assumed that the high-to-low transition was simply a noise pulse and is therefore ignored. Once a valid start bit has been detected and verified, the verification circuit samples the

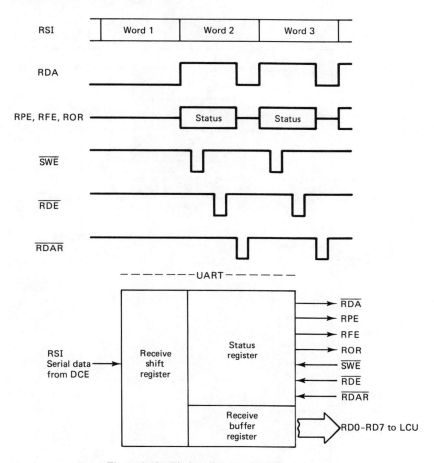

Figure 2-10 Timing diagram: UART receiver.

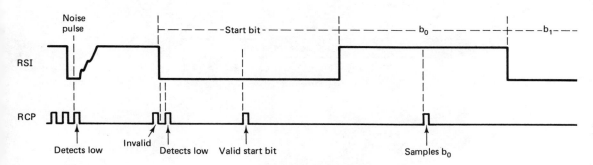

Figure 2-11 Start-bit verification.

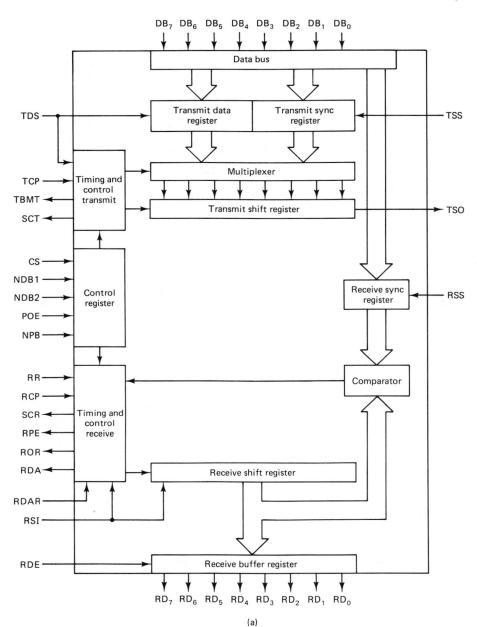

(a)

NPB	1 = no parity bit (RPE disabled)
	0 = parity bit
POE	1 = parity even
	0 = parity odd

NDB2	NDB1	Bits/word
0	0	5
0	1	6
1	0	7
1	1	8

(b)

Figure 2-12 USRT transceiver: (a) block diagram; (b) control word.

incoming data once every 16 clock cycles, which is equal to the data rate. Sampling at 16 times the bit rate also establishes the sample time to within $\frac{1}{16}$ of a bit time from the center of a bit.

Universal synchronous receiver/transmitter (USRT). The USRT is used for synchronous data transmission between the DTE and the DCE. Synchronous transmission means that there is clocking information transferred between the USRT and the modem and each transmission begins with a unique SYN character. The primary functions of the USRT are:

1. To perform serial-to-parallel and parallel-to-serial conversion of data
2. To perform error detection by inserting and checking parity bits
3. Insert and detect SYN characters

The block diagram of the USRT is shown in Figure 2-12a. The USRT operates very similarly to the UART, and therefore only the differences are explained. With the USRT, start and stop bits are not allowed. Instead, unique SYN characters are loaded into the transmit and receive SYN registers prior to transferring data. The programming information for the control word is shown in Figure 2-12b.

USRT transmitter. The transmit clock signal (TCP) is set at the desired bit rate and the desired SYN character is loaded from the parallel input pins (DB1–DB8) into the transmit SYN register by pulsing transmit SYN strobe (TSS). Data are loaded into the transmit data register from DB1–DB8 by pulsing the transmit data strobe (TDS). The next character transmitted is extracted from the transmit data register provided that the TDS pulse occurs during the presently transmitted character. If TDS is not pulsed, the next transmitted character is extracted from the transmit SYN register and the SYN character transmitted (SCT) signal is set. The transmit buffer empty (TBMT) signal is used to request the next character from the DTE. The serial output data appears on the transmit serial output (TSO) pin.

USRT receiver. The receive clock signal (RCP) is set at the desired bit rate and the desired SYN character is loaded into the receive SYN register from DB1–

DB8 by pulsing receive SYN strobe (RSS). On a high-to-low transition of the receiver rest input (RR), the receiver is placed in the search (bit phase) mode. In the search mode, serially received data are examined on a bit-by-bit basis until a SYN character is found. After each bit is clocked into the receive shift register, its contents are compared to the contents of the receive SYN register. If they are identical, a SYN character has been found and the SYN character receive (SCR) output is set. This character is transferred into the receive buffer register and the receiver is placed into the character mode. In the character mode, receive data are examined on a character-by-character basis and receiver flags for receive data available (RDA), receiver overrun (ROR), receive parity error (RPE), and SYN character received are provided to the status word register. Parallel receive data are outputted to the DTE on RB1–RB8.

SERIAL INTERFACES

To ensure an orderly flow of data between the line control unit and the modem, a *serial interface* is placed between them. This interface coordinates the flow of data, control signals, and timing information between the DTE and the DCE.

Before serial interfaces were standardized, every company that manufactured data communications equipment used a different interface configuration. More specifically, the cabling arrangement between the DTE and the DCE, the type and size of the connectors used, and the voltage levels varied considerably from vender to vender. To interconnect equipment manufactured by different companies, special level converters, cables, and connectors had to be built. The Electronic Industries Association (EIA), in an effort to standardize interface equipment between the data terminal equipment and data communications equipment, agreed on a set of standards which are called the RS-232C specifications. The RS-232C specifications identify the mechanical, electrical, and functional description for the interface between the DTE and the DCE. The RS-232C interface is similar to the combined CCITT standards V.28 (electrical specifications) and V.24 (functional description) and is designed for serial transmission of data up to 20,000 bps for a distance of approximately 50 ft. The EIA has recently adopted a new set of standards called the RS-449A, which when used in conjunction with the RS-422A or RS-423A standard, can operate at data rates up to 10 Mbps and span distances up to 1200 m.

RS-232C Interface

The RS-232C interface specifies a 25-wire cable with a DB25P/DB25S-compatible connector. Figure 2-13 shows the electrical characteristics of the RS-232C interface. The terminal load capacitance of the cable is specified as 2500 pF, which includes cable capacitance. The impedance at the terminating end must be between 3000 and 7000 Ω, and the output impedance is specified as greater than 300 Ω. With these

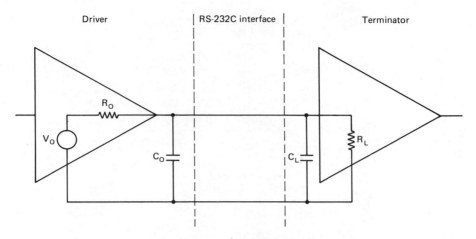

Figure 2-13 RS-232C electrical specifications.

electrical specifications and for a maximum bit rate of 20,000 bps, the nominal maximum length of the RS-232C interface is approximately 50 ft.

Although the RS-232C interface is simply a cable and two connectors, the standard also specifies limitations on the voltage levels that the DTE and DCE can output onto or receive from the cable. In both the DTE and DCE, there are circuits that convert the internal logic level to RS-232C values. For example, a DTE uses TTL logic and is interfaced to a DCE which uses ECL logic; they are not compatible. Voltage-leveling circuits convert the internal voltage values of the DTE and DCE to RS-232C values. If both the DCE and DCE output and input RS-232C levels, they are electrically compatible regardless of which logic family they use internally. A leveler is called a *driver* if it outputs a signal voltage to the cable and a *terminator* if it accepts a signal voltage from the cable. Table 2-5 lists the voltage limits for both drivers and terminators. Note that the data lines use negative logic and the control lines use positive logic.

From Table 2-5 it can be seen that the limits for a driver are more inclusive than those for a terminator. The driver can output any voltage between +5 and +15 or −5 and −15 V dc, and a terminator will accept any voltage between +3 and +25 and −3 and −25 V dc. The difference in the voltage levels between a driver and a terminator is called *noise margin*. The noise margin reduces the susceptibility of the interface to noise transients on the cable. Typical voltages used for data and control signals are ±7 V dc and ±10 V dc.

The pins on the RS-232C interface cable are functionally categorized as either ground, data, control (handshaking), or timing pins. All the pins are unidirectional (signals are propagated only from the DTE to the DCE, or vice versa). Table 2-6 lists the 25 pins of the RS-232C interface, their designations, and the direction of signal propagation (i.e., either toward the DTE or toward the DCE). The RS-232C specifications designate the ground, data, control, and timing pins as A, B,

**TABLE 2-5   RS232C VOLTAGE
SPECIFICATIONS (V DC)**

	Data pins	
	Logic 1	*Logic 0*
Driver	−5 to −15	+5 to +15
Terminator	−3 to −25	+3 to +25
	Control pins	
	Enable "on"	*Disable "off"*
Driver	+5 to +15	−5 to −15
Terminator	+3 to +25	−3 to −25

C, and D, respectively. These are nondescriptive designations. It is more practical and useful to use acronyms to designate the pins that reflect the pin functions. Table 2-6 lists the CCITT and EIA designations and the nomenclature more commonly used by industry in the United States.

EIA RS-232C pin functions.   Twenty of the 25 pins of the RS-232C interface are designated for specific purposes or functions. Pins 9, 10, 11, 18, and 25 are unassigned; pins 1 and 7 are grounds; pins 2, 3, 14, and 16 are data pins; pins 15, 17, and 24 are timing pins; and all the other assigned pins are reserved for control or handshaking signals. There are two full-duplex data channels available with the RS-232C interface; one channel is for primary data (actual information) and the second channel is for secondary data (diagnostic information and handshaking signals). The functions of the 20 assigned pins are summarized below.

Pin 1—protective ground. This pin is frame ground and is used for protection against electrical shock. Pin 1 should be connected to the third-wire ground of the ac electrical system at one end of the cable (either at the DTE or the DCE, but not at both ends).

Pin 2—transmit data (TD). Serial data on the primary channel from the DTE to the DCE are transmitted on this pin. TD is enabled by an active condition on the CS pin.

Pin 3—received data (RD). Serial data on the primary channel are transferred from the DCE to the DTE on this pin. RD is enabled by an active condition on the RLSD pin.

Pin 4—request to send (RS). The DTE bids for the primary communications channel from the DCE on this pin. An active condition on RS turns on the modem's analog carrier. The analog carrier is modulated by a unique bit pattern called a training sequence which is used to initialize the communications channel and synchronize the receive modem. RS cannot go active unless pin 6 (DSR) is active.

TABLE 2-6   EIA RS-232C PIN DESIGNATIONS

Pin number	EIA nomenclature	Common acronym	Direction
1	Protective ground (AA)	GND	None
2	Transmitted data (BA)	TD, SD	DTE to DCE
3	Received data (BB)	RD	DCE to DTE
4	Request to send (CA)	RS, RTS	DTE to DCE
5	Clear to send (CB)	CS, CTS	DCE to DTE
6	Data set ready (CC)	DSR, MR	DCE to DTE
7	Signal ground (AB)	GND	None
8	Received line signal detect (CF)	RLSD	DCE to DTE
9	Unassigned		
10	Unassigned		
11	Unassigned		
12	Secondary received line signal detect (SCF)	SRLSD	DCE to DTE
13	Secondary clear to send (SCB)	SCS	DCE to DTE
14	Secondary transmitted data (SBA)	STD	DTE to DCE
15	Transmission signal element timing (DB)	SCT	DCE to DTE
16	Secondary received data (SBB)	SRD	DCE to DTE
17	Receiver signal element timing (DD)	SCR	DCE to DTE
18	Unassigned		
19	Secondary request to send (SCA)	SRS	DTE to DCE
20	Data terminal ready (CD)	DTR	DTE to DCE
21	Signal quality detector (CG)	SQD	DCE to DTE
22	Ring indicator (CE)	RI	DCE to DTE
23	Data signal rate selector (CH)	DSRS	DTE to DCE
24	Transmit signal element timing (DA)	SCTE	DTE to DCE
25	Unassigned		

Pin 5—clear to send (CS). This signal is a handshake from the DCE to the DTE in response to an active condition on request to send. CS enables the TD pin.

Pin 6—data set ready (DSR). On this pin the DCE indicates the availability of the communications channel. DSR is active as long as the DCE is connected to the communications channel (i.e., the modem or the communications channel is not being tested or is not in the voice mode).

Pin 7—signal ground. This pin is the signal reference for all the data, control, and timing pins. Usually, this pin is strapped to frame ground (pin 1).

Pin 8—receive line signal detect (RLSD). The DCE uses this pin to signal the DTE when the DCE is receiving an analog carrier on the primary data channel. RSLD enables the RD pin.

Pin 9. Unassigned.

Pin 10. Unassigned.

Pin 11. Unassigned.

Pin 12—secondary receive line signal detect (SRLSD). This pin is active when the DCE is receiving an analog carrier on the secondary channel. SRLSD enables the SRD pin.

Pin 13—secondary clear to send (SCS). This pin is used by the DCE to send a handshake to the DTE in response to an active condition on the secondary request to send pin. SCS enables the STD pin.

Pin 14—secondary transmit data (STD). Diagnostic data are transferred from the DTE to the DCE on this pin. STD is enabled by an active condition on the SCS pin.

Pin 15—transmission signal element timing (SCT). Transmit clocking signals are sent from the DCE to the DTE on this pin.

Pin 16—secondary received data (SRD). Diagnostic data are transferred from the DCE to the DTE on this pin. SRD is enabled by an active condition on the SCS pin.

Pin 17—receive signal element timing (SCR). Receive clocking signals are sent from the DCE to the DTE on this pin. The clock frequency is equal to the bit rate of the primary data channel.

Pin 18. Unassigned.

Pin 19—secondary request to send (SRS). The DTE bids for the secondary communications channel from the DCE on this pin.

Pin 20—data terminal ready (DTR). The DTE sends information to the DCE on this pin concerning the availability of the data terminal equipment (i.e., access to the mainframe at the primary station or status of the computer terminal at the secondary station). DTR is used primarily with dial-up data communications circuits to handshake with RI.

Pin 21—signal quality detector (SQD). The DCE sends signals to the DTE on this pin that reflect the quality of the received analog carrier.

Pin 22—ring indicator (RI). This pin is used with dial-up lines for the DCE to signal the DTE that there is an incoming call.

Pin 23—data signal rate selector (DSRS). The DTE uses this pin to select the transmission bit rate (clock frequency) of the DCE.

Pin 24—transmit signal element timing (SCTE). Transmit clocking signals are sent from the DTE to the DCE on this pin when the master clock oscillator is located in the DTE.

Pin 25. Unassigned.

Pins 1 through 8 are used with both asynchronous and synchronous modems. Pins 15, 17, and 24 are used for only synchronous modems. Pins 12, 13, 14, 16, and 19 are used only when the DCE is equipped with a secondary channel. Pins 19 and 22 are used exclusively for dial-up telephone connections.

The basic operation of the RS-232C interface is shown in Figure 2-14 and described as follows. When the DTE has primary data to send, it enables request

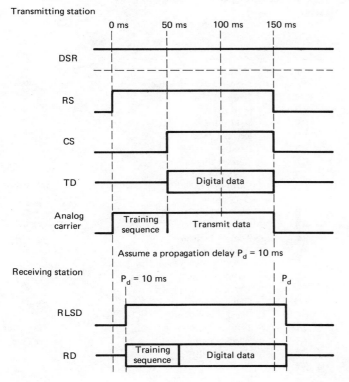

Figure 2-14 Timing diagram: basic operation of the RS-232C interface.

to send ($t = 0$ ms). After a predetermined time delay (50 ms), CS goes active. During the RS/CS delay the modem is outputting an analog carrier that is modulated by a unique bit pattern called a *training sequence*. The training sequence is used to initialize the communications line and synchronize the carrier and clock recovery circuits in the receive modem. After the RS/CS delay, TD is enabled and the DTE begins to transmit data. After the receive DTE detects an analog carrier, RD is enabled. When the transmission is complete ($t = 150$ ms), RS goes low turning off the analog carrier and shutting off CS. For a more detailed explanation, timing diagrams, and illustrative examples, see V. Alisouskas and W. Tomasi, *Digital and Data Communications* (Englewood Cliffs, N.J.: Prentice-Hall, 1985).

RS-449A Interface

Contemporary data rates have exceeded the capabilities of the RS-232C interface. Therefore, it was necessary to adopt and implement a new standard that allows higher bit rates to be transmitted for longer distances. The RS-232C has a maximum

TABLE 2-7   EIA RS-449A PRIMARY CHANNEL PIN DESIGNATIONS

Pin number	Mnemonic	Circuit name
1	None	Shield
2	SI	Signaling rate indicator
3,21	None	Spare
4,22	SD	Send data
5,23	ST	Send timing
6,24	RD	Receive data
7,25	RS	Request to send
8,26	RT	Receive timing
9,27	CS	Clear to send
10	LL	Local loopback
11,29	DM	Data mode
12,30	TR	Terminal ready
13,31	RR	Receiver ready
14	RL	Remote loopback
15	IC	Incoming call
16	SF/SR	Select frequency/signaling rate
17,23	TT	Terminal timing
18	TM	Test mode
19	SG	Signal ground
20	RC	Receive common
28	IS	Terminal in service
32	SS	Select standby
33	SQ	Signal quality
34	NS	New signal
36	SB	Standby indicator
37	SC	Send common

**TABLE 2-8 EIA RS-449A SECONDARY
DIAGNOSTIC CHANNEL PIN DESIGNATIONS**

Pin number	Mnemonic	Circuit name
1	None	Shield
2	SRR	Secondary receiver ready
3	SSD	Secondary send data
4	SRD	Secondary receive data
5	SG	Signal ground
6	RC	Receive common
7	SRS	Secondary request to send
8	SCS	Secondary clear to send
9	SC	Send common

bit rate of 20,000 bps and a maximum distance of approximately 50 ft. Consequently, the EIA has adopted a new standard: the RS-449A interface. The RS-449A is essentially an updated version of the RS-232C except that the RS-449A outlines only the mechanical and functional specifications of the cable and connectors.

The RS-449A specifies two cables: one with 37 wires that is used for serial data transmission and one with 9 wires that is used for secondary diagnostic information. Table 2-7 lists the 37 pins of the RS-449A primary cable and their designations, and Table 2-8 lists the 9 pins of the diagnostic cable and their designations. Note that the acronyms used with the RS-449A are more descriptive than those recommended by the EIA for the RS-232C. The functions specified by the RS-449A are very similar to the RS-232C. The major difference between the two standards is the separation of the primary data and secondary diagnostic channels onto two cables.

The electrical specifications used with the RS-449A are specified by either the RS-422A or the RS-423A standard. The RS-422A standard specifies a balanced interface cable that will operate at bit rates up to 10 Mbps and span distances up to 1200 m. This does not mean that 10 Mbps can be transmitted 1200 m. At 10 Mbps the maximum distance is 15 m, and 90 kbps is the maximum bit rate that can be transmitted 1200 m. The RS-423A standard specifies an unbalanced interface cable that will operate at a maximum line speed of 100 kbps and span a maximum distance of 90 m.

Figure 2-15 shows the *balanced* digital interface circuit for the RS-422A, and Figure 2-16 shows the *unbalanced* digital interface circuit for the RS-423A.

TRANSMISSION MEDIA AND DATA MODEMS

In its simplest form, data communication is the transmittal of digital information between two DTEs. The DTEs may be separated by a few feet or several thousand miles. At the present time, there is an insufficient number of transmission media to carry digital information from source to destination in digital form. Therefore, the

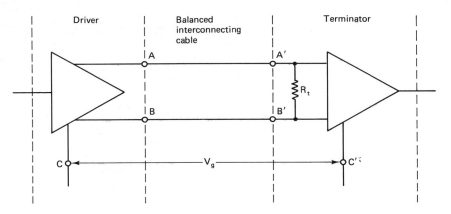

Figure 2-15 RS-422A interface circuit. R_t, Optional cable termination resistance; V_g, ground potential difference; A, B, driver interface points; A', B', terminator interface points; C, driver circuit ground; C', terminator circuit ground; A-B, balanced driver output; A'-B', balanced terminator input.

most convenient alternative is to use the existing public telephone network (PTN) as the transmission media for data communications circuits. Unfortunately, the PTN was designed (and most of it constructed) long before the advent of large-scale data communications. The PTN was intended to be used for transferring voice telephone communications signals, not digital data. Therefore, to use the PTN for data communications, the data must be converted to a form more suitable for transmission over analog carrier systems.

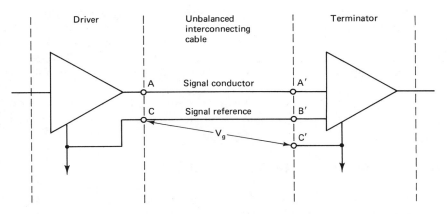

Figure 2-16 RS-423A interface circuit. A, C, Driver interface; A', B', terminator interface; V_g, ground potential difference; C, driver circuit ground; C', terminator circuit ground.

Transmission Media

As stated previously, the public telephone network is a convenient alternative to constructing alternate digital facilities (at a tremendous cost) for carrying only digital data. The public telephone network comprises over 2000 local telephone companies and several long-distance common carriers such as Microwave Communications Incorporated (MCI), GTE Sprint, and the American Telephone and Telegraph Company (AT&T). Local telephone companies provide voice and data services for relatively small geographic areas, whereas long-distance common carriers provide voice and data services for relatively large geographic areas.

Essentially, there are two types of circuits available from the public telephone network: *direct distance dialing* (DDD) and *private line*. The DDD network is commonly called the *dial-up network*. Anyone who has a telephone number subscribes to the DDD network. With the DDD network, data links are established and disconnected in the same manner as normal voice calls are established and disconnected—with a standard telephone or some kind of an automatic dial/answer machine. Data links that are established through the DDD network use *common usage* equipment and facilities. Common usage means that a subscriber uses the equipment and transmission medium for the duration of the call, then they are relinquished to the network for other subscribers to use. With private line circuits, a subscriber has a permanent dedicated communications link 24 hours a day.

Figure 2-17 shows a simplified block diagram of a telephone communications link. Each subscriber has a dedicated cable facility between his station and the nearest telephone office called a *local loop*. The local loop is used by the subscriber to access

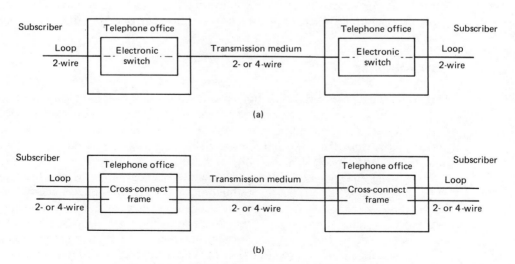

(a)

(b)

Figure 2-17 Telephone communications link: (a) direct distance dialing; (b) dedicated private line.

the PTN. The facilities used to interconnect telephone offices are called *trunk* circuits and can be a metallic cable, a digital carrier system, a microwave radio, a fiber optic link, or a satellite radio system, depending on the distance between the two offices. For temporary connections using the DDD network, telephone offices are interconnected through sophisticated electronic switching systems (ESS) and use intricate switching arrangements. With private line circuits, data links are permanently hardwired through telephone offices without going through a switch. Dial-up data links are preferred when there are a large number of subscribers in a network or if there is a small volume of data traffic. Private line circuits are preferred for limited-access networks when there is a large volume of data throughput.

The quality of a dial-up circuit is guaranteed to meet the minimum requirements for a *voice band* (VB) communications circuit. With a private line circuit, the communications link can be improved by adding amplifiers and equalizers to the circuit. This is called *conditioning* the line. A voice-grade circuit using the PTN has an ideal passband from 0 to 4 kHz, although the usable passband is limited to approximately 300 to 3000 Hz. The minimum-quality circuit available using the PTN is called a basic voice grade (VG) circuit. The quality of a dial-up circuit is guaranteed to meet basic requirements and can be as good as a private line circuit. However, with the DDD network, the transmission characteristics of the data link vary from call to call, while in a private line circuit they remain relatively constant. With the DDD network, *contention* can be a problem; each subscriber must contend for a connection through the network with every other subscriber in the network. With private line circuits, there is no contention because each circuit has only one subscriber. Consequently, there are several advantages that private line circuits have over dial-up networks: increased availability, more consistent performance, greater reliability, and lower costs for moderate to high volumes of data. Dial-up circuits are limited to two-wire operation, whereas private line circuits can operate either two- or four-wire.

Data Modems

The primary purpose of the data modem is to interface the digital terminal equipment to an analog communications channel. The data modem is also called a DCE, a *dataset*, a *dataphone*, or simply a *modem*. At the transmit end, the modem converts digital pulses from the serial interface to analog signals, and at the receive end, the modem converts analog signals to digital pulses.

Modems are generally classified as either asynchronous or synchronous and use either FSK, PSK, or QAM modulation. With synchronous modems, clocking information is recovered in the receive modem; with asynchronous modems, it is not. Asynchronous modems use FSK modulation and are restricted to low-speed applications (below 2000 bps). Synchronous modems use PSK and QAM modulation and are used for medium-speed (2400 to 4800 bps) and high-speed (9600 bps) applications.

Asynchronous modems. Asynchronous modems are used primarily for low-speed dial-up circuits. There are several standard modem designs commonly used for asynchronous data transmission. For half-duplex operation using the two-wire DDD network or full-duplex operation with four-wire private line circuit, the Bell System 202T/S or equivalent is a popular modem. The 202T is a four-wire, full-duplex modem and the 202S is a two-wire, half-duplex modem. The 202T/S uses FSK modulation. The mark and space frequencies are 1200 and 2200 Hz, respectively. With a 1200-bps data rate, the modulation index for the 202T/S is 0.83 and requires approximately 2400 Hz of bandwidth. The output spectrum for the 202T/S modem is shown in Figure 2-18.

To operate full duplex with a two-wire dial-up circuit, it is necessary to divide the usable bandwidth of a voice band circuit in half, creating two equal-capacity data channels. A popular modem that does this is the Bell System 103 or equivalent. The 103 modem is capable of full-duplex operation over a two-wire line at bit rates up to 300 bps. With the 103 modem, there are two data channels each with separate mark and space frequencies. One channel is the *low-band channel* and occupies a passband from 300 to 1650 Hz. The second channel is the *high-band channel* and occupies a passband from 1650 to 3000 Hz. The mark and space frequencies for the low-band channel are 1270 and 1070 Hz, respectively. The mark and space frequencies for the high-band channel are 2225 and 2025 Hz, respectively. For a bit rate of 300 bps, the modulation index for the 103 modem is 0.67. The output spectrum for the 103 modem is shown in Figure 2-19. The high- and low-band data channels occupy different frequency bands and can therefore use the same two-wire facility without interferring with each other. This is called *frequency-division multiplexing* and is explained in detail in Chapter 6.

The low-band channel is commonly called the *originate channel* and the high-band channel is called the *answer channel*. It is standard procedure on a dial-up circuit for the station that originates the call to transmit on the low-band frequencies and receive on the high-band frequencies, and the station that answers the call to transmit on the high-band frequencies and receive on the low-band frequencies.

Synchronous modems. Synchronous modems are used for medium- and high-speed data transmission and use either PSK or QAM modulation. With synchronous

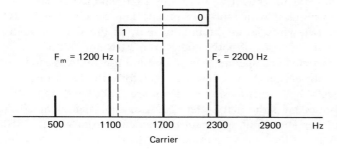

Figure 2-18 Output spectrum for a 202T/S modem. Carrier frequency = 1700 Hz, input data = 1200 bps alternating 1/0 pattern, modulation index = 0.83.

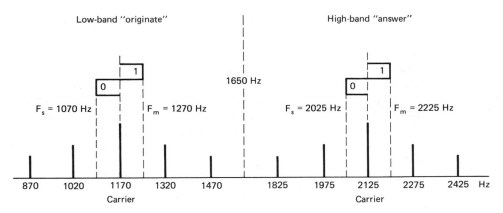

Figure 2-19 Output spectrum for a 103 modem. Carrier frequency: low band = 1170, high band = 2125; input data = 300 bps alternating 1/0 sequence; modulation index = 0.67.

modems the transmit clock, together with the data, digitally modulate an analog carrier. The modulated carrier is transmitted to the receive modem, where a coherent carrier is recovered and used to demodulate the data. The transmit clock is recovered from the data and used to clock the received data into the DTE. Because of the clock and carrier recovery circuits, a synchronous modem is more complicated and thus more expensive than its asynchronous counterpart.

PSK modulation is used for medium-speed (2400 to 4800 bps) synchronous modems. More specifically, QPSK is used with 2400-bps modems and 8PSK is used with 4800-bps modems. QPSK has a bandwidth efficiency of 2 bps/Hz; therefore, the baud rate and minimum bandwidth for a 2400-bps synchronous modem are 1200 baud and 1200 Hz. The standard 2400-bps synchronous modem is the Bell System 201C or equivalent. The 201C uses a 1600-Hz carrier and has an output spectrum that extends from 1000 to 2200 Hz. 8PSK has a bandwidth efficiency of 3 bps/Hz; therefore, the baud rate and minimum bandwidth for 4800-bps synchronous modems are 1600 baud and 1600 Hz. The standard 4800-bps synchronous modem is the Bell System 208A or equivalent. The 208A also uses a 1600-Hz carrier but has an output spectrum that extends from 800 to 2400 Hz. Both the 201C and 208A are full-duplex modems designed to be used with four-wire private line circuits. The 201C and 208A can operate over two-wire dial-up circuits but only in the simplex mode. There are half-duplex two-wire versions of both models: the 201B and 208B.

High-speed synchronous modems operate at 9600 bps and use 16QAM modulation. 16QAM has a bandwidth efficiency of 4 bps/Hz; therefore, the baud rate and minimum bandwidth for 9600-bps synchronous modems are 2400 baud and 2400 Hz. The standard 9600-bps modem is the Bell System 209A or equivalent. The 209A uses a 1650-Hz carrier and has an output spectrum that extends from 450 to 2850 Hz. The Bell System 209A is a four-wire synchronous modem designed to be used on full-duplex private line circuits. The 209B is the two-wire version designed for half-duplex dial-up circuits.

Normally, an asynchronous data format is used with asynchronous modems and a synchronous data format is used with synchronous modems. However, asynchronous data are occasionally used with synchronous modems; this is called *isochronous transmission*. Synchronous data are never used with asynchronous modems.

Table 2-9 summarizes the standard Bell System modems.

TABLE 2-9   MODEM SUMMARY

Bell designation	Line facility	Operating mode	Synchronization	Type of modulation	Maximum data rate (bps)
103	Dial-up	FDX	Asynchronous	FSK	300
113A	Dial-up	Simplex	Asynchronous	FSK	300
113B	Dial-up	Simplex	Asynchronous	FSK	300
201B	Dial-up	HDX/FDX	Synchronous	QPSK	2400
201C	Private	HDX/FDX	Synchronous	QPSK	2400
202S	Dial-up	HDX	Asynchronous	FSK	1200
202T	Private	HDX/FDX	Asynchronous	FSK	1200 (basic)
					1800 (Cl conditioning)
208A	Private	HDX/FDX	Synchronous	8PSK	4800
208B	Dial-up	HDX	Synchronous	8PSK	4800
209A	Private	HDX/FDX	Synchronous	16QAM	9600
209B	Dial-up	HDX	Synchronous	16QAM	9600

QUESTIONS

2-1. Define *data communications*.

2-2. What was the significance of the Carterfone decision?

2-3. Explain the difference between a two-point and a multipoint circuit.

2-4. What is a data communications topology?

2-5. Define the four transmission modes for data communications circuits.

2-6. Which of the four transmission modes can be used only with multipoint circuits?

2-7. Explain the differences between two-wire and four-wire circuits.

2-8. What is a data communications code? What are some of the other names for data communications codes?

2-9. What are the three types of characters used in data communications codes?

2-10. Which data communications code is the most powerful? Why?

2-11. What are the two general categories of error control? What is the difference between them?

2-12. Explain the following error detection techniques: redundancy, exact-count encoding, parity, vertical redundancy checking, longitudinal redundancy checking, and cyclic redundancy checking.

2-13. Which error detection technique is the simplest?

2-14. Which error detection technique is the most reliable?

2-15. Explain the following error correction techniques: symbol substitution, retransmission, and forward error correction.

2-16. Which error correction technique is designed to be used in a human environment?

2-17. Which error correction technique is the most reliable?

2-18. Define *character synchronization*.

2-19. Describe the asynchronous data format.

2-20. Describe the synchronous data format.

2-21. Which data format is best suited to long messages? Why?

2-22. What is a cluster?

2-23. Describe the functions of a control unit.

2-24. What is the purpose of the data modem?

2-25. What are the primary functions of the UART?

2-26. What is the maximum number of bits that can make up a single character with a UART?

2-27. What do the status signals RPE, RFE, and ROR indicate?

2-28. Why does the receive clock for a UART operate 16 times faster than the receive bit rate?

2-29. What are the major differences between a UART and a USRT?

2-30. What is the purpose of the serial interface?

2-31. What is the most prominent serial interface in the United States?

2-32. Why did the EIA establish the RS-232C interface?

2-33. What is the nominal maximum length for the RS-232C interface?

2-34. What are the four general classifications of pins on the RS-232C interface?

2-35. What is the maximum positive voltage that a driver will output?

2-36. Which classification of pins uses negative logic?

2-37. What is the primary difference between the RS-449A interface and the RS-232C interface?

2-38. Higher bit rates are possible with a (balanced, unbalanced) interface cable.

2-39. Who provides the most commonly used transmission medium for data communications circuits? Why?

2-40. Explain the differences between DDD circuits and private line circuits.

2-41. Define the following terms: local loop, trunk, common usage, and dial switch.

2-42. What is a DCE?

2-43. What is the primary difference between a synchronous and an asynchronous modem?

2-44. What is necessary for full-duplex operation using a two-wire circuit?

2-45. What do *originate* and *answer mode* mean?

2-46. What modulation scheme is used for low-speed applications? For medium-speed applications? For high-speed applications?

2-47. Why are synchronous modems required for medium- and high-speed applications?

PROBLEMS

2-1. Determine the LRC and VCR for the following message (use even parity for LRC and odd parity for VRC).

```
P S S S                                              E B P
A Y Y T D A T A sp C O M M U N I C A T I O N S T C A
D N N X                                              X S D
```

2-2. Determine the BCS for the following data- and CRC-generating polynomials:

$$G(x) = x^7 + x^4 + x^2 + x^0 = 1\ 0\ 0\ 1\ 0\ 1\ 0\ 1$$

$$P(x) = x^5 + x^4 + x^1 + x^0 = 1\ 1\ 0\ 0\ 1\ 1$$

2-3. How many Hamming bits are required for a single ASCII character?

2-4. Determine the Hamming bits for the ASCII character "B." Insert the Hamming bits into every other location starting at the left.

Chapter 3

DATA COMMUNICATIONS PROTOCOLS

INTRODUCTION

In essence, a *protocol* is a set of customs or regulations dealing with formality or precedence such as diplomatic or military protocol. A data communications protocol is a set of rules governing the orderly exchange of data information. As stated previously, the function of a line control unit is to control the flow of data between the applications program and the remote terminals. Therefore, there must be a set of rules that govern how an LCU reacts to or initiates different types of transmissions. This set of rules is called a *data link protocol*. Essentially, a data link protocol is a set of procedures, including precise character sequences, that ensure an orderly exchange of data between two LCUs.

In a data communications circuit, the station that is presently transmitting is called the *master* and the receiving station is called the *slave*. In a centralized network, the primary station controls when each secondary station can transmit. When a secondary station is transmitting, it is the master and the primary station is now the slave. The role of master is temporary and which station is master is delegated by the primary. Initially, the primary is master. The primary station solicits each secondary station, in turn, by *polling* it. A poll is an invitation from the primary to a secondary to transmit a message. Secondaries cannot poll a primary. When a primary polls a secondary, the primary is initiating a *line turnaround*; the polled secondary has been designated the master and must respond. If the primary *selects* a secondary, the secondary is identified as a receiver. A selection is an interrogation by the primary of a secondary to determine the secondary's status (i.e., ready to receive or not ready to receive a message). Secondary stations cannot select the primary. Transmissions from the primary go to all the secondaries; it is up to the secondary stations to individually decode each transmission and determine if it is intended for them. When a secondary transmits, it sends only to the primary.

Data link protocols are generally categorized as either asynchronous or synchronous. As a rule, asynchronous protocols use an asynchronous data format and asynchronous modems, whereas synchronous protocols use a synchronous data format and synchronous modems.

Asynchronous Protocols

Two of the most commonly used asynchronous data protocols are the Bell System's *selective calling system* (8A1/8B1) and IBM's *asynchronous data link protocol* (83B). In essence, these two protocols are the same set of procedures.

Asynchronous protocols are *character oriented*. That is, unique data link control characters such as end of transmission (EOT) and start of text (STX), no matter where they occur in a transmission, warrant the same action or perform the same function. For example, the end-of-transmission character used with ASCII is 04H. No matter when 04H is received by a secondary, the LCU is cleared and placed in the line monitor mode. Consequently, care must be taken to ensure that the bit sequences for data link control characters do not occur within a message unless they are intended to perform their designated data link functions. Vertical redundancy checking (parity) is the only type of error detection used with asynchronous protocols, and symbol substitution and ARQ (retransmission) are used for error correction. With asynchronous protocols, each secondary station is generally limited to a single terminal/printer pair. This station arrangement is called a *stand alone*. With the stand-alone configuration, all messages transmitted from or received on the terminal CRT are also written on the printer. Thus the printer simply generates a hard copy of all transmissions.

In addition to the line monitoring mode, a remote station can be in any one of three operating modes: *transmit*, *receive*, and *local*. A secondary station is in the transmit mode whenever it has been designated master. In the transmit mode, the secondary can send formatted messages or acknowledgments. A secondary is in the receive mode whenever it has been selected by the primary. In the receive mode, the secondary can receive formatted messages from the primary. For a terminal operator to enter information into his or her computer terminal, the terminal must be in the local mode. A terminal can be placed in the local mode through software sent from the primary or the operator can do it manually from the keyboard.

The polling sequence for most asynchronous protocols is quite simple and usually encompasses sending one or two data link control characters, then a *station polling address*. A typical polling sequence is

```
E   D
O   C   A
T   3
```

The EOT character is the *clearing* character and always precedes the polling sequence. EOT places all the secondaries in the line monitor mode. When in the line monitor mode, a secondary station listens to the line for its polling or selection address. When DC3 immediately follows EOT, it indicates that the next character is a station polling address. For this example, the station polling address is the single ASCII character "A." Station "A" has been designated the master and must respond with either a formatted message or an acknowledgement. There are two acknowledgment sequences that may be transmitted in response to a poll. They are listed below together with their functions.

Acknowledgement	Function
A \C K	No message to transmit, ready to receive
\\	No message to transmit, not ready to receive

The selection sequence, which is very similar to the polling sequence, is

$$
\begin{matrix} E \\ O \\ T \end{matrix} \quad X \quad Y
$$

Again, the EOT character is transmitted first to ensure that all the secondary stations are in the line monitor mode. Following the EOT is a two-character selection address "XY." Station XY has been selected by the primary and designated as a receiver. Once selected, a secondary station must respond with one of three acknowledgment sequences indicating its status. They are listed below together with their functions.

Acknowledgment	Function
A \C K	Ready to receive
\\	Not ready to receive, terminal in local, or printer out of paper
**	Not ready to receive, have a formatted message to transmit

More than one station can be selected simultaneously with *group* or *broadcast* addresses. Group addresses are used when the primary desires to select more than one but not all of the remote stations. There is a single broadcast address that is used to select simultaneously all the remote stations. With asynchronous protocols, acknowledgment procedures for group and broadcast selections are somewhat involved and for this reason are seldom used.

Messages transmitted from the primary and secondary use exactly the same data format. The format is as follows:

```
S                    E
T  message data  O
X                    T
```

The preceding format is used by the secondary to transmit data to the primary in response to a poll. The STX and EOT characters frame the message. STX precedes the data and indicates that the message begins with the character that immediately follows it. The EOT character signals the end of the message and relinquishes the role of master to the primary. The same format is used when the primary transmits a message except that the STX and EOT characters have an additional function. The STX is a *blinding* character. Upon receipt of the STX character, all previously unselected stations are "blinded," which means that they ignore all transmissions except EOT. Consequently, the subsequent message transmitted by the primary is received only by the previously selected station. The unselected secondaries remain blinded until they receive an EOT character, at which time they will return to the line monitor mode and again listen to the line for their polling or selection addresses. STX and EOT are not part of the message; they are data link control characters and are inserted and deleted by the LCU.

Sometimes it is necessary or desirable to transmit coded data in addition to the message that are used only for data link management, such as date, time of message, message number, message priority, routing information, and so on. This bookkeeping information is not part of the message; it is overhead and is transmitted as *heading* information. To identify the heading, the message begins with a start-of-heading character (SOH). SOH is transmitted first, followed by the heading information, STX, then the message. The entire sequence is terminated with an EOT character. When a heading is included, STX terminates the heading and also indicates the beginning of the message. The format for transmitting heading information together with message data is

```
S                S                    E
O  heading  T  message data  O
H                X                    T
```

Synchronous Protocols

With synchronous protocols, a secondary station can have more than a single terminal/printer pair. The group of devices is commonly called a *cluster*. A single LCU can serve a cluster with as many as 50 devices (terminals and printers). Synchronous protocols can be either character or bit oriented. The most commonly used character-oriented synchronous protocol is IBM's 3270 binary synchronous communications

(BSC or bisync), and the most popular bit-oriented protocol (BPO) is IBM's synchronous data link communications (SDLC).

IBM's bisync protocol.   With bisync, each transmission is preceded by a unique SYN character: 16H for ASCII and 32H for EBCDIC. The SYN character places the receive USRT in the character or byte mode and prepares it to receive data in 8-bit groupings. With bisync, SYN characters are always transmitted in pairs (hence the name "bisync"). Therefore, if 8 successive bits are received in the middle of a message that are equivalent to a SYN character, they are ignored. For example, the characters "A" and "b" have the following hex and binary codes:

```
A = 41H = 0 1 0 0 0 0 0 1
b = 62H = 0 1 1 0 0 0 1 0
```

If the ASCII characters A and b occur successively during a message or heading, the following bit sequence occurs:

```
    A (41H)              b (62H)
 ‾‾‾‾‾‾‾‾‾‾‾‾‾‾       ‾‾‾‾‾‾‾‾‾‾‾‾‾‾
 0 1 0 0 0 0 0 1 0 1 1 0 0 0 1 0
         ‾‾‾‾‾‾‾‾‾‾‾‾‾‾‾
            SYN (16H)
```

As you can see, it appears that a SYN character has been transmitted when actually it has not. To avoid this situation, SYN characters are always transmitted in pairs, and consequently, if only one is received, it is ignored. The likelihood of two false SYN characters occurring one immediately after the other is remote.

With synchronous protocols, the concepts of polling, selecting, and acknowledging are identical to those used with asynchronous protocols except, with bisync, group, and broadcast selections are not allowed. There are two polling formats used with bisync: general and specific. The format for a general poll is

```
P S S E P S S S     E P
A Y Y O A Y Y P P " " N A
D N N T D N N A A     Q D
```

The PAD character at the beginning of the sequence is called a *leading* pad and is either a 55H or an AAH (01010101 or 10101010 binary). As you can see, a leading pad is simply a string of alternating 1's and 0's. The purpose of the leading pad is to ensure that transitions occur in the data prior to the actual message. The transitions are needed for clock recovery in the receive modem to maintain bit synchronization. Next, there are two SYN characters to establish character synchronization. The EOT character is again used as a clearing character and places all the secondary stations into the line monitor mode. The PAD character immediately following the second SYN character is simply a string of successive logic 1's that is used for a time fill, giving each of the secondary stations time to clear. The number of

1's transmitted during this time fill may not be a multiple of 8 bits. Consequently, the two SYN characters are repeated to reestablish character synchronization. The SPA is not an ASCII or EBCDIC character. The letters SPA stand for *station polling address*. Each secondary station has a unique SPA. Two SPAs are transmitted for the purpose of error detection (redundancy). A secondary will not respond to a poll unless its SPA appears twice. The two quotation marks signify that the poll is for any device at that station that is in the send mode. If two or more devices are in the send mode when a general poll is received, the LCU determines which device's message is transmitted. The enquiry (ENQ) character is sometimes called a *format* or *line turnaround character* because it completes the polling format and initiates a line turnaround (i.e., the secondary station identified by the SPA is designated master and must respond).

The PAD character at the end of the polling sequence is called a *trailing* pad and is simply a 7FH (DEL or delete character). The purpose of the trailing pad is to ensure that the RLSD signal in the receive modem is held active long enough for the entire received message to be demodulated. If the carrier were shut off immediately at the end of the message, RLSD would go inactive and disable the receive data pin. If the last character of the message were not completely demodulated, the end of it would be cut off.

The format for a specific poll is

```
P S S E P S S S     D D   E P
A Y Y O A Y Y P P         N A
D N N T D N N A A   A A   Q D
```

The character sequence for a specific poll is similar to that of a general poll except that two DAs (*device addresses*) are substituted for the two quotation marks. With a specific poll, both the station and device addresses are included. Therefore, a specific poll is an invitation to transmit to a specific device at a given station. Again, two DAs are transmitted for redundancy error detection.

The character sequence for a selection is

```
P S S E P S S S     D D   E P
A Y Y O A Y Y S S         N A
D N N T D N N A A   A A   Q D
```

The sequence for a selection is similar to that of a specific poll except that two SSA characters are substituted for the two SPAs. SSA stands for "station select address." All selections are specific; they are for a specific device (device DA). Table 3-1 lists the SPAs, SSAs, and DAs for a network that can have a maximum of 32 stations and the LCU at each station can serve a 32-device cluster.

EXAMPLE 3-1

Determine the character sequences for (a) a general poll for station 8, (b) a specific poll for device 6 at station 8, and (c) a selection of device 6 at station 8.

TABLE 3-1   STATION AND DEVICE ADDRESSES

Station or device number	SPA	SSA	DA	Station or device number	SPA	SSA	DA
0	sp	-	sp	16	&	Ø	&
1	A	/	A	17	J	1	J
2	B	S	B	18	K	2	K
3	C	T	C	19	L	3	L
4	D	U	D	20	M	4	M
5	E	V	E	21	N	5	N
6	F	W	F	22	O	6	O
7	G	X	G	23	P	7	P
8	H	Y	H	24	Q	8	Q
9	I	Z	I	25	R	9	R
10	[	¦	[	26	]	:	]
11	.	,	.	27	$	#	$
12	<	%	<	28	*	@	*
13	(	—	(	29	)	'	)
14	+	>	+	30	;	=	;
15	!	?	!	31	∧	"	∧

Solution   (a) From Table 3-1 the SPA for station 8 is H; therefore, the sequence for a general poll is

```
P S S E P S S       E P
A Y Y O A Y Y H H " " N A
D N N T D N N       Q D
```

(b) From Table 3-1 the DA for device 6 is F; therefore, the sequence for a specific poll is

```
P S S E P S S       E P
A Y Y O A Y Y H H F F N A
D N N T D N N       Q D
```

(c) From Table 3-1 the SSA for station 8 is Y; therefore, the sequence for a selection is

```
P S S E P S S       E P
A Y Y O A Y Y Y Y F F N A
D N N T D N N       Q D
```

With bisync, there are only two ways in which a secondary can respond to a poll: with a formatted message or with a *handshake*. A handshake is simply a response from the secondary that indicates it has no formatted messages to transmit (i.e., a handshake is a negative acknowledgment to a poll). The character sequence for a handshake is

```
P S S E P
A Y Y O A
D N N T D
```

A secondary can respond to a selection with either a positive or a negative acknowledgement. A positive acknowledgment to a selection indicates that the device selected is ready to receive. The character sequence for a positive acknowledgment is

```
P S S D   P
A Y Y L θ A
D N N E   D
```

A negative acknowledgment to a selection indicates that the device selected is not ready to receive. A negative acknowledgment is called a *reverse interrupt* (RVI). The character sequence for an RVI is

```
P S S D   P
A Y Y L < A
D N N E   D
```

With bisync, formatted messages are sent from a secondary to the primary in response to a poll and sent from the primary to a secondary after the secondary has been selected. Formated messages use the following format:

```
P S S S         S         E B P
A Y Y O heading T message T C A
D N N H         X         X C D
```

Note: If CRC-16 is used for error detection, there are two block check characters.

Longitudinal redundancy checking (LRC) is used for error detection with ASCII-coded messages, and cyclic redundancy checking (CRC) is used for EBCDIC. The BCC is computed beginning with the first character after SOH and continues through and includes ETX. (If there is no heading, the BCC is computed beginning with the first character after STX.) With synchronous protocols, data are transmitted in blocks. Blocks of data are generally limited to 256 characters. ETX is used to terminate the last block of a message. ETB is used for multiple block messages to terminate all message blocks except the last one. The last block of a message is always terminated with ETX. All BCCs must be acknowledged by the receiving station. A positive acknowledgment indicates that the BCC was good and a negative acknowledgment means that the BCC was bad. A negative acknowledgement is an automatic request for retransmission. The character sequences for positive and negative acknowledgments are as follows:

Positive acknowledgment:

```
P S S D   P       P S S D   P
A Y Y L θ A  or  A Y Y L 1 A
D N N E   D       D N N E   D
```

  even-numbered   odd-numbered
     blocks   blocks

Negative acknowledgment:

```
P S S N P
A Y Y A A
D N N K D
```

Examples of dialogue using bisync protocol

```
P S S E P S S       E P
A Y Y O A Y Y A A " " N A ─────────────────────→
D N N T D N N       Q D
```

Primary station sends a general poll for station 1.

```
              P S S E P
←──────────── A Y Y O A ────
              D N N T D
```

Station 1 responds with a negative acknowledgment—no messages
to transmit.

```
┌──── P S S E P S S       E P
      A Y Y O A Y Y B B " " N A ────────→
      D N N T D N N       Q D
```

Primary station sends a general poll for station 2.

```
┌──── P S S S       S  message  E B P
   ←─ A Y Y O heading T  block 1 T C A ────
      D Y Y H       X           B C D
```

Station 2 responds with the first block of a multiblock message.

```
┌──── P S S D   P
      A Y Y L 1 A ────────→
      D N N E   D
```

Primary sends a positive acknowledgment indicating that block 1 was
received without any errors—because block 1 is an odd-numbered
block, DLE 1 is used.

```
┌──── P S S S  message  E B P
   ←─ A Y Y T  block 2  T C A ────
│     D N N X           X C D
```

Station 2 sends the second and final block of the message—note that there is no heading at the beginning of the second block—a heading is transmitted only with the first block of a message.

```
            P S S N P
          — A Y Y A A ——————————————→
            D N N K D
```

Primary sends a negative acknowledgment to station 2 indicating that block 2 was received with an error and must be retransmitted.

```
        P S S S               E B P
    ←—— A Y Y T  message      T C A ———————
        D N N X  block 2      X C D
```

Station 2 resends block 2.

```
            P S S D   P
          — A Y Y L θ A ——————————→
            D N N E   D
```

Primary sends a positive acknowledgment to station 2 indicating that block 2 was received without any errors—because block 2 is an even-numbered block, DLE θ is used.

```
        P S S E P
    ←—— A Y Y O A ———————————————————
        D N N T D
```

Secondary responds with a handshake—a secondary sends a handshake whenever it is its turn to transmit but it has nothing to say.

```
    P S S E P S S           E P
  — A Y Y O A Y Y T T E E N A ——————→
    D N N T D N N           Q D
```

Primary selects station 3, device 5.

```
            P S S D   P
    ←—————— A Y Y L θ A ————————————————
            D N N E   D
```

Station 3 sends a positive acknowledgment to the selection; device 5 is ready to receive.

```
        P S S S         S            E B P
    ——— A Y Y O heading T  message   T C A ———→
        D N N H         X  block 1   X C D
```

Primary sends a single block message to station 3.

```
        P  S  S  D      P
 ◄───── A  Y  Y  L  1   A ──────
        D  N  N  E      D
```

Station 3 responds with a positive acknowledgment indicating the block of data was received without any errors.

Transparency.   It is possible that a device that is attached to one of the ports of a station LCU is not a computer terminal or a printer. For example, a microprocessor-controlled monitor system that is used to monitor environmental conditions (temperature, humidity, etc.) or a security alarm system. If so, the data transferred between it and the applications program are not ASCII- or EBCDIC-encoded characters; they are microprocessor op-codes or binary-encoded data. Consequently, it is possible that an 8-bit sequence could occur in the message that is equivalent to a data link control character. For example, if the binary code 00000011 (03H) occurred in a message, the LCU would misinterpret it as the ASCII code for ETX. Consequently, the receive LCU would prematurely terminate the message and interpret the next 8-bit sequence as a BCC. To prevent this from occurring, the LCU is made *transparent* to the data. With bisync, a *data link escape* character (DLE) is used to achieve transparency. To place an LCU in the transparent mode, STX is preceded by a DLE (i.e., the LCU simply transfers the data to the selected device without searching through the message for data link control characters). To come out of the transparent mode, DLE ETX is transmitted. To transmit a DLE as part of the text, it must be preceded by DLE (i.e., DLE DLE). Actually, there are only five characters that it is necessary to precede with DLE:

1. *DLE STX*: places the receive LCU into the transparent mode.
2. *DLE ETX*: used to terminate the last block of transparent text and take the LCU out of the transparent mode.
3. *DLE ETB*: used to terminate blocks of transparent text other than the final block.
4. *DLE ITB*: used to terminate blocks of transparent text other than the final block when ITB is used for a block terminating character.
5. *DLE SYN*: used only with transparent messages that are more than 1 s long. With bisync, two SYN characters are inserted in the text every 1 s to ensure that the receive LCU does not lose character synchronization. In a multipoint circuit with a polling environment, it is highly unlikely that any blocks of data would exceed 1 s in duration. SYN character insertion is used almost exclusively for two-point circuits.

Synchronous Data Link Communications

Synchronous data link communications (SDLC) is a synchronous *bit-oriented* protocol developed by IBM. A bit-oriented protocol (BOP) is a discipline for serial-by-bit information transfer over a data communication channel. With a BOP, data link control information is transferred and interpreted on a bit-by-bit basis rather than with unique data link control characters. SDLC can transfer data either simplex, half-duplex, or full duplex. With a BOP, there is a single control field that performs essentially all the data link control functions. The character language used with SDLC is EBCDIC and data are transferred in groups called *frames*. Frames are generally limited to 256 characters in length. There are two types of stations in SDLC: primary stations and secondary stations. The *primary station* controls data exchange on the communications channel and issues *commands*. The *secondary station* receives commands and returns *responses* to the primary.

There are three transmission states with SDLC: transient, idle, and active. The *transient state* exists before and after the initial transmission and after each line turnaround. An *idle state* is presumed after 15 or more consecutive 1's have been received. The *active state* exists whenever either the primary or a secondary station is transmitting information or control signals.

Figure 3-1 shows the frame format used with SDLC. The frames sent from the primary and the frames sent from a secondary use exactly the same format. There are five fields used with SDLC: the flag field, the address field, the control field, the text or information field, and the frame check field.

Information field.   All information transmitted in an SDLC frame must be in the information field (I field), and the number of bits in the I field must be a multiple of 8. An I field is not allowed with all SDLC frames. The types of frames that allow an I field are discussed later.

Flag field.   There are two flag fields per frame: the beginning flag and the ending flag. The flags are used for the *delimiting sequence* and to achieve character synchronization. The delimiting sequence sets the limits of the frame (i.e., when the frame begins and when it ends). The flag is used with SDLC in the same manner that SYN characters are used with bisync, to achieve character synchronization.

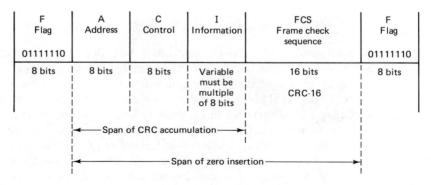

Figure 3-1   SDLC frame format.

The sequence for a flag is 7EH, 01111110 binary, or the EBCDIC character "=." There are several variations of how flags are used. They are:

1. One beginning and one ending flag for each frame.

```
        beginning flag                          ending flag
    . . . 01111110 address control text FCC 01111110 . . .
```

2. The ending flag from one frame can be used for the beginning flag for the next frame.

```
                                    ←──────── frame N + 1 ────────→
    ──────── frame N ────────→
    . . . text FCC 01111110 address control text FCC 01111110 . . .
                          ╱   ╲
                  ending flag  beginning flag
                    frame N      frame N+1
```

3. The last zero of an ending flag is also the first zero of the beginning flag of the next frame.

```
                                    ←──────── frame N+1 ────────
                          shared 0
                  frame N ────→  ╱
    . . . text FCC 0111111101111110 address control FCC . . .
                      ╱       ╲
              ending flag   beginning flag
                frame N        frame N+1
```

4. Flags are transmitted in lieu of idle line 1's.

```
    011111101111110111111101111110 address control text . . .
          ╱     ╱     ╱     ╰──┬──╯
    idle line flags  beginning flag
```

Address field. The address field has 8 bits; thus 256 addresses are possible with SDLC. The address 00H (00000000) is called the *null* or *void address* and is never assigned to a secondary. The null address is used for network testing. The address FFH (11111111) is the *broadcast address* and is common to all secondaries. The remaining 254 addresses can be used as *unique* station addresses or as *group* addresses. In frames sent from the primary, the address field contains the address of the destination station (a secondary). In frames sent from a secondary, the address field contains the address of that secondary. Therefore, the address is always that of a secondary. The primary station has no address because all transmissions from secondary stations go to the primary.

Control field. The control field is an 8-bit field that identifies the type of frame it is. The control field is used for polling, confirming previously received information frames, and several other data link management functions. There are three frame formats used with SDLC: information, supervisory, and unnumbered.

Information frame. With an information frame there must be an information field. Information frames are used for transmitting sequenced information. The bit pattern for the control field of an information frame is

Bit:	b_0	b_1	b_2	b_3	b_4	b_5	b_6	b_7
Function:	$\longleftarrow$	nr	$\longrightarrow$	P or F $\overline{P}$ or $\overline{F}$	$\longleftarrow$	ns	$\longrightarrow$	0

An information frame is identified by a 0 in the least significant bit position (b_7 with EBCDIC code). Bits b_4, b_5, and b_6 are used for numbering transmitted frames (ns = number sent). With 3 bits, the binary numbers 000 through 111 (0–7) can be represented. The first frame transmitted is designated frame 000, the second frame 001, and so on up to frame 111 (the eighth frame); then the count cycles back to 000 and repeats.

Bits b_0, b_1, and b_2 are used to confirm correctly received information frames (nr = number received) and to automatically request retransmission of incorrectly received information frames. The nr is the number of the next frame that the transmitting station expects to receive, or the number of the next frame that the receiving station will transmit. The nr confirms received frames through nr-1. Frame nr-1 is the last frame received without a transmission error. Any transmitted I frame not confirmed must be retransmitted. Together, the ns and nr bits are used for error correction (ARQ). The primary must keep track of an ns and nr for each secondary. Each secondary must keep track of only its ns and nr. After all frames have been confirmed, the primary's ns must agree with the secondary's nr, and vice versa. For the example shown next, the primary and secondary stations begin with their ns and nr counters reset to 000. The primary sends three numbered information frames (ns = 0, 1, and 2). At the same time the primary sends nr = 0 because the next frame it expects to receive is frame 0, which is the secondary's present ns. The secondary responds with two information frames (ns = 0 and 1). The secondary received all three frames from the primary without any errors, so the nr transmitted in the secondary's control field is 3 (which is the number of the next frame that the primary will send). The primary now sends information frames 3 and 4 with an nr = 2. The nr = 2 confirms the correct reception of frames 0 and 1. The secondary responds with frames ns = 2, 3, and 4 with an nr = 4. The nr = 4 confirms reception of only frame 3 from the primary (nr-1). Consequently, the primary must retransmit frame 4. Frame 4 is retransmitted together with four additional frames (ns = 5, 6, 7, and 0). The primary's nr = 5, which confirms frames 2, 3, and 4 from the secondary. Finally, the secondary sends information frame 5 with an nr = 1. The nr = 1 confirms frames 4, 5, 6, 7 and 0 from the primary. At this point, all the frames transmitted have been confirmed except frame 5 from the secondary.

Primary's ns:	0 1 2		3 4		4 5 6 7 0
Primary's nr:	0 0 0		2 2		5 5 5 5 5
Secondary's ns:		0 1		2 3 4	5
Secondary's nr:		3 3		4 4 4	1

With SDLC, a station can never send more than seven numbered frames without receiving a confirmation. For example, if the primary sent eight frames (ns = 0, 1, 2, 3, 4, 5, 6, and 7) and the secondary responded with an nr = 0, it is ambiguous which frames are being confirmed. Does nr = 0 mean that all eight frames were received correctly, or that frame 0 had an error in it and all eight frames must be retransmitted? (With SDLC, all previously transmitted frames beginning with frame nr-1 must be retransmitted.)

Bit b_3 is the *poll* (P) or *not-a-poll* ($\overline{P}$) bit when sent from the primary and the *final* (F) or *not-a-final* ($\overline{F}$) bit when sent by a secondary. In a frame sent by the primary, if the primary desires to poll the secondary, the P bit is set (1). If the primary does not wish to poll the secondary, the P bit is reset (0). A secondary cannot transmit unless it receives a frame addressed to it with the P bit set. In a frame sent from a secondary, if it is the last (final) frame of the message, the F bit is set (1). If it is not the final frame, the F bit is reset (0). With I frames, the primary can select a secondary station, send formatted information, confirm previously received I frames, and poll with a single transmission.

EXAMPLE 3-2

Determine the bit pattern for the control field of a frame sent from the primary to a secondary station for the following conditions: primary is sending information frame 3, it is a poll, and the primary is confirming the correct reception of frames 2, 3, and 4 from the secondary.

Solution

$b_7 = 0$ because it is an information frame.

b_4, b_5, and b_6 are 011 (binary 3 for ns =3).

$b_3 = 1$, it is a polling frame.

b_0, b_1, and b_2 are 101 (binary 5 for nr = 5).

control field = B6H.

b_0	b_1	b_2	b_3	b_4	b_5	b_6	b_7
1	0	1	1	0	1	1	0

Supervisory frame. An information field is not allowed with a supervisory frame. Consequently, supervisory frames cannot be used to transfer information; they are used to assist in the transfer of information. Supervisory frames are used to confirm previously received information frames, convey ready or busy conditions, and to report frame numbering errors. The bit pattern for the control field of a supervisory frame is

Bit:	b_0	b_1	b_2	b_3	b_4	b_5	b_6	b_7
Function:	–	–	nr	– –	P or F $\overline{P}$ or $\overline{F}$	X	X	0	1

A supervisory frame is identified by a 01 in bit positions b_6 and b_7, respectively, of the control field. With the supervisory format, bit b_3 is again the poll/not-a-poll or final/not-a-final bit and b_0, b_1, and b_2 are the nr bits. However, with a supervisory format, b_5 and b_6 are used to indicate either the receive status of the station transmitting the frame or to request transmission or retransmission of sequenced information frames. With two bits, there are four combinations possible. The four combinations and their functions are as follows:

b_4	b_5	Receiver status
0	0	Ready to receive (RR)
0	1	Ready not to receive (RNR)
1	0	Reject (REJ)
1	1	Not used with SDLC

When the primary sends a supervisory frame with the P bit set and a status of ready to receive, it is equivalent to a general poll with bisync. Supervisory frames are used by the primary for polling and for confirming previously received information frames when there is no information to send. A secondary uses the supervisory format for confirming previously received information frames and for reporting its receive status to the primary. If a secondary sends a supervisory frame with RNR status, the primary cannot send it numbered information frames until that status is cleared. RNR is cleared when a secondary sends an information frame with the F bit = 1 or a RR or REJ frame with the F bit = 0. The REJ command/response is used to confirm information frames through nr-1 and to request retransmission of numbered information frames beginning with the frame number identified in the REJ frame. An information field is prohibited with a supervisory frame and the REJ command/ response is used only with full-duplex operation.

EXAMPLE 3-3

Determine the bit pattern for the control field of a frame sent from a secondary station to the primary for the following conditions: the secondary is ready to receive, it is the final frame, and the secondary station is confirming frames 3, 4, and 5.

Solution

b_6 and $b_7 = 01$ because it is a supervisory frame.

b_4 and $b_5 = 00$ (ready to receive).

$b_3 = 1$ (it is the final frame).

b_0, b_1, and $b_2 = 110$ (binary 6 for nr = 6).

control field = D1H

b_0	b_1	b_2	b_3	b_4	b_5	b_6	b_7
1	1	0	1	0	0	0	1

Unnumbered frame. An unnumbered frame is identified by making bits b_6 and b_7 in the control field 11. The bit pattern for the control field of an unnumbered frame is

Bit:	b_0	b_1	b_2	b_3	b_4	b_5	b_6	b_7
Function:	X	X	X	P or F	X	X	1	1
				$\overline{P}$ or $\overline{F}$				

With an unnumbered frame, bit b_3 is again either the $P/\overline{P}$ or $F/\overline{F}$ bit. Bits b_0, b_1, b_2, b_4, and b_5 are used for various unnumbered commands and responses. With 5 bits available, 32 unnumbered commands/responses are possible. The control field in an unnumbered frame sent by the primary is a command. The control field in an unnumbered frame sent by a secondary is a response. With unnumbered frames, there are no ns or nr bits. Therefore, numbered information frames cannot be sent or confirmed with the unnumbered format. Unnumbered frames are used to send network control and status information. Two examples of control functions are (1) placing secondary stations on-line and off-line and (2) LCU initialization. Table 3-2 lists several of the more commonly used unnumbered commands and responses. An information field is prohibited with all the unnumbered commands/responses except UI, FRMR, CFGR, TEST and XID.

A secondary station must be in one of three modes: the initialization mode, the normal response mode, or the normal disconnect mode. The procedures for the *initialization mode* are system specified and vary considerably. A secondary in the *normal response mode* cannot initiate unsolicited transmissions; it can transmit only in response to a frame received with the P bit set. When in the *normal disconnect mode*, a secondary is off-line. In this mode, a secondary can receive only a TEST,

TABLE 3-2   UNNUMBERED COMMANDS AND RESPONSES

Binary configuration						I Field	Resets
b_0		b_7	Acronym	Command	Response	prohibited	ns and nr
000	P/F	0011	UI	Yes	Yes	No	No
000	F	0111	RIM	No	Yes	Yes	No
000	P	0111	SIM	Yes	No	Yes	Yes
100	P	0011	SNRM	Yes	No	Yes	Yes
000	F	1111	DM	No	Yes	Yes	No
010	P	0011	DISC	Yes	No	Yes	No
011	F	0011	UA	No	Yes	Yes	No
100	F	0111	FRMR	No	Yes	No	No
111	F	1111	BCN	No	Yes	Yes	No
110	P/F	0111	CFGR	Yes	Yes	No	No
010	F	0011	RD	No	Yes	Yes	No
101	P/F	1111	XID	Yes	Yes	No	No
001	P	0011	UP	Yes	No	Yes	No
111	P/F	0011	TEST	Yes	Yes	No	No

XID, CFGR, SNRM, or SIM command from the primary and can respond only if the P bit is set.

The unnumbered commands and responses are summarized below.

Unnumbered information (UI). UI is a command/response that is used to send unnumbered information. Unnumbered information transmitted in the I field is not confirmed.

Set initialization mode (SIM). SIM is a command that places the secondary station into the initialization mode. The initialization procedure is system specified and varies from a simple self-test of the station controller to executing a complete IPL (initial program logic) program. SIM resets the ns and nr counters at the primary and secondary stations. A secondary is expected to respond to a SIM command with a UA response.

Request initialization mode (RIM). RIM is a response sent by a secondary station to request the primary to send an SIM command.

Set normal response mode (SNRM). SNRM is a command that places a secondary station in the normal response mode (NRM). A secondary station cannot send or receive numbered information frames unless it is in the normal response mode. Essentially, SNRM places a secondary station on-line. SNRM resets the ns and nr counters at the primary and secondary stations. UA is the normal response to an SNRM command. Unsolicited responses are not allowed when the secondary is in the NRM. A secondary remains in the NRM until it receives a DISC or SIM command.

Disconnect mode (DM). DM is a response that is sent from a secondary station if the primary attempts to send numbered information frames to it when the secondary is in the normal disconnect mode.

Request disconnect (RD). RD is a response sent when a secondary wishes to be placed in the disconnect mode.

Disconnect (DISC). DISC is a command that places a secondary station in the normal disconnect mode (NDM). A secondary cannot send or receive numbered information frames when it is in the normal disconnect mode. When in the normal disconnect mode, a secondary can receive only an SIM or SNRM command and can transmit only a DM response. The expected response to a DISC command is UA.

Unnumbered acknowledgment (UA). UA is an affirmative response that indicates compliance to a SIM, SNRM, or DISC command. UA is also used to acknowledge unnumbered information frames.

Frame reject (FRMR). FRMR is for reporting procedual errors. The FRMR sequence is a response transmitted when the secondary has received an invalid frame from the primary. A received frame may be invalid for any one of the following reasons:

1. The control field contains an invalid or unassigned command.
2. The amount of data in the information field exceeds the buffer space at the secondary.
3. An information field is received in a frame that does not allow information.
4. The nr received is incongruous with the secondary's ns. For example, if the secondary transmitted ns frames 2, 3, and 4 and then the primary responded with an nr of 7.

A secondary cannot release itself from the FRMR condition, nor does it act on the frame that caused the condition. The secondary repeats the FRMR response until it receives one of the following *mode-setting* commands: SNRM, DISC, or SIM. The information field for a FRMR response always contains three bytes (24 bits) and has the following format:

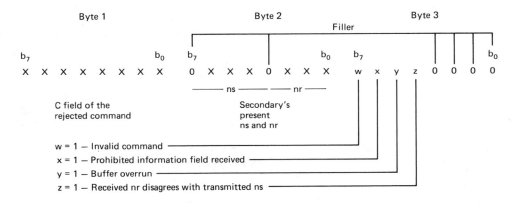

TEST. TEST is a command that can be sent in any mode to solicit a TEST response. If an information field is included with the command, the secondary returns it with the response. The TEST command/response is exchanged for link testing purposes.

Exchange station identification (XID). As a command, XID solicits the identification of the secondary station. An information field can be included in the frame to convey the identification data of either the primary or secondary station. For dial-up circuits, it is often necessary that the secondary station identify itself before the primary will exchange information frames with it, although XID is not restricted to only dial-up data circuits.

Frame check sequence field.   The FCS field contains the error detection mechanism for SDLC. The FCS is equivalent to the BCC used with bisync. SDLC uses CRC-16 and the following generating polynomial: $x^{16} + x^{12} + x^5 + x^1$.

SDLC Loop Operation

An SDLC *loop* is operated in the half-duplex mode. The primary difference between the loop and bus configurations is that in a loop, all transmissions travel in the same direction on the communications channel. In a loop configuration, only one station transmits at a time. The primary transmits first, then each secondary station responds sequentially. In an SDLC loop, the transmit port of the primary station controller is connected to one or more secondary stations in a serial fashion; then the loop is terminated back at the receive port of the primary. Figure 3-2 shows an SDLC loop configuration.

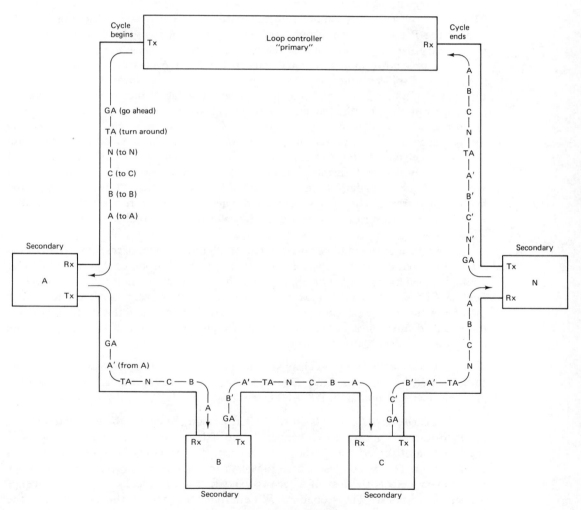

Figure 3-2   SDLC loop configuration.

In an SDLC loop, the primary transmits frames that are addressed to any or all of the secondary stations. Each frame transmitted by the primary contains an address of the secondary station to which that frame is directed. Each secondary station, in turn, decodes the address field of every frame, then serves as a repeater for all stations that are down-loop from it. If a secondary detects a frame with its address, it accepts the frame, then passes it on to the next down-loop station. All frames transmitted by the primary are returned to the primary. When the primary has completed transmitting, it follows the last flag with eight consecutive 0's. A flag followed by eight consecutive 0's is called a *turnaround* sequence which signals the end of the primary's transmission. Immediately following the turnaround sequence, the primary transmits continuous 1's, which generates a *go-ahead* sequence (01111111). A secondary cannot transmit until it has received a frame addressed to it with the P bit set, a turnaround sequence, and then a go-ahead sequence. Once the primary has begun transmitting 1's, it goes into the receive mode.

The first down-loop secondary station that has received a frame addressed to it with the P bit set, changes the seventh 1 bit in the go-ahead sequence to a 0, thus creating a flag. That flag becomes the beginning flag of the secondary's response frame or frames. After the secondary has transmitted its last frame, it again becomes a repeater for the idle line 1's from the primary. These idle line 1's again become the go-ahead sequence for the next secondary station. The next down-loop station that has received a frame addressed to it with the P bit set detects the turnaround sequence, any frames transmitted from up-loop secondaries, and then the go-ahead sequence. Each secondary station inserts its response frames immediately after the last repeated frame. The cycle is completed when the primary receives its own turnaround sequence, a series of response frames, and then the go-ahead sequence.

Configure command/response.   The configure command/response (CFGR) is an unnumbered command/response that is used only in a loop configuration. CFGR contains a one-byte *function descriptor* (essentially a subcommand) in the information field. A CFGR command is acknowledged with a CFGR response. If the low-order bit of the function descriptor is set, a specified function is initiated. If it is reset, the specified function is cleared. There are six subcommands that can appear in the configure command's function field.

1. *Clear—00000000.* A clear subcommand causes all previously set functions to be cleared by the secondary. The secondary's response to a clear subcommand is another clear subcommand, 00000000.

2. *Beacon test (BCN)—0000000X.* The beacon test causes the secondary receiving it to turn on or turn off its carrier. If the X bit is set, the secondary suppresses transmission of the carrier. If the X bit is reset, the secondary resumes transmission of the carrier. The beacon test is used to isolate an open-loop problem. Also, whenever a secondary detects the loss of a receive carrier, it automatically begins to transmit its beacon response. The secondary will continue transmiting the beacon until the loop resumes normal status.

3. *Monitor mode—0000010X*. The monitor command causes the addressed secondary to place itself into a monitor (receive only) mode. Once in the monitor mode, a secondary cannot transmit until it receives a monitor mode clear (00000100) or a clear (00000000) subcommand.

4. *Wrap—0000100X*. The wrap command causes the secondary station to loop its transmissions directly to its receiver input. The wrap command places the secondary effectively off-line for the duration of the test. A secondary station does not send the results of a wrap test to the primary.

5. *Self-test—0000101X*. The self-test subcommand causes the addressed secondary to initiate a series of internal diagnostic tests. When the tests are completed, the secondary will respond. If the P bit in the configure command is set, the secondary will respond following completion of the self-test at its earliest opportunity. If the P bit is reset, the secondary will respond following completion of the test to the next poll-type frame it receives. All other transmissions are ignored by the secondary while it is performing the self-tests. The secondary indicates the results of the self-test by setting or resetting the low-order bit (X) of its self-test response. A 1 indicates that the tests were unsuccessful, and a 0 indicates that they were successful.

6. *Modified link test—0000110X*. If the modified link test function is set (X bit set), the secondary station will respond to a TEST command with a TEST response that has an information field containing the first byte of the TEST command information field repeated *n* times. The number *n* is system implementation dependent. If the X bit is reset, the secondary station will respond to a TEST command, with or without an information field, with a TEST response with a zero-length information field. The modified link test is an optional subcommand and is only used to provide an alternative form of link test to that previously described for the TEST command.

Transparency

The transparency mechanism used with SDLC is called *zero-bit insertion* or *zero stuffing*. The flag bit sequence (01111110) can occur in a frame where this pattern is not intended to be a flag. For example, any time that 7EH occurs in the address, control, information, or FCS field it would be interpreted as a flag and disrupt character synchronization. Therefore, 7EH must be prohibited from occurring except when it is intended to be a flag. To prevent a 7EH sequence from occurring, a zero is automatically inserted after any occurrence of five consecutive 1's except in a designated flag sequence (i.e., flags are not zero inserted). When five consecutive 1's are received and the next bit is a 0, the 0 is deleted or removed. If the next bit is a 1, it must be a valid flag. An example of zero insertion/deletion is shown below.

Original frame bits at the transmit station:

```
01111110   01101111   11010011   1110001100110101   01111110
  flag      address    control          FCS            flag
```

After zero insertion but prior to transmission:

```
01111110  01101111  101010011  11100001100110101  01111110
  flag     address    control        /      FCS         flag
                         \
                      inserted zeros
```

After zero deletion at the receive end:

```
01111110  01101111  11010011  1110001100110101  01111110
  flag     address    control        FCS            flag
```

Message Abort

Message abort is used to prematurely terminate a frame. Generally, this is only done to accommodate high-priority messages such as emergency link recovery procedures, and so on. A message abort is any occurrence of 7 to 14 consecutive 1's. Zeros are not inserted in an abort sequence. A message abort terminates an existing frame and immediately begins the higher-priority frame. If more than 14 consecutive 1's occur in succession, it is considered an idle line condition. Therefore, 15 or more successive 1's place the circuit into the idle state.

Invert-on-Zero Encoding

A binary synchronous transmission such as SDLC is time synchronized to enable identification of sequential binary digits. Synchronous data communications assumes that bit or clock synchronization is provided by either the DCE or the DTE. With synchronous transmissions, a receiver samples incoming data at the same rate that they were transmitted. Although minor variations in timing can exist, synchronous modems provide received data clock recovery and dynamically adjusted sample timing to keep sample times midway between bits. For a DTE or a DCE to recover the clock, it is necessary that transitions occur in the data. *Invert-on-zero coding* is an encoding scheme that guarantees at least one transition in the data for every 7 bits transmitted. Invert-on-zero coding is also called NRZI (*nonreturn-to-zero inverted*).

   With NRZI encoding, the data are encoded in the transmitter, then decoded in the receiver. Figure 3-3 shows an example of NRZI encoding. 1's are unaffected by the NRZI encoder. However, 0's invert the encoded transmission level. Consequently, consecutive 0's generate an alternating high/low sequence. With SDLC, there can never be more than six 1's in succession (a flag). Therefore, a high-to-low transition is guaranteed to occur at least once for every 7 bits transmitted except during a message abort or an idle line condition. In a NRZI decoder, whenever a high/low transition occurs in the received data, a 0 is generated. The absence of a transition simply generates a 1. In Figure 3-3, a high level is assumed prior to encoding the incoming data.

   NRZI encoding was intended to be used with asynchronous modems which

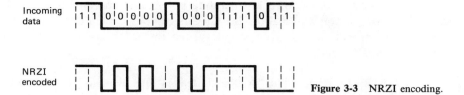

Figure 3-3 NRZI encoding.

do not have clock recovery capabilities. Consequently, the DTE must provide time synchronization which is aided by using NRZI-encoded data. Synchronous modems have built in scramblers and descramblers which ensure that transitions occur in the data, and thus NRZI encoding is unnecessary. The NRZI encoder/decoder is placed between the DTE and the DCE.

High-Level Data Link Control

In 1975, the International Standards Organization (ISO) defined several sets of sub-standards that, when combined, are called *high-level data link control* (HDLC). Since HDLC is a superset of SDLC, only the added capabilities are explained.

HDLC comprises three standards (subdivisions) that, when combined, outline the frame structure, control standards, and class of operation for a bit-oriented data link control (DLC).

ISO 3309–1976(E). This standard defines the frame structure, delimiting, sequence, and transparency mechanism used with HDLC. These are essentially the same as with SDLC except that HDLC has extended addressing capabilities and checks the FCS in a slightly different manner. The delimiting sequence used with HDLC is identical to SDLC: a 01111110 sequence.

HDLC can use either the *basic* 8-bit address field or an *extended* addressing format. With extended addressing the address field may be extended recursively. If b_0 in the address byte is a logic 1, the 7 remaining bits are the secondary's address (the ISO defines the low-order bit as b_0, whereas SDLC designates the high-order bit as b_0). If b_0 is a logic 0, the next byte is also part of the address. If b_0 of the second byte is a 0, a third address byte follows, and so on, until an address byte with a logic 1 for the low-order bit is encountered. Essentially, there are 7 bits available in each address byte for address encoding. An example of a three-byte extended addressing scheme is shown below, b_0 in the first two bytes of the address field are 0's, indicating that additional address bytes follow and b_0 in the third address byte is a logic 1, which terminates the address field.

```
            b₀ = 0       b₀ = 0       b₀ = 1
             |            |            |
 01111110  0XXXXXXX    0XXXXXXX    1XXXXXXX  . . .
   flag       three-byte address field   control field, etc.
```

HDLC uses CRC-16 with a generating polynomial specified by CCITT V.41 as the FCS. At the transmit station, the CRC is computed such that if it is included in the FCS computation at the receive end, the remainder for an errorless transmission is always F0BBH.

ISO 4335–1979(E).   This standard defines the elements of procedure for HDLC. The control field, information field, and supervisory format have increased capabilities over SDLC.

Control field. With HDLC, the control field can be extended to 16 bits. Seven bits are for the ns and 7 bits are for the nr. Therefore, with the extended control format, there can be a maximum of 127 outstanding (unconfirmed) frames at any given time.

Information field. HDLC permits any number of bits in the information field of an information command or response (SDLC is limited to 8-bit bytes). With HDLC any number of bits may be used for a character in the I field as long as all characters have the same number of bits.

Supervisory format. With HDLC, the supervisory format includes a fourth status condition: selective reject (SREJ). SREJ is identified by a 11 in bit position b_4 and b_5 of a supervisory control field. With a SREJ, a single frame can be rejected. A SREJ calls for the retransmission of only the frame identified by nr, whereas a REJ calls for the retransmission of all frames beginning with nr. For example, the primary sends I frames ns = 2, 3, 4, and 5. Frame 3 was received in error. A REJ would call for a retransmission of frames 3, 4, and 5; a SREJ would call for the retransmission of only frame 3. SREJ can be used to call for the retransmission of any number of frames except that only one is identified at a time.

Operational modes. HDLC has two operational modes not specified in SDLC: asynchronous response mode and asynchronous disconnect mode.

1. *Asynchronous response mode (ARM).* With the ARM, secondary stations are allowed to send unsolicited responses. To transmit, a secondary does not need to have received a frame from the primary with the P bit set. However, if a secondary receives a frame with the P bit set, it must respond with a frame with the F bit set.
2. *Asynchronous disconnect mode (ADM).* An ADM is identical to the normal disconnect mode except that the secondary can initiate a DM or RIM response at any time.

ISO 7809–1985(E).   This standard combines previous standards 6159(E) (unbalanced) and 6256(E) (balanced) and outlines the class of operation necessary to establish the link-level protocol.

Unbalanced operation. This class of operation is logically equivalent to a multipoint private line circuit with a polling environment. There is a single primary station responsible for central control of the network. Data transmission may be either half- or full-duplex.

Balanced operation. This class of operation is logically equivalent to a two-point private line circuit. Each station has equal data link responsibilities, and channel access is through contention using the asynchronous response mode. Data transmission may be half- or full-duplex.

PUBLIC DATA NETWORK

A *public data network* (PDN) is a switched data communications network similar to the public telephone network except that a PDN is designed for transferring data only. Public data networks combine the concepts of both *value-added networks* (VANs) and *packet-switching networks*.

Value-Added Network

A value-added network *"adds value"* to the services or facilities provided by a common carrier to provide new types of communication services. Examples of added values are error control, enhanced connection reliability, dynamic routing, failure protection, logical multiplexing, and data format conversions. A VAN comprises an organization that leases communications lines from common carriers such as AT&T and MCI and adds new types of communications services to those lines. Examples of value-added networks are GTE Telnet, DATAPAC, TRANSPAC, and Tymnet Inc.

Packet-Switching Network

Packet switching involves dividing data messages into small bundles of information and transmitting them through communications networks to their intended destinations using computer-controlled switches. Three common switching techniques are used with public data networks: *circuit switching*, *message switching*, and *packet switching*.

Circuit switching. Circuit switching is used for making a standard telephone call on the public telephone network. The call is established, information is transferred, and then the call is disconnected. The time required to establish the call is called the *setup* time. Once the call has been established, the circuits interconnected by the network switches are allocated to a single user for the duration of the call. After a call has been established, information is transferred in *real time*. When a call is terminated, the circuits and switches are once again available for another user. Because there are a limited number of circuits and switching paths available, *blocking* can occur. Blocking is when a call cannot be completed because there are no facilities or switching paths available between the source and destination locations. When circuit switching is used for data transfer, the terminal equipment at the source and destination must be compatible; they must use compatible modems and the same bit rate, character set, and protocol.

A circuit switch is a *transparent* switch. The switch is transparent to the data; it does nothing more than interconnect the source and destination terminal equipment. A circuit switch adds no value to the circuit.

Message switching.   Message switching is a form of *store-and-forward* network. Data, including source and destination identification codes, are transmitted into the network and stored in a switch. Each switch within the network has message storage capabilities. The network transfers the data from switch to switch when it is convenient to do so. Consequently, data are not transferred in real time; there can be a delay at each switch. With message switching, blocking cannot occur. However, the delay time from message transmission to reception varies from call to call and can be quite long (possibly as long as 24 hours). With message switching, once the information has entered the network, it is converted to a more suitable format for transmission through the network. At the receive end, the data are converted to a format compatible with the receiving data terminal equipment. Therefore, with message switching, the source and destination data terminal equipment do not need to be compatible. Message switching is more efficient than circuit switching because data that enter the network during busy times can be held and transmitted later when the load has decreased.

A message switch is a *transactional* switch because it does more than simply transfer the data from the source to the destination. A message switch can store data or change its format and bit rate, then convert the data back to their original form or an entirely different form at the receive end. Message switching multiplexes data from different sources onto a common facility.

Packet switching.   With packet switching, data are divided into smaller segments called *packets* prior to transmission through the network. Because a packet can be held in memory at a switch for a short period of time, packet switching is sometimes called a *hold-and-forward* network. With packet switching, a message is divided into packets and each packet can take a different path through the network. Consequently, all packets do not necessarily arrive at the receive end at the same time or in the same order in which they were transmitted. Because packets are small, the hold time is generally quite short and message transfer is near real time and blocking cannot occur. However, packet-switching networks require complex and expensive switching arrangements and complicated protocols. A packet switch is also a transactional switch. Circuit, message, and packet switching techniques are summarized in Table 3-3.

CCITT X.1 International User Class of Service

The CCITT X.1 standard divides the various classes of service into three basic modes of transmission for a public data network. The three modes are: *start/stop*, *synchronous*, and *packet*.

TABLE 3-3   SWITCHING TECHNIQUE SUMMARY

Circuit switching	Message switching	Packet switching
Dedicated transmission path	No dedicated transmission path	No dedicated transmission path
Continuous transmission of data	Transmission of messages	Transmission of packets
Operates in real time	Not real time	Near real time
Messages not stored	Messages stored	Messages held for short time
Path established for entire message	Route established for each message	Route established for each packet
Call setup delay	Message transmission delay	Packet transmission delay
Busy signal if called party busy	No busy signal	No busy signal
Blocking may occur	Blocking cannot occur	Blocking cannot occur
User responsible for message-loss protection	Network responsible for lost messages	Network may be responsible for each packet but not for entire message
No speed or code conversion	Speed and code conversion	Speed and code conversion
Fixed bandwidth transmission (i.e., fixed information capacity)	Dynamic use of bandwidth	Dynamic use of bandwidth
No overhead bits after initial setup delay	Overhead bits in each message	Overhead bits in each packet

Start/stop mode.   With the start/stop mode, data are transferred from the source to the network and from the network to the destination in an asynchronous data format (i.e., each character is framed within a start and stop bit). Call control signaling is done in International Alphabet No. 5 (ASCII-77). Two common protocols used for start/stop transmission are IBM's 83B protocol and AT&T's 8A1/B1 selective calling arrangement.

Synchronous mode.   With the synchronous mode, data are transferred from the source to the network and from the network to the destination in a synchronous data format (i.e., each message is preceded by a unique synchronizing character). Call control signaling is identical to that used with private line data circuits and common protocols used for synchronous transmission are IBM's 3270 bisync, Burrough's BASIC, and UNIVAC's UNISCOPE.

Packet mode.   With the packet mode, data are transferred from the source to the network and from the network to the destination in a frame format. The ISO HDLC frame format is the standard data link protocol used with the packet mode. Within the network, data are divided into smaller packets and transferred in accordance with the CCITT X.25 user to network interface protocol.

Figure 3-4 illustrates a typical layout for a public data network showing each of the three modes of operation. The packet assembler/disassembler (PAD) interfaces user data to X.25 format when the user's data are in either the asynchronous or synchronous mode of operation. A PAD is unnecessary when the user is operating

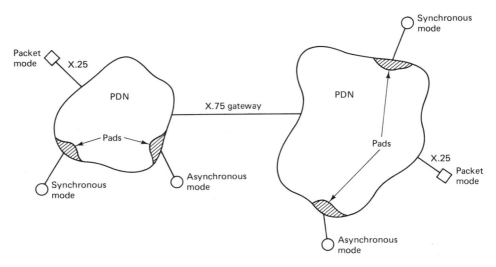

Figure 3-4 Public data network.

in the packet mode. X.75 is recommended by the CCITT for the *gateway* protocol. A gateway is used to interface two public data networks.

ISO PROTOCOL HIERARCHY

The ISO international protocol hierarchy is shown in Figure 3-5. This hierarchy was developed to facilitate the intercommunications of data processing equipment by separating network responsibilities into seven *levels* or *layers*. The basic concept of layering responsibilities is that each layer adds value to services provided by the sets of lower layers. In this way, the highest level is offered the full set of services needed to run a distributed data application.

From Figure 3-5 it can be seen that each level of the hierarchy adds overhead to the data. In fact, if all seven levels are addressed, less than 15% of the transmitted message is source information. The basic services provided by each layer of the hierarchy are summarized below.

1. *Physical layer.* The physical layer is the lowest level of the hierarchy and specifies the physical, electrical, functional, and procedural standards for accessing the data communications network. The specifications outlined by the physical layer are similar to those specified by the EIA RS-232C interface standard.

2. *Data link layer.* The data link layer is responsible for communications between primary and secondary nodes within the network. The data link layer provides a means to activate, maintain, and deactivate the data link. The data link layer provides the final framing of the information envelope, facilitates the orderly flow of data between nodes, and allows for error detection and correction.

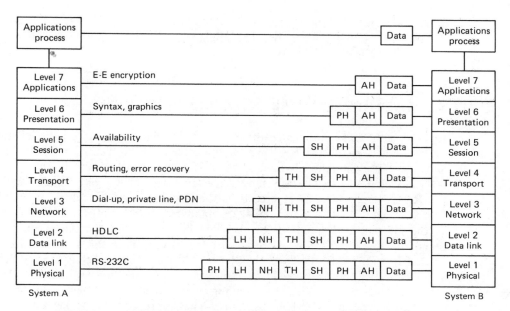

Figure 3-5 ISO international protocol hierarchy. AH, Applications header; PH, presentation header; SH, session header; TH, transport header; NH, network header; LH, link header; PH, physical header.

3. *Network layer.* The network layer determines which network configuration (dial-up, leased, or packet) is most appropriate for the function provided by the network.

4. *Transport layer.* The transport layer controls the end-to-end integrity of the message, which includes message routing, segmenting, and error recovery. The transport layer acts as the interface between the network and session layers.

5. *Session layer.* The session layer is responsible for network availability (i.e., buffer storage and processor capacity). Session responsibilities includes network log-on and log-off procedures and user authentication. A session is the temporary condition that exists when data are actually in the process of being transferred and does not include procedures such as call establishment, setup, or disconnect procedures. The session layer determines the type of dialogue available (i.e., simplex, half-duplex, or full-duplex).

6. *Presentation layer.* The presentation layer is concerned with syntax or representation. Presentation functions include data formatting, encoding, the encryption/decryption of messages, dialogue procedures, synchronization, interruption, and termination. The presentation layer performs code and character set translation and determines the display mechanism for messages.

7. *Application layer.* The application layer is analogous to the general manager of the network. The application layer controls the sequence of activities within an

application and also the sequence of events between the computer application and the user of another application. The application layer communicates directly with the user's application program.

CCITT X.25 USER-TO-NETWORK INTERFACE PROTOCOL

In 1976, the CCITT designated the X.25 user interface as the international standard for packet network access. Keep in mind that X.25 is strictly a *user-to-network* interface and addresses only the physical, data link, and network layers in the ISO seven layer model. X.25 uses existing standards whenever possible. For example, X.25 specifies X.21, X.26, and X.27 standards as the physical interface, which correspond to EIA RS-232C, RS-423A, and RS-422A standards, respectively. X.25 defines HDLC as the international standard for the data link layer and the American National Standards Institute (ANSI) 3.66 *advanced data communications control procedures* (ADCCP) as the U.S. standard. ANSI 3.66 and ISO HDLC specify exactly the same set of data link control procedures. However, ANSI 3.66 and HDLC were designed for private line data circuits with a polling environment. Consequently, the addressing and control procedures outlined by them are not appropriate for packet data networks. ANSI 3.66 and HDLC were selected for the data link layer because of their frame format, delimiting sequence, transparency mechanism, and error detection method.

The network layer of X.25 specifies three switching services offered in a switched data network: permanent virtual circuit, virtual call, and datagram.

Permanent Virtual Circuit

A *permanent virtual circuit* (PVC) is logically equivalent to a two-point dedicated private line circuit except slower. A PVC is slower because a hardwired connection is not provided. Each time a connection is requested, the appropriate switches and circuits must be established through the network to provide the interconnection. A PVC establishes a connection between two predetermined subscribers of the network on demand, but not permanently. With a PVC, a source and destination address are unnecessary because the two users are fixed.

Virtual Call

A *virtual call* (VC) is logically equivalent to making a telephone call through the DDD network. A VC is a one-to-many arrangement. Any VC subscriber can access any other VC subscriber through a network of switches and communication channels. Virtual calls are temporary connections that use common usage equipment and circuits. The source must provide its address and the address of the destination before a VC can be completed.

Datagram

A *datagram* (DG) is, at best, vaguely defined by X.25 and, until it is completely outlined, has very limited usefulness. With a DG, users send small packets of data into the network. The network does not acknowledge packets nor does it guarantee successful transmission. However, if a message will fit into a single packet, a DG is somewhat reliable. This is called a *single-packet-per-segment* protocol.

X.25 Packet Format

A virtual call is the most efficient service offered for a packet network. There are two packet formats used with virtual calls: a call request packet and a data transfer packet.

Call request packet.   Figure 3-6 shows the field format for a call request packet. The delimiting sequence is 01111110 (an HDLC flag), and the error detection/correction mechanism is CRC-16 with ARQ. The link address field and the control field have little use and are therefore seldom used with packet networks. The rest of the fields are defined in sequence.

Format identifier. The format identifier identifies whether the packet is a new call request or a previously established call. The format identifier also identifies the packet numbering sequence (either 0–7 or 0–127).

Logical channel identifier (LCI). The LCI is a 12-bit binary number that identifies the source and destination users for a given virtual call. After a source user has gained access to the network and has identified the destination user, they are assigned an LCI. In subsequent packets, the source and destination addresses are unnecessary; only the LCI is needed. When two users disconnect, the LCI is relinquished and can be reassigned two new users. There are 4096 LCIs available. Therefore, there may be as many as 4096 virtual calls established at any given time.

Packet type. This field is used to identify the function and the content of the packet (i.e., new request, call clear, call reset, etc.).

Calling address length. This 4-bit field gives the number of digits (in binary) that appear in the calling address field. With 4 bits, up to 15 digits can be specified.

Called address length. This field is the same as the calling address field except that it identifies the number of digits that appear in the called address field.

Called address. This field contains the destination address. Up to 15 BCD digits can be assigned to a destination user.

Calling address. This field is the same as the called address field except that it contains up to 15 BCD digits that can be assigned to a source user.

Flag	Link address field	Link control field	Format identifier	Logical channel identifier	Packet type	Calling address length	Called address length	Called address	Calling address	0	Facilities field length	Facilities field	Protocol ID	User data	Frame check sequence	Flag
8	8	8	4	12	8	4	4	To 60	To 60	2	6	To 512	32	To 96	16	8

Bits:

Figure 3-6 Call request packet format.

130

Facilities length field. This field identifies (in binary) the number of 8-bit octets present in the facilities field.

Facilities field. This field contains up to 512 bits of optional network facility information, such as reverse billing information, closed user groups, and whether it is a simplex transmit or simplex receive connection.

Protocol identifier. This 32-bit field is reserved for the subscriber to insert user-level protocol functions such as log-on procedures and user identification practices.

User data field. Up to 96 bits of user data can be transmitted with a call request packet. These are unnumbered data which are not confirmed. This field is generally used for user passwords.

Data transfer packet. Figure 3-7 shows the field format for a data transfer packet. A data transfer packet is similar to a call request packet except that a data transfer packet has considerably less overhead and can accommodate a much larger user data field. The data transfer packet contains a send and receive packet sequence field that were not included with the call request format.

The flag, link address, link control, format identifier, LCI, and FCS fields are identical to those used with the call request packet. The send and receive packet sequence fields are described as follows.

Send packet sequence field. This field is used in the same manner that the ns and nr sequences are used with SDLC and HDLC. P(s) is analogous to ns, and P(r) is analogous to nr. Each successive data transfer packet is assigned the next P(s) number in sequence. The P(s) can be a 3- or a 7-bit binary number and thus number packets from either 0–7 or 0–127. The numbering sequence is identified in the format identifier. The send packet field always contains 8 bits and the unused bits are reset.

Receive packet sequence field. P(r) is used to confirm received packets and call for retransmission of packets received in error (ARQ). The I field in a data transfer packet can have considerably more source information than an I field in a call request packet.

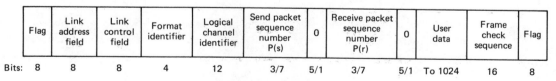

Flag	Link address field	Link control field	Format identifier	Logical channel identifier	Send packet sequence number P(s)	0	Receive packet sequence number P(r)	0	User data	Frame check sequence	Flag
Bits: 8	8	8	4	12	3/7	5/1	3/7	5/1	To 1024	16	8

Figure 3-7 Data transfer packet format.

LOCAL AREA NETWORKS

A *local area network* (LAN) is a data communications network that is designed to provide two-way communications between a large variety of data communications terminal equipment within a relatively small geographic area. LANs are privately owned and operated and are used to interconnect data terminal equipment in the same building or building complex.

Local Area Network System Considerations

Topology. The topology or physical architecture of a LAN identifies how the stations are interconnected. The most common configurations used with LANs are the star, bus, ring, and mesh topologies.

Connecting medium. Presently, all LANs use coaxial cable as the transmission medium, although in the near future it is likely that fiber optic cables will also be used. Fiber cables can operate at higher bit rates and, consequently, have a larger capacity to transfer information than coaxial cables. LANs that use a coaxial cable are limited to an overall length of approximately 1500 m. Fiber links are expected to far exceed this distance.

Transmission format. There are two basic approaches to transmission format for LANs: *baseband* and *broadband*. Baseband transmission uses the connecting medium as a single-channel device. Only one station can transmit at a time and all stations must transmit and receive the same types of signals (encoding schemes and bit rates). Essentially, a baseband format time division multiplexes signals onto the transmission medium. Broadband transmission uses the connecting medium as a multichannel device. Each channel occupies a different frequency band (i.e., frequency-division multiplexing). Consequently, each channel can contain different encoding schemes and operate at different bit rates. A broadband network permits voice, digital data, and video to be transmitted simultaneously over the same transmission medium. However, baseband systems require RF modems, amplifiers, and more complicated transceivers than baseband systems. For this reason, baseband systems are more prevalent. Table 3-4 summarizes baseband and broadband transmission techniques.

Channel Accessing

Channel accessing describes the mechanism used by a station to gain access to a local area network. There are essentially two methods used for channel accessing with LANs: carrier sense, multiple access with collision detection (CSMA/CD) and token passing.

Carrier sense, multiple access with collision detection. With CSMA/CD, a station monitors (listens to) the line to determine if the line is busy. If a station has a message to transmit but the line is busy, it waits for an idle condition before it transmits its message. If two stations begin transmitting at the same time, a *collision* occurs. When this happens, both stations cease transmitting (*back off*) and each station waits a random period of time before attempting a retransmission. The random

TABLE 3-4   TRANSMISSION FORMAT SUMMARY

Baseband	Broadband
Characteristics	
Digital signaling	Analog signaling (requires RF modem)
Entire bandwidth used by signal	FDM possible (i.e., multiple data channels)
Bidirectional	Unidirectional
Bus topology	Bus topology
Maximum length approximately 1500 meters	Maximum length up to tens of kilometers
Advantages	
Less expensive	High capacity
Simpler technology	Multiple traffic types
Easy and quick to install	More flexible circuit configurations
	Larger area covered
Disadvantages	
Single channel	Modem required
Limited capacity	Complex installation and maintenance
Grounding problems	Double propagation delay
Limited distance	

delay time for each station is different and therefore allows for prioritizing the stations on the network. With CSMA/CD, stations must contend for the network. A station is not guaranteed access to the network. To detect the occurrence of a collision, a station must be capable of transmitting and receiving simultaneously. CSMA/CD is used by most baseband LANs in the bus configuration. Ethernet is a popular local area network that uses baseband transmission with CSMA/CD. The transmission rate with Ethernet is 10 Mbps over a coaxial cable. Collision detection is accomplished by monitoring the line for phase violations in a Manchester-encoded (biphase) digital encoding scheme.

Token passing.   Token passing is a channel-accessing arrangement that is best suited for a ring topology with either a baseband or a broadband network. With token passing, an electrical *token* (*code*) is circulated around the ring from station to station. Each station, in turn, acquires the token. In order to transmit, a station must first possess the token; then the station removes the token and places its message on the line. After a station transmits, it passes the token on to the next sequential station. With token passing, each station has equal access to the transmission medium.

The Cambridge ring is a popular local area network that uses baseband transmission with token passing. The transmission rate with a Cambridge ring is 10 Mbps. Table 3-5 lists several local area networks and some of their characteristics.

TABLE 3-5   LOCAL AREA NETWORK SUMMARY

Ethernet:	Developed by Xerox Corporation in conjunction with Digital Equipment Corporation and Intel Corporation; baseband system using CSMA/CD; 10 Mbps
Wangnet:	Developed by Wang Computer Corporation; broadband system using CSMA/CD
Localnet:	Developed by Sytek Corporation; broadband system using CSMA/CD
Domain:	Developed by Apollo Computer Corporation; broadband network using token passing
Cambridge ring:	Developed by the University of Cambridge; baseband system using CSMA/CD; 10 Mbps

In 1980, the IEEE local area network committee was established to standardize the means of connecting digital computer equipment and peripherals with the local area network environment. In 1983, the committee established IEEE standards 802.3 (CSMA/CD) and 802.4 (token passing) for a bus topology. IEEE standard 802.5, which is still pending approval, addresses token passing with a ring topology.

QUESTIONS

3-1. Define *data communications protocol*.

3-2. What is a master station? A slave station?

3-3. Define *polling* and *selecting*.

3-4. What is the difference between a synchronous and an asynchronous protocol?

3-5. What is the difference between a character-oriented protocol and a bit-oriented protocol?

3-6. Define the three operating modes used with data communications circuits.

3-7. What is the function of the clearing character?

3-8. What is a unique address? A group address? A broadcast address?

3-9. What does a negative acknowledgment to a poll indicate?

3-10. What is the purpose of a heading?

3-11. Why is IBM's 3270 synchronous protocol called bisync?

3-12. Why are SYN characters always transmitted in pairs?

3-13. What is an SPA? An SSA? A DA?

3-14. What is the purpose of a leading pad? A trailing pad?

3-15. What is the difference between a general poll and a specific poll?

3-16. What is a handshake?

3-17. (Primary, secondary) stations transmit polls.

3-18. What does a negative acknowledgment to a poll indicate?

3-19. What is a positive acknowledgment to a poll?

3-20. What is the difference between ETX, ETB, and ITB?

3-21. What character is used to terminate a heading and begin a block of text?

3-22. What is transparency? When is it necessary? Why?

3-23. What is the difference between a command and a response with SDLC?

3-24. What are the three transmission states used with SDLC? Explain them.

3-25. What are the five fields used with an SDLC frame? Briefly explain each.

3-26. What is the delimiting sequence used with SDLC?

3-27. What is the null address in SDLC? When is it used?

3-28. What are the three frame formats used with SDLC? Explain what each format is used for.

3-29. How is an information frame identified in SDLC? A supervisory frame? An unnumbered frame?

3-30. What are the purposes of the nr and ns sequences in SDLC?

3-31. With SDLC, when is the P bit set? The F bit?

3-32. What is the maximum number of unconfirmed frames that can be outstanding at any one time with SDLC? Why?

3-33. With SDLC, which frame formats can have an information field?

3-34. With SLDC, which frame formats can be used to confirm previously received frames?

3-35. What command/response is used for reporting procedual errors with SDLC?

3-36. Explain the three modes in SDLC that a secondary station can be in.

3-37. When is the configure command/response used with SDLC?

3-38. What is a go-ahead sequence? A turnaround sequence?

3-39. What is the transparency mechanism used with SDLC?

3-40. What is a message abort? When is it transmitted?

3-41. Explain invert-on-zero encoding. Why is it used?

3-42. What supervisory condition exists with HDLC that is not included with SDLC?

3-43. What is the delimiting sequence used with HDLC? The transparency mechanism?

3-44. Explain extended addressing as it is used with HDLC.

3-45. What is the difference between the basic control format and the extended control format with HDLC?

3-46. What is the difference in the information fields used with SDLC and HDLC?

3-47. What operational modes are included with HDLC that are not included with SDLC?

3-48. What is a public data network?

3-49. Describe a value-added network.

3-50. Explain the differences in circuit-, message-, and packet-switching techniques.

3-51. What is blocking? With which switching techniques is blocking possible?

3-52. What is a transparent switch? A transactional switch?

3-53. What is a packet?

3-54. What is the difference between a store-and-forward and a hold-and-forward network?

3-55. Explain the three modes of transmission for public data networks.

3-56. What is the user-to-network protocol designated by CCITT?

3-57. What is the user-to-network protocol designated by ANSI?

3-58. Which layers of the ISO protocol hierarchy are addressed by X.25?

3-59. Explain the following terms: permanent virtual circuit, virtual call, and datagram.

3-60. Why was HDLC selected as the link-level protocol for X.25?

3-61. Briefly explain the fields that make up an X.25 call request packet.

3-62. Describe a local area network.

3-63. What is the connecting medium used with local area networks?

3-64. Explain the two transmission formats used with local area networks.

3-65. Explain CSMA/CD.

3-66. Explain token passing.

PROBLEMS

3-1. Determine the hex code for the control field in an SDLC frame for the following conditions: information frame, poll, transmitting frame 4, and confirming reception of frames 2, 3, and 4.

3-2 Determine the hex code for the control field in an SDLC frame for the following conditions: supervisory frame, ready to receive, final, confirming reception of frames 6, 7, and 0.

3-3. Insert 0's into the following SDLC data stream.

1110010000111111111001111101001110101111111111001011

3-4. Delete 0's from the following SDLC data stream.

010111110100011011111011101101011111010111000111100

3-5. Sketch the NRZI levels for the following data stream (start with a high condition).

        1   0   0   1   1   1   0   0   1   0   1   0

Chapter 4

DIGITAL TRANSMISSION

INTRODUCTION

As stated previously, digital transmission is the transmittal of digital pulses between two points in a communications system. The original source information may already be in digital form or it may be analog signals that must be converted to digital pulses prior to transmission and converted back to analog form at the receive end. With digital transmission systems, a physical facility such as a metallic wire pair, a coaxial cable, or a fiber optic link is required to interconnect the two points in the system. The pulses are contained in and propagate down the facility.

Advantages of Digital Transmission

1. The primary advantage of digital transmission is noise immunity. Analog signals are more susceptible than digital pulses to undesired amplitude, frequency, and phase variations. This is because with digital transmission, it is not necessary to evaluate these parameters as precisely as with analog transmission. Instead, the received pulses are evaluated during a sample interval, and a simple determination is made whether the pulse is above or below a certain threshold.

2. Digital pulses are better suited to processing and multiplexing than analog signals. Digital pulses can be stored easily, whereas analog signals cannot. Also, the transmission rate of a digital system can easily be changed to adapt to different environments and to interface with different types of equipment. Multiplexing is explained in detail in Chapter 5.

Disadvantages of Digital Transmission

1. The transmission of digitally encoded analog signals requires more bandwidth than simply transmitting the analog signal.
2. Analog signals must be converted to digital codes prior to transmission and converted back to analog at the receiver.
3. Digital transmission requires precise time synchronization between transmitter and receiver clocks.
4. Digital transmission systems are incompatible with existing analog facilities.

PULSE MODULATION

Pulse modulation includes many different methods of transferring pulses from a source to a destination. The four predominant methods are *pulse width modulation* (PWM), *pulse position modulation* (PPM), *pulse amplitude modulation* (PAM), and *pulse code modulation* (PCM). The four most common methods of pulse modulation are summarized below and shown in Figure 4-1.

1. *PWM*. This method is sometimes called pulse duration modulation (PDM) or pulse length modulation (PLM). The pulse width (active portion of the duty cycle) is proportional to the amplitude of the analog signal.
2. *PPM*. The position of a constant-width pulse within a prescribed time slot is varied according to the amplitude of the analog signal.
3. *PAM*. The amplitude of a constant-width, constant-position pulse is varied according to the amplitude of the analog signal.

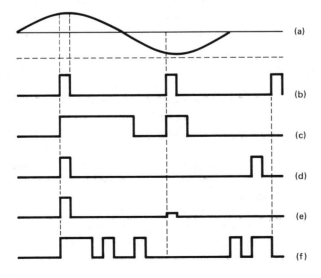

Figure 4-1 Pulse modulation: (a) analog signal; (b) sample pulse; (c) PWM; (d) PPM; (e) PAM; (f) PCM.

4. *PCM*. The analog signal is sampled and converted to a fixed-length, serial binary number for transmission. The binary number varies according to the amplitude of the analog signal.

PAM is used as an intermediate form of modulation with PSK, QAM, and PCM, although it is seldom used by itself. PWM and PPM are used in special-purpose communications systems (usually for the military) but are seldom used for commercial systems. PCM is by far the most prevalent method of pulse transmission and consequently, will be the topic of discussion for the remainder of this chapter.

PULSE CODE MODULATION

Pulse code modulation (PCM) is the only one of the pulse modulation techniques previously mentioned that is a digital transmission system. With PCM, the pulses are of fixed length and fixed amplitude. PCM is a binary system; a pulse or lack of a pulse within a prescribed time slot represents either a logic 1 or a logic 0 condition. With PWM, PPM, or PAM; a single pulse does not represent a single binary digit (bit).

Figure 4-2 shows a simplified block diagram of a single-channel, *simplex* (*one-way-only*) PCM system. The bandpass filter limits the input analog signal to the standard voice band frequency range 300 to 3000 Hz. The *sample-and-hold* circuit periodically samples the analog input and converts those samples to a multilevel PAM signal. The *analog-to-digital converter* (ADC) converts the PAM samples to a serial binary data stream for transmission. The transmission medium is generally a metallic wire pair.

At the receive end, the *digital-to-analog converter* (DAC) converts the serial binary data stream to a multilevel PAM signal. The sample-and-hold circuit and

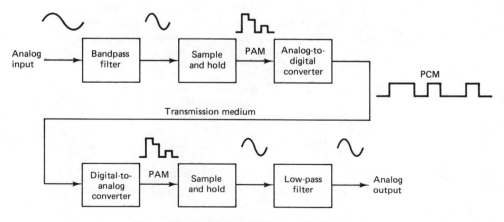

Figure 4-2 Simplified PCM system block diagram.

low-pass filter convert the PAM signal back to its original analog form. An integrated circuit that performs the PCM encoding and decoding is called a *codec* (*co*der/ *dec*oder). The codec is explained in detail in Chapter 5.

Sample-and-Hold Circuit

The purpose of the sample-and-hold circuit is to sample periodically the continually changing analog input signal and convert the sample to a series of constant-amplitude PAM levels. For the ADC to accurately convert a signal to a digital code, the signal must be relatively constant. If not, before the ADC can complete the conversion, the input would change. Therefore, the ADC would continually be attempting to following the analog changes and never stabilize on any PCM code.

Figure 4-3 shows the schematic diagram of a sample-and-hold circuit. The FET acts like a simple switch. When turned "on," it provides a low-impedance path to deposit the analog sample across capacitor C1. The time that Q1 is "on" is called the *aperture* or *acquisition time*. Essentially, C1 is the hold circuit. When Q1 is "off," the capacitor does not have a complete path to discharge through and therefore stores the sampled voltage. The *storage time* of the capacitor is also called the A/D *conversion time* because it is during this time that the ADC converts the sample voltage to a digital code. The acquisition time should be very short. This assures that a minimum change occurs in the analog signal while it is being deposited across C1. If the input to the ADC is changing while it is performing the conversion, distortion results. This distortion is called *aperture distortion*. Thus, by having a short aperture time and keeping the input to the ADC relatively constant, the sample-and-hold circuit reduces aperture distortion. If the analog signal is sampled for a short period of time and the sample voltage is held at a constant amplitude during the A/D conversion time, this is called *flat-top sampling*. If the sample time is made longer and the analog-to-digital conversion takes place with a changing analog signal, this is called *natural sampling*. Natural sampling introduces more aperture distortion than flat-top sampling and requires a faster A/D converter.

Figure 4-4 shows the input analog signal, the sampling pulse, and the waveform developed across C1. It is important that the output impedance of voltage follower

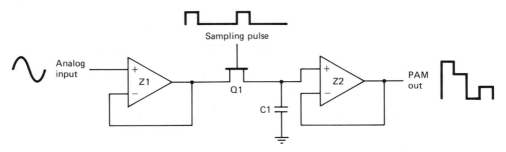

Figure 4-3   Sample-and-hold circuit.

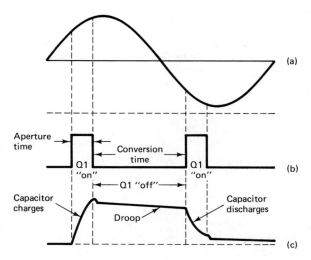

Aperture time

Conversion time

Q1 "on"

Q1 "on"

Q1 "off"

Capacitor charges

Droop

Capacitor discharges

(a)

(b)

(c)

Figure 4-4 Sample-and-hold waveforms: (a) analog in; (b) sample pulse; (c) capacitor voltage.

Z1 and the "on" resistance of Q1 be as small as possible. This assures that the *RC* charging time constant of the capacitor is kept very short, allowing the capacitor to charge or discharge rapidly during the short acquisition time. The rapid drop in the capacitor voltage immediately following each sample pulse is due to the redistribution of the charge across C1. The interelectrode capacitance between the gate and drain of the FET is placed in series with C1 when the FET is "off," thus acting like a capacitive voltage-divider network. Also, note the gradual discharge across the capacitor during the conversion time. This is called *droop* and is caused by the capacitor discharging through its own leakage resistance and the input impedance of voltage follower Z2. Therefore, it is important that the input impedance of Z2 and the leakage resistance of C1 be as high as possible. Essentially, voltage followers Z1 and Z2 isolate the sample-and-hold circuit (Q1 and C1) from the input and output circuitry.

EXAMPLE 4-1

For the sample-and-hold circuit shown in Figure 4-3, determine the largest-value capacitor that can be used. Use an output impedance for Z1 of 10 Ω, an "on" resistance for Q1 of 10 Ω, an acquisition time of 10 μs, a maximum peak-to-peak input voltage of 10 V, a maximum output current from Z1 of 10 mA, and an accuracy of 1%.

Solution The expression for the current through a capacitor is

$$i = C\frac{dv}{dt}$$

Rearranging and solving for *C* yields

$$C = i\frac{dt}{dv}$$

where

C = maximum capacitance
i = maximum output current from Z1, 10 mA
dv = maximum change in voltage across C1, which equals 10 V
dt = charge time, which equals the aperture time, 10 μs

Therefore,

$$C_{max} = \frac{(10 \text{ mA}) (10 \text{ μs})}{10 \text{ V}} = 10 \text{ nF}$$

The charge time constant for C when Q1 is "on" is

$$\tau = RC$$

where

τ = one charge time constant
R = output impedance of Z1 plus the "on" resistance of Q1
C = capacitance value of C1

Rearranging and solving for C gives us

$$C_{max} = \frac{\tau}{R}$$

The charge time of capacitor C1 is also dependent on the accuracy desired from the device. The percent accuracy and its required RC time constant are summarized as follows:

Accuracy (%)	Charge time
10	3τ
1	4τ
0.1	7τ
0.01	9τ

For an accuracy of 1%,

$$C = \frac{10 \text{ μs}}{4 (20)} = 125 \text{ nF}$$

To satisfy the output current limitations of Z1, a maximum capacitance of 10 nF was required. To satisfy the accuracy requirements, 125 nF was required. To satisfy both requirements, the smaller-value capacitor must be used. Therefore, C1 can be no larger than 10 nF.

Sampling Rate

The Nyquist sampling theorem establishes the *minimum sampling rate* (F_s) that can be used for a given PCM system. For a sample to be reproduced accurately at the receiver, each cycle of the analog input signal (F_a) must be sampled at least twice. Consequently, the minimum sampling rate is equal to twice the highest audio input frequency. If F_s is less than two times F_a, distortion will result. This distortion is called *aliasing* or *foldover distortion*. Mathematically, the minimum Nyquist sample rate is

$$F_s \geq 2F_a \qquad (4\text{-}1)$$

where

F_s = minimum Nyquist sample rate
F_a = highest frequency to be sampled

Essentially, a sample-and-hold circuit is an AM modulator. The switch is a nonlinear device that has two inputs: the sampling pulse and the input analog signal. Consequently, *nonlinear mixing* (*heterodyning*) occurs between these two signals. Figure 4-5a shows the frequency-domain representation of the output spectrum from a sample-and-hold circuit. The output includes the two original inputs (the audio and the fundamental frequency of the sampling pulse), their sum and difference frequencies ($F_s \pm F_a$), all the harmonics of F_s and F_a ($2F_s$, $2F_a$, $3F_s$, $3F_a$, etc.), and their associated sidebands ($2F_s \pm F_a$, $3F_s \pm F_a$, etc.).

Because the sampling pulse is a repetitive waveform, it is made up of a series of harmonically related sine waves. Each of these sine waves is amplitude modulated by the analog signal and produces sum and difference frequencies symetrical around each of the harmonics of F_s. Each sum and difference frequency generated is separated from its respective center frequency by F_a. As long as F_s is at least twice F_a, none of the side frequencies from one harmonic will spill into the sidebands of another harmonic and aliasing does not occur. Figure 4-5b shows the results when an analog input frequency greater than $F_s/2$ modulates F_s. The side frequencies from one harmonic foldover into the sideband of another harmonic. The frequency that folds over is an alias of the input signal (hence the names "aliasing" or "foldover distortion"). If an alias side frequency from the first harmonic folds over into the input audio spectrum, it cannot be removed through filtering or any other technique.

EXAMPLE 4-2

For a PCM system with a maximum audio input frequency of 4 kHz, determine the minimum sample rate and the alias frequency produced if a 5-kHz audio signal were allowed to enter the sample-and-hold circuit.

Solution Using Nyquist's sampling theorem (Equation 4-1), we have

$$F_s \geq 2F_a \qquad \text{therefore, } F_s \geq 8 \text{ kHz}$$

If a 5-kHz audio frequency entered the sample-and-hold circuit, the output spectrum shown in Figure 4-6 is produced. It can be seen that the 5-kHz signal produces an alias frequency of 3 kHz that folds over into the original audio spectrum.

The input bandpass filter shown in Figure 4-2 is called an *antialiasing* or *antifoldover filter*. Its upper cutoff frequency is chosen such that no frequency greater than

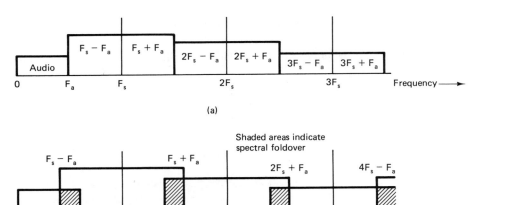

Figure 4-5 Output spectrum for a sample-and-hold circuit: (a) no aliasing; (b) aliasing distortion.

one-half of the sampling rate is allowed to enter the sample-and-hold circuit, thus eliminating the possibility of foldover distortion occurring.

PCM CODES

With PCM, the analog input signal is sampled, then converted to a serial binary code. The binary code is transmitted to the receiver, where it is converted back to the original analog signal. The binary codes used for PCM are n-bit codes, where n may be any positive whole number greater than 1. The codes currently used for PCM are *sign-magnitude codes*, where the *most significant bit* (MSB) is the sign bit and the remaining bits are used for magnitude. Table 4-1 shows an n-bit PCM code where n equals 3. The most significant bit is used to represent the sign of the sample (logic 1 = positive and logic 0 = negative). The two remaining bits represent the magnitude. With 2 magnitude bits, there are four codes possible for positive numbers and four possible for negative numbers. Consequently, there is a total of eight possible codes ($2^3 = 8$).

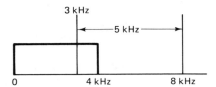

Figure 4-6 Output spectrum for Example 4-2.

TABLE 4-1 3-BIT PCM CODE

Sign	Magnitude		Level	Decimal
1	1	1	———————	+3
1	1	0	———————	+2
1	0	1	———————	+1
1	0	0	———————	+0
0	0	0	———————	−0
0	0	1	———————	−1
0	1	0	———————	−2
0	1	1	———————	−3

Folded Binary Code

The PCM code shown in Table 4-1 is called a *folded binary code*. Except for the sign bit, the codes on the bottom half of the table are a mirror image of the codes on the top half. (If the negative codes were folded over on top of the positive codes, they would match perfectly.) Also, with folded binary there are two codes assigned to zero volts: 100 (+0) and 000 (−0). For this example, the magnitude of the minimum step size is 1 V. Therefore, the maximum voltage that may be encoded with this scheme is +3 V (111) or −3 V (011). If the magnitude of a sample exceeds the highest quantization interval, *overload distortion* (also called *peak limiting*) occurs. Assigning PCM codes to absolute magnitudes is called *quantizing*. The magnitude of the minimum step size is called *resolution*, which is equal in magnitude to the voltage of the least significant bit (V_{lsb} or the magnitude of the minimum step size of the DAC). The resolution is the minimum voltage other than 0 V that can be decoded by the DAC at the receiver. The smaller the magnitude of the minimum step size, the better (smaller) the resolution and the more accurately the quantization interval will resemble the actual analog sample.

In Table 4-1, each 3-bit code has a range of input voltages that will be converted to that code. For example, any voltage between +0.5 and +1.5 will be converted to the code 101. Any voltage between +1.5 and +2.5 will be encoded as 110. Each code has a *quantization range* equal to + or − one-half the resolution except the codes for +0 V and −0 V. The 0-V codes each have an input range equal to only one-half the resolution, but because there are two 0-V codes, the range for 0 V is also + or − one-half the resolution. Consequently, the maximum input voltage to the system is equal to the voltage of the highest magnitude code plus one-half of the voltage of the least significant bit.

Figure 4-7 shows an analog input signal, the sampling pulse, the corresponding PAM signal, and the PCM code. The analog signal is sampled three times. The first sample occurs at time t_1 when the analog voltage is +2 V. The PCM code that corresponds to sample 1 is 110. Sample 2 occurs at time t_2 when the analog

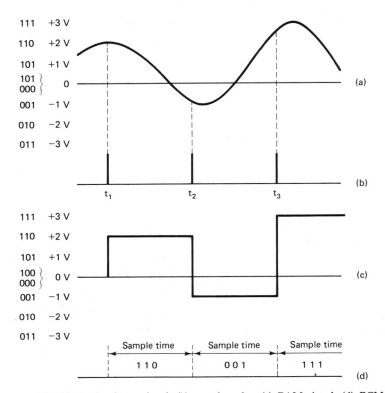

Figure 4-7 (a) Analog input signal; (b) sample pulse; (c) PAM signal; (d) PCM code.

voltage is −1 V. The corresponding PCM code is 001. To determine the PCM code for a particular sample, simply divide the voltage of the sample by the resoltuion, convert it to an n-bit binary code, and add the sign bit to it. For sample 1, the sign bit is 1, indicating a positive voltage. the magnitude code (10) corresponds to a binary 2. Two times 1 V equals 2 V, the magnitude of the sample.

Sample 3 occurs at time t_3. The voltage at this time is +2.6 V. The binary PCM code for +2.6 V is 2.6/1 = 2.6. There is no code for this magnitude. If successive approximation ADCs are used, the magnitude of the sample is rounded off to the nearest valid code (111 or +3 V for this example). This results in an error when the code is converted back to analog by the DAC at the receive end. This error is called *quantization error* (Qe). The quantization error is equivalent to additive noise (it alters the signal amplitude). Like noise, the quantization error may add to or subtract from the actual signal. Consequently, quantization error is also called *quantization noise* (Qn) and its maximum magnitude is one-half the voltage of the minimum step size ($V_{lsb}/2$). For this example, Qe = 1 V/2 or 0.5 V.

Dynamic Range

The number of PCM bits transmitted per sample is determined by several variables, which include maximum allowable input amplitude, resolution, and dynamic range. *Dynamic range* (DR) is the ratio of the largest possible magnitude to the smallest possible magnitude that can be decoded by the DAC. Mathematically, dynamic range is

$$DR = \frac{V_{max}}{V_{min}} \tag{4-2a}$$

where V_{min} is equal to the resolution and V_{max} is the maximum voltage that can be decoded by the DACs. Thus

$$DR = \frac{V_{max}}{resolution}$$

For the system shown in Table 4-1,

$$DR = \frac{3 \text{ V}}{1 \text{ V}} = 3$$

It is common to represent dynamic range in decibels; therefore,

$$DR = 20 \log \frac{V_{max}}{V_{min}} = 20 \log \frac{3}{1} = 9.54 \text{ dB} \tag{4-2b}$$

A dynamic range of 3 indicates that the ratio of the largest to the smallest decoded receive signal is 3.

If a smaller resolution is desired, such as 0.5 V, to maintain a dynamic range of 3, the maximum allowable input voltage must be reduced by the same factor, one-half.

$$DR = \frac{1.5}{0.5} = 3$$

Therefore, dynamic range is independent of resolution. If the resolution were reduced by a factor of 2 (0.25 V), to maintain the same maximum input amplitude, the dynamic range must double:

$$DR = \frac{1.5}{0.25} = 6$$

The number of bits used for a PCM code depends on the dynamic range. With a 2-bit PCM code, the minimum decodable magnitude has a binary code of 01. The maximum magnitude is 11. The ratio of the maximum binary code to the minimum binary code is 3, the same as the dynamic range. Because the minimum binary code is always 1, DR is simply the maximum binary number for a system. Consequently, to determine the number of bits required for a PCM code the following mathematical relationship is used:

$$2^n - 1 \geq DR$$

and for a minimum value of n,

$$2^n - 1 = DR \qquad (4\text{-}3a)$$

where

$$n = \text{number of PCM bits}$$
$$DR = \text{absolute value of DR}$$

Why $2^n - 1$? One PCM code is used for 0 V, which is not considered for dynamic range. Therefore,

$$2^n = DR + 1 \qquad (4\text{-}3b)$$

To solve for n, convert to logs:

$$\log 2^n = \log (DR + 1)$$
$$n \log 2 = \log (DR + 1) \qquad (4\text{-}3c)$$
$$n = \frac{\log (3 + 1)}{\log 2} = \frac{0.602}{0.301} = 2$$

For a dynamic range of 3, a PCM code with 2 bits is required.

EXAMPLE 4-3

A PCM system has the following parameters: a maximum analog input frequency of 4 kHz, a maximum decoded voltage at the receiver of $\pm 2.55 V_p$, and a minimum dynamic range of 46 dB. Determine the following: minimum sample rate, minimum number of bits used in the PCM code, resolution, and quantization error.

Solution   Substituting into Equation 4-1, the minimum sample rate is

$$F_s = 2F_a = 2\,(4 \text{ kHz}) = 8 \text{ kHz}$$

To determine the absolute value of the dynamic range, substitute into Equation 4-2b:

$$46 \text{ dB} = 20 \log \frac{V_{\text{max}}}{V_{\text{min}}}$$

$$2.3 = \log \frac{V_{\text{max}}}{V_{\text{min}}}$$

$$10^{2.3} = \frac{V_{\text{max}}}{V_{\text{min}}} = DR$$

$$199.5 = DR$$

Substitute into Equation 4-3b and solve for n:

$$n = \frac{\log(199.5 + 1)}{\log 2} = 7.63$$

The closest whole number greater than 7.63 is 8; therefore, 8 bits must be used for the magnitude.

Because the input amplitude range is $\pm 2.55 V_p$, one additional bit, the sign bit, is required. Therefore, the total number of PCM bits is 9 and the total number of PCM codes is 2^9 or 512. (There are 255 positive codes, 255 negative codes, and 2 zero codes.)

To determine the actual dynamic range, substitute into Equation 4-3c:

$$DR = 20 \log 255 = 48.13 \text{ dB}$$

To determine the resolution, divide the maximum $+$ or $-$ magnitude by the number of positive or negative nonzero PCM codes.

$$\text{resolution} = \frac{V_{\max}}{2^n - 1}$$

$$= \frac{2.55}{2^8 - 1} = \frac{2.55}{256 - 1} = 0.01 \text{ V}$$

The maximum quantization error is

$$Qe = \frac{\text{resolution}}{2} = 0.005 \text{ V}$$

Coding Efficiency

Coding efficiency is a numerical indication of how efficiently a PCM code is utilized. Coding efficiency is the ratio of the minimum number of bits required to achieve a certain dynamic range to the actual number of PCM bits used. Mathematically, coding efficiency is

$$\text{coding efficiency} = \frac{\text{minimum number of bits}}{\text{actual number of bits}} \times 100 \qquad (4\text{-}4)$$

The coding efficiency for Example 4-3 is

$$\text{coding efficiency} = \frac{8.63}{9} \times 100 = 95.89\%$$

Signal-to-Quantization Noise Ratio

The 3-bit PCM coding scheme described in the preceding section is a linear code. That is, the magnitude change between any two successive codes is uniform. Consequently, the magnitude of their quantization error is also equal. The maximum quantization noise is the voltage of the least significant bit divided by 2. Therefore, the worst possible *signal-to-quantization noise ratio* (SQR) occurs when the input signal is at its minimum amplitude (101 or 001). Mathematically, the worst-case SQR is

$$\text{SQR} = \frac{\text{minimum voltage}}{\text{quantization noise}} = \frac{V_{lsb}}{V_{lsb}/2} = 2$$

For a maximum amplitude input signal of 3 V (either 111 or 011), the maximum quantization noise is also the voltage of the least significant bit divided by 2. Therefore, the SQR for a maximum input signal condition is

$$\text{SQR} = \frac{\text{minimum voltage}}{\text{quantization noise}} = \frac{V_{max}}{V_{lsb}/2} = \frac{3}{0.5} = 6$$

From the preceding example it can be seen that even though the magnitude of error remains constant throughout the entire PCM code, the percentage of error does not; it decreases as the magnitude or the input signal increases. As a result, the SQR is not constant.

The preceding expression for SQR is for voltage and presumes the maximum quantization error and a constant-amplitude analog signal; therefore, it is of little practical use and is shown only for comparison purposes. In reality and as shown in Figure 4-7, the difference between the PAM waveform and the analog input waveform varies in magnitude. Therefore, the signal-to-quantization noise ratio is not constant. Generally, the quantization error or distortion caused by digitizing an analog sample is expressed as an average noise power-to-average signal power ratio. For linear PCM codes (all quantization intervals have equal magnitudes), the signal-to-quantizing noise ratio (also called *signal-to-distortion ratio* or *signal-to-noise ratio*) is determined as follows:

$$\text{SQR (dB)} = 10 \log \frac{v^2/R}{(q^2/12)/R}$$

where

$$R = \text{resistance}$$

$$v = \text{rms signal voltage}$$

$$q = \text{rms noise voltage}$$

$$\frac{v^2}{R} = \text{rms signal power}$$

$$\frac{q^2/12}{R} = \text{average quantization noise power}$$

If the resistances are assumed to be equal,

$$\text{SQR (dB)} = 10 \log \frac{v^2}{q^2/12} \tag{4-5a}$$

$$= 10.8 + 20 \log \frac{v}{q} \tag{4-5b}$$

Linear versus Nonlinear PCM Codes

Early PCM systems used *linear codes* (i.e., the magnitude change between any two successive steps is uniform). With linear encoding, the accuracy (resolution) for the higher-magnitude analog signals is the same as for the lower-magnitude signals, and the SQR for the lower-magnitude signals is less than for the higher-magnitude signals. With voice transmission, low-amplitude signals are more likely to occur than large-amplitude signals. Therefore, if there were more codes for the lower magnitudes, it would increase the accuracy where the accuracy is needed. As a result, there would be fewer codes available for the higher amplitudes, which would increase the quantization error for the larger-amplitude signals (thus decreasing the SQR). Such a coding technique is called *nonlinear* or *nonuniform encoding*. With nonlinear encoding, the resolution increases with the magnitude of the input signal. Figure 4-8 shows the step outputs from a linear and a nonlinear ADC.

Note, with nonlinear encoding, there are more codes at the bottom of the scale than there are at the top, thus increasing the accuracy for the smaller signals. Also note that the distance between successive codes is greater for the higher-amplitude signals, thus increasing the quantization error and reducing the SQR. Also, because the ratio of V_{max} to V_{min} is increased with nonlinear encoding, the dynamic range is larger than with a uniform code. It is evident that nonlinear encoding is a compromise; SQR is sacrificed for the high-amplitude signals to achieve more accuracy for the low-amplitude signals and to achieve a larger dynamic range.

It is difficult to fabricate nonlinear ADCs; consequently, alternative methods of achieving the same results have been devised.

Idle Channel Noise

During times when there is no analog input signal, the only input to the PAM sampler is random, thermal noise. This noise is called *idle channel noise* and is converted to a PAM sample just as if it were a signal. Consequently, even input noise is quantized by the ADC. Figure 4-9 shows a way to reduce idle channel noise by a method called *midtread quantization*. With midtread quantizing, the first quantization interval

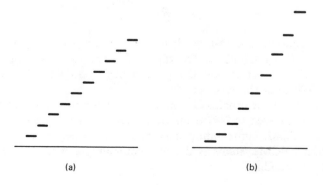

(a)   (b)

Figure 4-8 (a) Linear versus (b) nonlinear encoding.

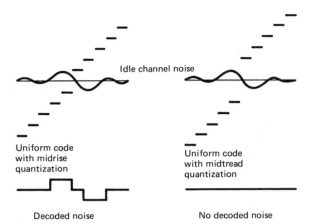

Idle channel noise

Uniform code
with midrise
quantization

Uniform code
with midtread
quantization

Decoded noise

No decoded noise

Figure 4-9   Idle channel noise.

is made larger in amplitude then the rest of the steps. Consequently, input noise can be quite large and still be quantized as a positive or negative zero code. As a result, the noise is suppressed during the encoding process.

In the PCM codes described thus far, the lowest-magnitude positive and negative codes have the same voltage range as all the other codes (+ or − one-half the resolution). This is called *midrise quantization*. Figure 4-9 contrasts the idle channel noise transmitted with a midrise PCM code to the idle channel noise transmitted when midtread quantization is used. The advantage of midtread quantization is less idle channel noise. The disadvantage is a larger possible magnitude for Qe in the lowest quantization interval.

With a folded binary PCM code, residual noise that fluctuates slightly above and below 0 V is converted to either a + or − zero PCM code and is consequently eliminated. In systems that do not use the two 0-V assignments, the residual noise could cause the PCM encoder to alternate between the zero code and the minimum + or − code. Consequently, the decoder would reproduce the encoded noise. With a folded binary code, most of the residual noise is inherently eliminated by the encoder.

Companding

Companding is the process of *compressing*, then *expanding*. With companded systems, the higher-amplitude analog signals are compressed (amplified less than the lower-amplitude signals) prior to transmission, then expanded (amplified more than the smaller-amplitude signals) at the receiver.

Figure 4-10 illustrates the process of companding. A input signal with a dynamic range of 120 dB is compressed to 60 dB for transmission, then expanded to 120 dB at the receiver. With PCM, companding may be accomplished through analog or digital techniques. Early PCM systems used analog companding, whereas more modern systems use digital companding.

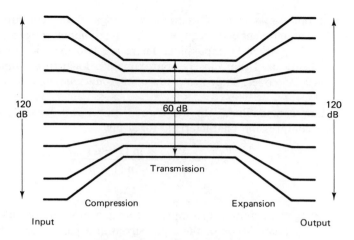

Figure 4-10 Basic companding process.

Analog Companding

Historically, analog compression was implemented using specially designed diodes inserted in the analog signal path in the PCM transmitter prior to the sample-and-hold circuit. Analog expansion was also implemented with diodes that were placed just after the receive low-pass filter. Figure 4-11 shows the basic process of analog compression. In the transmitter, the analog signal is compressed, sampled, then converted to a linear PCM code. In the receiver, the PCM code is converted to a PAM signal, filtered, then expanded back to its original input amplitude characteristics.

Different signal distributions require different companding characteristics. For instance, voice signals require relatively constant SQR performance over a wide dynamic range, which means that the distortion must be proportional to signal amplitude

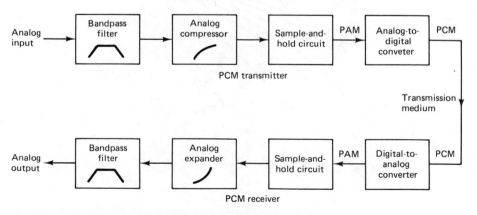

Figure 4-11 PCM system with analog companding.

for any input signal level. This requires a logarithmic compression ratio. A truly logarithmic assignment code requires an infinite dynamic range and an infinite number of PCM codes, which is impossible. There are two methods of analog companding currently being used that closely approximate a logarithmic function and are often called *log-PCM* codes. They are *μ-law* and *A-law companding*.

μ-Law companding.   In the United States and Japan, μ-law companding is used. The compression characteristic for μ-law

$$V_{out} = \frac{V_{max} \times \ln(1 + \mu V_{in}/V_{max})}{\ln(1 + \mu)} \qquad (4\text{-}6)$$

where

V_{max} = maximum uncompressed analog input amplitude
V_{in} = amplitude of the input signal at a particular instant of time
μ = parameter used to define the amount of compression
V_{out} = compressed output amplitude

Figure 4-12 shows the compression for several values of μ. Note that the higher the μ, the more compression. Also note that for a $\mu = 0$, the curve is linear (no compression).

The parameter μ determines the range of signal power in which the SQR is relatively constant. Voice transmission requires a minimum dynamic range of 40 dB and a 7-bit PCM code. For a relatively constant SQR and a 40-dB dynamic range, $\mu = 100$ or larger is required. The early Bell System digital transmission systems used a 7-bit PCM code with $\mu = 100$. The most recent digital transmission systems use 8-bit PCM codes and $\mu = 255$.

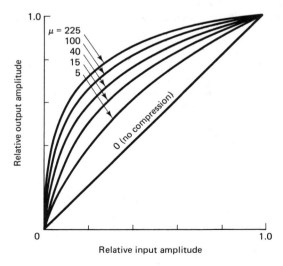

Figure 4-12   μ-Law compression characteristics.

EXAMPLE 4-4

For a compressor with $\mu = 255$, determine the gain for the following values of V_{in}: V_{max}, $0.75\ V_{max}$, $0.5\ V_{max}$, and $0.25\ V_{max}$.

Solution Substituting into Equation 4-6, the following gains are achieved for various input magnitudes:

V_{in}	Gain
V_{max}	1
$0.75V_{max}$	1.26
$0.5V_{max}$	1.75
$0.25V_{max}$	3

It can be seen that as the input signal amplitude increases, the gain decreases or is compressed.

A-Law companding. In Europe, the CCITT has established A-law companding to be used to approximate true logarithmic companding. For an intended dynamic range, A-law companding has a slightly flatter SQR than μ-law. A-law companding, however, is inferior to μ-law in terms of small-signal quality (idle channel noise). The compression characteristic for A-law companding is

$$V_{out} = V_{max}\frac{AV_{in}/V_{max}}{1 + \ln A} \qquad 0 \le \frac{V_{in}}{V_{max}} \le \frac{1}{A} \tag{4-7a}$$

$$= V_{max}\frac{1 + \ln (AV_{in}/V_{max})}{1 + \ln A} \qquad \frac{1}{A} \le \frac{V_{in}}{V_{max}} \le 1 \tag{4-7b}$$

Digital Companding

Digital companding involves compression at the transmit end after the input sample has been converted to a linear PCM code and expansion at the receive end prior to PCM decoding. Figure 4-13 shows the block diagram of a digitally companded PCM system.

With digital companding, the analog signal is first sampled and converted to a linear code, then the linear code is digitally compressed. At the receive end, the compressed PCM code is received, expanded, then decoded. The most recent digitally compressed PCM systems use a 12-bit linear code and an 8-bit compressed code. This companding process closely resembles a $\mu = 255$ analog compression curve by approximating the curve with a set of eight straight line *segments* (segments 0 through 7). The slope of each successive segment is exactly one-half that of the previous segment. Figure 4-14 shows the 12-bit-to-8-bit digital compression curve for positive values only. The curve for negative values is identical except the inverse. Although there are 16 segments (eight positive and eight negative) this scheme is often called

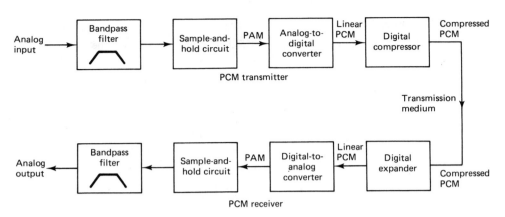

Figure 4-13 Digitally companded PCM system.

13-segment compression. This is because the curve for segments +0, +1, −0, and −1 is a straight line with a constant slope and is often considered as one segment.

The digital companding algorithm for a 12-bit-linear-to-8-bit-compressed code is actually quite simple. The 8-bit compressed code is comprised of a sign bit, a 3-

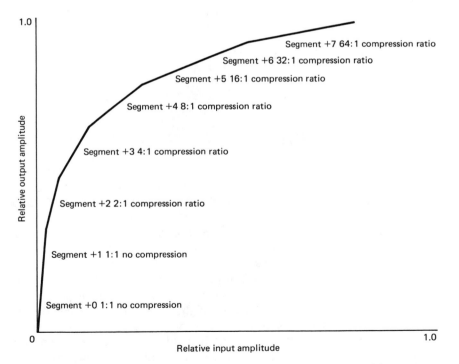

Figure 4-14 μ255 compression characteristics (positive values only).

bit segment identifier, and a 4-bit magnitude code which identifies the *quantization interval* within the specified segment (see Figure 4-15a).

In the $\mu255$ encoding table shown in Figure 4-15b, the bit positions designated with an X are truncated during compression and are consequently lost. Bits designated A, B, C, and D are transmitted as is. The sign bit (s) is also transmitted as is. Note that for segments 0 and 1, the original 12 bits are duplicated exactly at the output of the decoder (Figure 4-15c), whereas for segment 7, only the most significant 6 bits are recovered. With 11 magnitude bits, there are 2048 possible codes. There are 16 codes in segment 0 and in segment 1. In segment 2, there are 32 codes; segment 3 has 64. Each successive segment beginning with segment 3 has twice as many codes as the previous segment. In each of the eight segments, only sixteen 12-bit codes can be recovered. Consequently, in segments 0 and 1, there is no compression (of the 16 possible codes, all 16 can be recovered). In segment 2, there is a compression ratio of 2:1 (32 possible transmit codes and 16 possible recovered codes). In segment 3, there is a 4:1 compression ratio (64 possible transmit codes and 16 possible recovered codes). The compression ratio doubles with each successive segment. The compression ratio in segment 7 is 1024:16 or 64:1.

The compression process is as follows. The analog signal is sampled and converted to a linear 12-bit sign-magnitude code. The sign bit is transferred directly to the 8-bit code. The segment is determined by counting the number of leading 0's in the 11-bit magnitude portion of the code beginning with the MSB. Subtract the number of leading 0's (not to exceed 7) from 7. The result is the segment number, which is

Sign bit 1 = + 0 = −	3-Bit segment identifier	4-Bit quantization interval A B C D
	000 to 111	0000 to 1111

(a)

Segment	12-Bit linear code	8-Bit compressed code	8-Bit compressed code	12-Bit recovered code	Segment
0	s0000000ABCD	s000ABCD	s000ABCD	s0000000ABCD	0
1	s0000001ABCD	s001ABCD	s001ABCD	s0000001ABCD	1
2	s000001ABCDX	s010ABCD	s010ABCD	s000001ABCD1	2
3	s00001ABCDXX	s011ABCD	s011ABCD	s00001ABCD10	3
4	s0001ABCDXXX	s100ABCD	s100ABCD	s0001ABCD100	4
5	s001ABCDXXXX	s101ABCD	s101ABCD	s001ABCD1000	5
6	s01ABCDXXXXX	s110ABCD	s110ABCD	s01ABCD10000	6
7	s1ABCDXXXXXX	s111ABCD	s111ABCD	s1ABCD100000	7

(b)   (c)

Figure 4-15 12-bit to 8-bit digital companding: (a) 8-bit $\mu255$ compressed code format; (b) $\mu255$ encoding table; (c) $\mu255$ decoding table.

converted to a 3-bit binary number and substituted into the 8-bit code as the segment identifier. The four magnitude bits (A, B, C, and D) are the quantization interval and are substituted into the least significant 4 bits of the 8-bit compressed code.

Essentially, segments 2 through 7 are subdivided into smaller subsegments. Each segment has 16 subsegments, which correspond to the 16 conditions possible for the bits A, B, C, and D (0000–1111). In segment 2 there are two codes per subsegment. In segment 3 there are four. The number of codes per subsegment doubles with each subsequent segment. Consequently, in segment 7, each subsegment has 64 codes. Figure 4-16 shows the breakdown of segments versus subsegments for segments 2, 5, and 7. Note that in each subsegment, all 12-bit codes, once compressed and expanded, yield a single 12-bit code. This is shown in Figure 4-16.

From Figures 4-15 and 4-16, it can be seen that the most significant of the truncated bits is reinserted at the decoder as a 1. The remaining truncated bits are reinserted as 0's. This ensures that the maximum magnitude of error introduced by the compression and expansion process is minimized. Essentially, the decoder guesses what the truncated bits were prior to encoding. The most logical guess is halfway between the minimum- and maximum-magnitude codes. For example, in segment 5, the 5 least significant bits are truncated during compression. At the receiver, the decoder must determine what those bits were. The possibilities are any code between 00000 and 11111. The logical guess is 10000, approximately half the maximum magnitude. Consequently, the maximum compression error is slightly more than one-half the magnitude of that segment.

EXAMPLE 4-5

For a resolution of 0.01 V and analog sample voltages of (a) 0.05 V, (b) 0.32 V, and (c) 10.23 V, determine the 12-bit linear code, the 8-bit compressed code, and the recovered 12-bit code.

Solution   (a) To determine the 12-bit linear code for 0.05 V, simply divide the sample voltage by the resolution and convert the result to a 12-bit sign-magnitude binary number.

12-bit linear code:

$$\frac{0.05\,V}{0.01\,V} = 5 = \begin{matrix}1\ 0\ 0\ 0\ 0\ 0\ 0\ 0\ 0\ 1\ 0\ 1 \\ s\ \text{------magnitude-----}\end{matrix}$$

(11-bit binary number)

8-bit compressed code:

```
1   0   0   0   0   0   0     0   0   1   0   1
s       (7 − 7 = 0 or 000)        A   B   C   D
1           0   0   0         0   1   0   1
↑
sign bit         unit              quanti-
(+)           identifier            zation
            (segment 0)            interval
```

Segment	12-Bit linear code		12-Bit expanded code	Subsegment
7	s11111111111 .... s11111000000	} 64:1	s11111100000	15
7	s11110111111 .... s11110000000	} 64:1	s11110100000	14
7	s11101111111 .... s11101000000	} 64:1	s11101100000	13
7	s11100111111 .... s11100000000	} 64:1	s11100100000	12
7	s11011111111 .... s11011000000	} 64:1	s11011100000	11
7	s11010111111 .... s11010000000	} 64:1	s11010100000	10
7	s11001111111 .... s11001000000	} 64:1	s11001100000	9
7	s11000111111 .... s11000000000	} 64:1	s11000100000	8
7	s10111111111 .... s10111000000	} 64:1	s10111100000	7
7	s10110111111 .... s10110000000	} 64:1	s10110100000	6
7	s10101111111 .... s10101000000	} 64:1	s10101100000	5
7	s10100111111 .... s10100000000	} 64:1	s10100100000	4
7	s10011111111 .... s10011000000	} 64:1	s10011100000	3
7	s10010111111 .... s10010000000	} 64:1	s10010100000	2
7	s10001111111 .... s10010000000	} 64:1	s10001100000	1
7	s10001111111 .... s10000000000	} 64:1	s10000100000	0
	s1ABCD------			

(a)

Figure 4-16 12-bit segments divided into subsegments: (a) segment 7; (b) segment 5; (c) segment 2.

Segment	12-Bit linear code		12-Bit expanded code	Subsegment
5	s00111111111 .... s00111110000	16:1	s00111111000	15
5	s00111101111 .... s00111100000	16:1	s00111101000	14
5	s00111011111 .... s00111010000	16:1	s00111011000	13
5	s00111001111 .... s00111000000	16:1	s00111001000	12
5	s00110111111 .... s00110110000	16:1	s00110111000	11
5	s00110101111 .... s00110100000	16:1	s00110101000	10
5	s00110011111 .... s00110010000	16:1	s00110011000	9
5	s00110001111 .... s00110000000	16:1	s00110001000	8
5	s00101111111 .... s00101110000	16:1	s00101111000	7
5	s00101101111 .... s00101100000	16:1	s00101101000	6
5	s00101011111 .... s00101010000	16:1	s00101011000	5
5	s00101001111 .... s00101000000	16:1	s00101001000	4
5	s00100111111 .... s00100110000	16:1	s00100111000	3
5	s00100101111 .... s00100100000	16:1	s00100101000	2
5	s00100011111 .... s00100010000	16:1	s00100011000	1
5	s00100001111 .... s00100000000	16:1	s00100001000	0
	s001ABCD----			

(b)

Figure 4-16 (*continued*)

Segment	12-Bit linear code		12-Bit expanded code	Subsegment
2	s00000111111 .... s00000111110	2:1	s00000111111	15
2	s00000111101 .... s00000111100	2:1	s00000111101	14
2	s00000111011 .... s00000111010	2:1	s00000111011	13
2	s00000111001 .... s00000111000	2:1	s00000111001	12
2	s00000110111 .... s00000110110	2:1	s00000110111	11
2	s00000110101 .... s00000110100	2:1	s00000110101	10
2	s00000110011 .... s00000110010	2:1	s00000110011	9
2	s00000110001 .... s00000110000	2:1	s00000110001	8
2	s00000101111 .... s00000101110	2:1	s00000101111	7
2	s00000101101 .... s00000101100	2:1	s00000101101	6
2	s00000101011 .... s00000101010	2:1	s00000101011	5
2	s00000101001 .... s00000101000	2:1	s00000101001	4
2	s00000100111 .... s00000100110	2:1	s00000100111	3
2	s00000100101 .... s00000100100	2:1	s00000100101	2
2	s00000100011 .... s00000100010	2:1	s00000100011	1
2	s00000100001 .... s00000100000	2:1	s00000100001	0

s000001ABCD-

(c)

Figure 4-16 (*continued*)

12-bit recovered code:

```
1              0  0  0            0  1  0  1
s       (7 − 0 = 7 leading 0's)   A  B  C  D
1    0     0  0  0  0  0   0      0  1  0  1
     ↑
 sign bit       segment identifier      quantization
                determines the            interval
                number of leading
                     0's
```

As you can see, the recovered 12-bit code is exactly the same as the original 12-bit linear code. This is true for all codes in segment 0 and 1. Consequently, there is no compression error in these two segments.

(b) For the 0.32-V sample:

12-bit linear code:

$$\frac{0.32\,\mathrm{V}}{0.01\,\mathrm{V}} = 32 = \begin{array}{l} 1\ 0\ 0\ 0\ 0\ 0\ 1\ 0\ 0\ 0\ 0\ 0 \\ s\ \text{------magnitude-----} \end{array}$$

8-bit compressed code:

```
1   0  0  0  0  0  1   0  0  0  0  0
s   (7 − 5 = 2 or 010)  A  B  C  D  X
1       0  1  0        0  0  0  0  ↑
(+)     (segment 2)             truncated
```

12-bit recovered code:

```
1           0  1  0       0  0  0  0
s    (7 − 2 = 5 leading 0's)   A  B  C  D  X
1    0  0  0  0  0  1      0  0  0  0  1
              ↑                        ↑
          inserted                 inserted
```

Note the two inserted 1's in the decoded 12-bit code. The least significant bit is determined from the decoding table in Figure 4-15. The stuffed 1 in bit position 6 was dropped during the 12-bit-to-8-bit conversion. Transmission of this bit is redundant because if it were not a 1, the sample would not be in segment 3. Consequently, in all segments except 0, a 1 is automatically inserted after the reinserted zeros. For this sample, there is an error in the received voltage equal to the resolution, 0.01 V. In segment 2, for every two 12-bit codes possible, there is only one recovered 12-bit code. Thus a coding compression of 2:1 is realized.

(c) To determine the codes for 10.23 V, the process is the same:

12-bit linear code:

```
1  1  1111    111111
↑
s    ABCD  truncated
```

8-bit compressed code:

```
1     111     1111
↑
s   segment  ABCD
```

12-bit recovered code:

```
1  1  1111   100000
↑  │  \___/  \____/
s  │  ABCD   inserted
   │
inserted
```

The difference in the original 12-bit linear code and the recovered 12-bit code is

```
  111111111111
 −111111100000
  000000011111 = 31( 0.01 V ) = 0.31 V
```

Percentage Error

For comparison purposes, the following formula is used for computing the *percentage of error* introduced by digital compression:

$$\% \text{ error} = \frac{|\text{Tx voltage} - \text{Rx voltage}|}{\text{Rx voltage}} \times 100 \qquad (4\text{-}8)$$

EXAMPLE 4-6

The maximum percentage of error will occur for the smallest number in the lowest subsegment within any given segment. Because there is no compression error in segments 0 and 1, for segment 3 the maximum % error is computed as follows:

```
Transmit 12-bit code:  s00001000000
Receive 12-bit code:   s00001000010
Magnitude of error:     00000000010
```

$$\% \text{ error} = \frac{|1000000 - 1000010|}{1000010} \times 100$$

$$= \frac{|64 - 66|}{66} \times 100 = 3.03\%$$

For segment 7:

```
Transmit 12-bit code:  s10000000000
Receive 12-bit code:   s10000100000
Magnitude of error:     00000100000
```

$$\% \text{ error} = \frac{|10000000000 - 10000100000|}{10000100000} \times 100$$

$$= \frac{|1024 - 1056|}{1056} \times 100 = 3.03\%$$

Although the magnitude of error is higher for segment 7, the percentage of error is the same. The maximum percentage of error is the same for segments 3 through 7, and consequently, the SQR degradation is the same for each segment.

Although there are several ways in which the 12-bit-to-8-bit compression and the 8-bit-to-12-bit expansion can be accomplished with hardware, the simplest and most economical is with a look-up table in ROM (read-only memory).

Essentially every function performed by a PCM encoder and decoder is now accomplished with a single integrated-circuit chip called a *codec*. Most of the more recently developed codecs include an antialiasing (bandpass) filter, a sample-and-hold circuit, and an analog-to-digital converter in the transmit section and a digital-to-analog converter, a sample-and-hold circuit, and a bandpass filter in the receive section. The operation of a codec is explained in detail in Chapter 5.

DELTA MODULATION PCM

Delta modulation uses a single-bit PCM code to achieve digital transmission of analog signals. With conventional PCM, each code is a binary representation of both the sign and magnitude of a particular sample. Therefore, multiple-bit codes are required to represent the many values that the sample can be. With delta modulation, rather than transmit a coded representation of the sample, only a single bit is transmitted which simply indicates whether that sample is larger or smaller than the previous sample. The algorithm for a delta modulation system is quite simple. If the current sample is smaller than the previous sample, a logic 0 is transmitted. If the current sample is larger than the previous sample, a logic 1 is transmitted.

Delta Modulation Transmitter

Figure 4-17 shows a block diagram of a delta modulation transmitter. The input analog is sampled and converted to a PAM signal which is compared to the output of the DAC. The output of the DAC is a voltage equal to the magnitude of the previous sample, which was stored in the up-down counter as a binary number. The up-down counter is incremented or decremented depending on whether the previous sample is larger or smaller than the current sample. The up-down counter is clocked at a rate equal to the sample rate. Therefore, the up-down counter is updated after each comparison.

Figure 4-18 shows the ideal operation of a delta modulation encoder. Initially, the up-down counter is zeroed and the DAC is outputting 0 V. The first sample is taken, converted to a PAM signal, and compared to zero volts. The output of the comparator is a logic 1 condition (+V), indicating that the current sample is larger in amplitude than the previous sample. On the next clock pulse, the up-down counter is incremented to a count of 1. The DAC now outputs a voltage equal to the magnitude of the minimum step size (resolution). The steps change value at a rate equal to

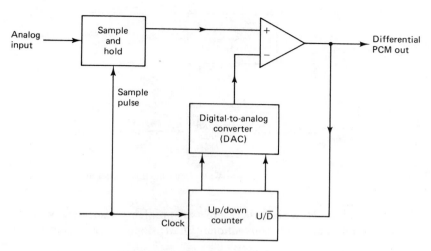

Figure 4-17 Delta modulation transmitter.

the clock frequency (sample rate). Consequently, with the input signal shown, the up-down counter follows the input analog signal up until the output of the DAC exceeds the analog sample; then the up-down counter will begin counting down until the output of the DAC drops below the sample amplitude. In the idealized situation (shown in Figure 4-18), the DAC output follows the input signal. Each time the up-down counter is incremented, a logic 1 is transmitted, and each time the up-down counter is decremented, a logic 0 is transmitted.

Delta Modulation Receiver

Figure 4-19 shows the block diagram of a delta modulation receiver. As you can see, the receiver is almost identical to the transmitter except for the comparator. As the logic 1's and 0's are received, the up-down counter is incremented or decremented accordingly. Consequently, the output of the DAC in the decoder is identical to the output of the DAC in the transmitter.

With delta modulation, each sample requires the transmission of only one bit; therefore, the bit rates associated with delta modulation are lower than conventional

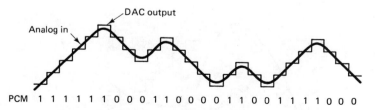

Figure 4-18 Ideal operation of a delta modulation encoder.

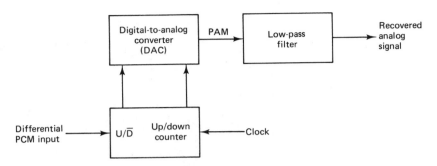

Figure 4-19 Delta modulation receiver.

PCM systems. However, there are two problems associated with delta modulation that do not occur with conventional PCM: slope overload and granular noise.

Slope overload.   Figure 4-20 shows what happens when the analog input signal changes at a faster rate than the DAC can keep up with. The slope of the analog signal is greater than the delta modulator can maintain. This is called *slope overload*. Increasing the clock frequency reduces the probability of slope overload occurring. Another way is to increase the magnitude of the minimum step size.

Granular noise.   Figure 4-21 contrasts the original and reconstructed signals associated with a delta modulation system. It can be seen that when the original analog input signal has a relatively constant amplitude, the reconstructed signal has variations that were not present in the original signal. This is called *granular noise*. Granular noise in delta modulation is analogous to quantization noise in conventional PCM.

Granular noise can be reduced by decreasing the step size. Therefore, to reduce the granular noise, a small resolution is needed, and to reduce the possibility of slope overload occurring, a large resolution is required. Obviously, a compromise is necessary.

Granular noise is more prevalent in analog signals that have gradual slopes and whose amplitudes vary only a small amount. Slope overload is more prevalent in analog signals that have steep slopes or whose amplitudes vary rapidly.

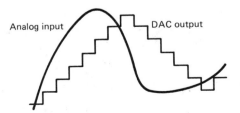

Figure 4-20 Slope overload distortion.

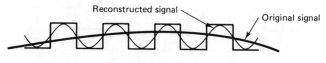

Figure 4-21 Granular noise.

ADAPTIVE DELTA MODULATION PCM

Adaptive delta modulation is a delta modulation system where the step size of the DAC is automatically varied depending on the amplitude characteristics of the analog input signal. Figure 4-22 shows how an adaptive delta modulator works. When the output of the transmitter is a string of consecutive 1's or 0's, this indicates that the slope of the DAC output is less than the slope of the analog signal in either the positive or negative direction. Essentially, the DAC has lost track of exactly where the analog samples are and the possibility of slope overload occurring is high. With an adaptive delta modulator, after a predetermined number of consecutive 1's or 0's, the step size is automatically increased. After the next sample, if the DAC output amplitude is still below the sample amplitude, the next step is increased even further until eventually the DAC catches up with the analog signal. When an alternating sequence of 1's and 0's is occurring, this indicates that the possibility of granular noise occurring is high. Consequently, the DAC will automatically revert to its minimum step size and thus reduce the magnitude of the noise error.

A common algorithm for an adaptive delta modulator is when three consecutive 1's or 0's occur, the step size of the DAC is increased or decreased by a factor of 1.5. Various other algorithms may be used for adaptive delta modulators, depending on particular system requirements.

DIFFERENTIAL PULSE CODE MODULATION

In a typical PCM-encoded speech waveform, there are often successive samples taken in which there is little difference between the amplitudes of the two samples. This

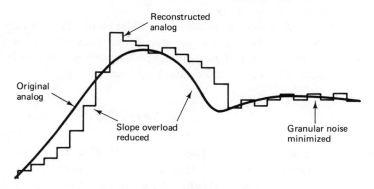

Figure 4-22 Adaptive delta modulation.

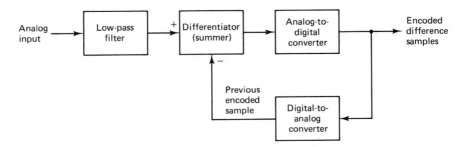

Figure 4-23   DPCM transmitter.

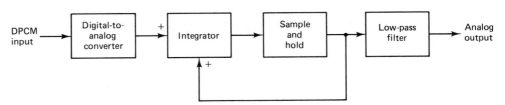

Figure 4-24   DPCM receiver.

facilitates transmitting several identical PCM codes, which is redundant. Differential pulse code modulation (DPCM) is designed specifically to take advantage of the sample-to-sample redundancies in typical speech waveforms. With DPCM, the difference in the amplitude of two successive samples is transmitted rather than the actual sample. Since the range of sample differences is typically less than the range of individual samples, fewer bits are required for DPCM than conventional PCM.

Figure 4-23 shows a simplified block diagram of a DPCM transmitter. The analog input signal is bandlimited to one-half of the sample rate, then compared to the preceding DPCM signal in the differentiator. The output of the differentiator is the difference between the two signals. The difference is PCM encoded and transmitted. The A/D converter operates the same as in a conventional PCM system except that it typically uses fewer bits per sample.

Figure 4-24 shows a simplified block diagram of a DPCM receiver. Each received sample is converted back to analog, stored, and then summed with the next sample received. In the receiver shown in Figure 4-24 the integration is performed on the analog signals, although it could also be performed digitally.

QUESTIONS

4-1. Contrast the advantages and disadvantages of digital transmission.

4-2. What are the four most common methods of pulse modulation?

4-3. Which method listed in Question 4.2 is the only form of pulse modulation that is a digital transmission system? Explain?

4-4. What is the purpose of the sample-and-hold circuit?

4-5. Define *aperture* and *acquisition time*.

4-6. What is the difference between natural and flat-top sampling?

4-7. Define *droop*. What causes it?

4-8. What is the Nyquist sampling rate?

4-9. Define and state the causes of foldover distortion.

4-10. Explain the difference between a magnitude-only code and a sign-magnitude code.

4-11. Explain overload distortion.

4-12. Explain quantizing.

4-13. What is quantization range? Quantization error?

4-14. Define *dynamic range*.

4-15. Explain the relationship between dynamic range, resolution, and the number of bits in a PCM code.

4-16. Explain coding efficiency.

4-17. What is SQR? What is the relationship between SQR, resolution, dynamic range, and the number of bits in a PCM code?

4-18. Contrast linear and nonlinear PCM codes.

4-19. Explain idle channel noise.

4-20. Contrast midtread and midrise quantization.

4-21. Define *companding*.

4-22. What does the parameter μ determine?

4-23. Briefly explain the process of digital companding.

4-24. What is the effect of digital compression on SQR, resolution, quantization interval, and quantization noise?

4-25. Contrast delta modulatin PCM and standard PCM.

4-26. Define *slope overload* and *granular noise*.

4-27. What is the difference between adaptive delta modulation and conventional delta modulation?

4-28. Contrast differential and conventional PCM.

PROBLEMS

4-1. Determine the Nyquist sample rate for a maximum analog input frequency of 4 kHz; 10 kHz.

4-2. For the sample-and-hold circuit shown in Figure 4-3, determine the largest-value capacitor that can be used. Use the following parameters: an output impedance for Z1 = 20 Ω, an "on" resistance of Q1 of 20 Ω, an acquisition time of 10 μs, a maximum output current from Z1 of 20 mA, and an accuracy of 1%.

4-3. For a sample rate of 20 kHz, determine the maximum analog input frequency.

4-4. Determine the alias frequency for a 4-kHz sample rate and an analog input frequency of 1.5 kHz.

4-5. Determine the dynamic range for a 10-bit sign-magnitude PCM code.

4-6. Determine the minimum number of bits required in a PCM code for a dynamic range of 80 dB. What is the coding efficiency?

4-7. For a resolution of 0.04 V, determine the voltages for the following linear 7-bit sign-magnitude PCM codes:

    (a) 0110101

    (b) 0000011

    (c) 1000001

    (d) 0111111

    (e) 1000000

4-8. Determine the SQR for a 2-V rms signal and a quantization noise magnitude of 0.2 V.

4-9. Determine the resolution and quantization noise for an 8-bit linear sign-magnitude PCM code for a maximum decoded voltage of $1.27V_p$.

4-10. A 12-bit linear PCM code is digitally compressed into 8 bits. The resolution = 0.03 V. Determine the following for an analog input voltage of 1.465 V.

    (a) 12-Bit linear PCM code.

    (b) 8-Bit compressed code.

    (c) Decoded 12-bit code.

    (d) Decoded voltage.

    (e) Percentage error.

4-11. For a 12-bit linear PCM code with a resolution of 0.02 V, determine the voltage range that would be converted to the following PCM codes.

    (a) 1 0 0 0 0 0 0 0 0 0 0 1

    (b) 0 0 0 0 0 0 0 0 0 0 0 0

    (c) 1 1 0 0 0 0 0 0 0 0 0 0

    (d) 0 1 0 0 0 0 0 0 0 0 0 0

    (e) 1 0 0 1 0 0 0 0 0 0 0 1

    (f) 1 0 1 0 1 0 1 0 1 0 1 0

4-12. For each of the following 12-bit linear PCM codes, determine their input voltage range and the 8-bit compressed code to which they would be converted.

    (a) 1 0 0 0 0 0 0 0 1 0 0 0

    (b) 1 0 0 0 0 0 0 0 1 0 0 1

    (c) 1 0 0 0 0 0 0 1 0 0 0 0

    (d) 0 0 0 0 0 0 1 0 0 0 0 0

    (e) 0 1 0 0 0 0 0 0 0 0 0 0

    (f) 0 1 0 0 0 0 1 0 0 0 0 0

Chapter 5

DIGITAL MULTIPLEXING

INTRODUCTION

Multiplexing is the transmission of information (either voice or data) from more than one source to more than one destination on the same transmission medium (*facility*). Transmissions occur on the same facility but not necessarily at the same time. The transmission medium may be a metallic wire pair, a coaxial cable, a microwave radio, a satellite radio, or a fiber optic cable. There are several ways in which multiplexing can be achieved, although the two most common methods are *frequency-division multiplexing* (FDM) and *time-division multiplexing* (TDM). Frequency-division multiplexing is discussed in Chapter 6.

TIME-DIVISION MULTIPLEXING

With TDM, transmissions from multiple sources occur on the same facility but not at the same time. Transmissions from various sources are *interleaved* in the time domain. The most common type of modulation used with TDM systems is PCM. With a PCM-TDM system, several voice band channels are sampled, converted to PCM codes, then time-division multiplexed onto a single metallic cable pair.

Figure 5-1a shows a simplified block diagram of a two-channel PCM-TDM carrier system. Each channel is alternately sampled and converted to a PCM code. While the PCM code for channel 1 is being transmitted, channel 2 is sampled and converted to a PCM code. While the PCM code from channel 2 is being transmitted, the next sample is taken from channel 1 and converted to a PCM code. This process

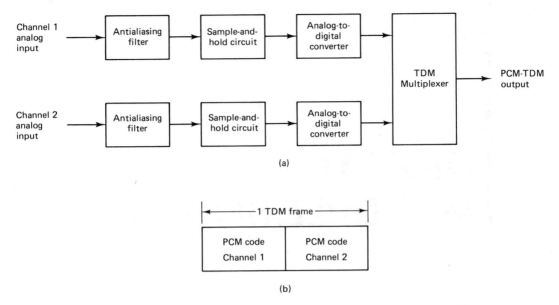

(a)

(b)

Figure 5-1   Two-channel PCM-TDM system: (a) block diagram; (b) TDM frame.

continues and samples are taken alternately from each channel, converted to PCM
codes, and transmitted. The multiplexer is simply a switch with two inputs and one
output. Channel 1 and channel 2 are alternately selected and connected to the multiplexer output. The time it takes to transmit one sample from each channel is called
the *frame time*.

The PCM code for each channel occupies a fixed time slot (epoch) within the
total TDM frame. With a two-channel system, the time allocated for each channel
is equal to one-half of the total frame time. A sample from each channel is taken
once during each frame. Therefore, the total frame time is equal to the reciprocal
of the sample rate ($1/F_s$). Figure 5-1b shows the TDM frame allocation for a two-
channel system.

T1 DIGITAL CARRIER SYSTEM

Figure 5-2 shows the block diagram of the Bell System T1 digital carrier system.
This system is the North American telephone standard. A T1 carrier time-division
multiplexes 24 PCM encoded samples for transmission over a single metallic wire
pair. Again, the multiplexer is simply a switch except that now it has 24 inputs
and one output. The 24 voice band channels are sequentially selected and connected
to the multiplexer output. Each voice band channel occupies a 300- to 3000-Hz
bandwidth.

Simply time-division multiplexing 24 voice band channels does not in itself

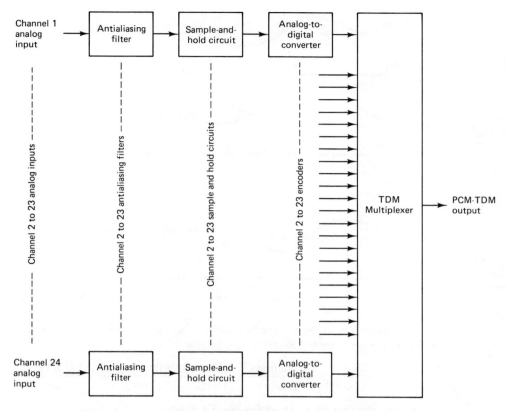

Figure 5-2 Bell system T1 PCM-TDM digital carrier system block diagram.

constitute a T1 carrier. At this point, the output of the multiplexer is simply a multiplexed digital signal (DS-1). It does not actually become a T1 carrier until it is line encoded and placed on special conditioned wire pairs called *T1 lines*. This is explained in more detail later in this chapter under the heading "North American Digital Hierarchy."

With the Bell System T1 carrier system, D-type (digital) channel banks perform the sampling, encoding, and multiplexing of 24 voice band channels. Each channel contains an 8-bit PCM code and is sampled 8000 times per second. (Each channel is sampled at the same rate but not at the same time; see Figure 5-3.) Therefore, a 64-kbps PCM encoded sample is transmitted for each voice band channel during each frame.

$$\frac{8 \text{ bits}}{\text{sample}} \times \frac{8000 \text{ samples}}{\text{second}} = 64 \text{ kbps}$$

Within each frame an additional bit called a *framing bit* is added. The framing bit occurs at an 8000-bps rate and is recovered in the receiver circuitry and used to

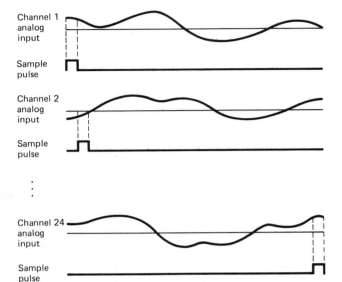

Figure 5-3 T1 sampling sequence.

maintain frame and sample synchronization between the TDM transmitter and receiver. As a result, each TDM frame contains 193 bits.

$$\frac{8 \text{ bits}}{\text{channel}} \times \frac{24 \text{ channels}}{\text{frame}} = \frac{192 \text{ bits}}{\text{frame}} + \frac{1 \text{ framing bit}}{\text{frame}} = \frac{193 \text{ bits}}{\text{frame}}$$

As a result, the line speed (bps) for the T1 carrier is

$$\text{line speed} = \frac{193 \text{ bits}}{\text{frame}} \times \frac{8000 \text{ frames}}{\text{second}} = 1.544 \text{ Mbps}$$

D-Type Channel Banks

The early T1 carrier systems were equipped with D1A channel banks which use a 7-bit magnitude-only PCM code with analog companding and $\mu = 100$. A later version of the D1 channel bank (D1D) used an 8-bit sign-magnitude PCM code. With D1A channel banks an eighth bit (the s bit) is added to each PCM code for the purpose of *signaling* (supervision: on-hook, off-hook, dial pulsing, etc.). Consequently, the signaling rate for D1 channel banks is 8 kbps. Also, with D1 channel banks, the framing bit sequence is simply an alternating 1/0 pattern. Figure 5-4 shows the frame and sample alignment for the T1 carrier system using D1A channel banks.

    Generically, the T1 carrier system has progressed through the D2, D3, and D4 channel banks, which use a digitally companded, 8-bit sign-magnitude compressed PCM code with $\mu = 255$. In the D1 channel bank, the compression and expansion

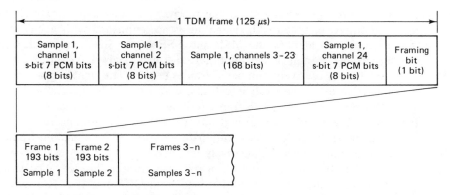

Figure 5-4 T1 carrier system frame and sample alignment using D1 channel banks.

characteristics were implemented in circuitry separate from the encoder and decoder. The D2, D3, and D4 channel banks incorporate the companding functions directly in the encoders and decoders. Although the D2 and D3 channel banks are functionally similar, the D3 channel banks were the first to incorporate integrated circuitry into their design, which reduced their size and power consumption. D4 channel banks were the first to incorporate separate customized LSI integrated circuits (codecs) for each voice band channel. With D1, D2, and D3 channel banks, common equipment performs the encoding and decoding functions. Consequently, a single equipment malfunction constitutes a total system failure.

D1A channel banks use a magnitude-only code; consequently, an error in the most significant bit (MSB) of a channel sample always produces a decoded error equal to one-half the total quantization range (one-half of V_{max}). Because D1D, D2, D3, and D4 channel banks use a sign-magnitude code, an error in the MSB (sign bit) causes a decoded error equal to twice the sample magnitude (from $+V$ to $-V$, or vice versa). The worst-case error is equal to twice the total quantization range. However, maximum amplitude samples occur rarely and most errors with D1D, D2, D3, and D4 coding are less than one-half the coding range. On the average, the error performance with a sign-magnitude code is less than with a magnitude-only code.

Superframe Format

The 8-kbps signaling rate used with D1 channel banks is excessive for voice transmission. Therefore, with D2 and D3 channel banks, a signaling bit is substituted only into the least significant bit (LSB) of every sixth frame. Therefore, five out of every six frames have 8-bit resolution, while one out of every six frames (the signaling frame) has only 7-bit resolution. Consequently, the signaling rate on each channel is 1.333 kbps (8000 bps/6) and the effective number of bits per sample is actually $7\frac{5}{6}$ bits and not 8.

Because only every sixth frame includes a signaling bit, it is necessary that all the frames are numbered so that the receiver knows when to extract the signaling information. Also, because the signaling is accomplished with a 2-bit binary word, it is necessary to identify the MSB and LSB of the signaling word. Consequently, the *superframe* format shown in Figure 5-5 was devised. Within each superframe, there are 12 consecutively numbered frames (1–12). The signaling bits are substituted in frames 6 and 12; the MSB into frame 6 and the LSB into frame 12. Frames 1–6 are called the A-highway with frame 6 designated as the A-channel signaling frame. Frames 7–12 are called the B-highway with frame 12 designated as the B-channel signaling frame. Therefore, in addition to identifying the signaling frames, the sixth and twelfth frames must be positively identified.

To identify frames 6 and 12, a different framing bit sequence is used for the odd- and even-numbered frames. The odd frames (frames 1, 3, 5, 7, 9, and 11) have an alternating 1/0 pattern, and the even frames (frames 2, 4, 6, 8, 10, and 12) have a 0 0 1 1 1 0 repetitive pattern. As a result, the combined bit pattern for the framing bits is a 1 0 0 0 1 1 0 1 1 1 0 0 repetitive pattern. The odd-numbered frames are used for frame and sample synchronization, while the even-numbered frames are used to identify the A and B channel signaling frames (6 and 12). Frame 6 is identified by a 0/1 transition in the framing bit between frames 4 and 6. Frame 12 is identified by a 1/0 transition in the framing bit between frames 10 and 12.

Figure 5-6 shows the frame, sample, and signaling alignment for the T1 carrier system using D2 or D3 channel banks.

In addition to *multiframe alignment* bits and PCM sample bits, certain time slots are used to indicate alarm conditions. For example, in the case of a transmit power supply failure, a common equipment failure, or loss of multiframe alignment; the second bit in each channel is made a 0 until the alarm condition has cleared. Also, the framing bit in frame 12 is complemented whenever multiframe alignment is lost (this is assumed whenever frame alignment is lost). In addition, there are special framing conditions that must be avoided in order to maintain clock and bit synchronization at the receive demultiplexing equipment. These special conditions are explained later in this chapter.

D4 Channel Bank

D4 channel banks time-division multiplex 48 voice band channels and operate at a transmission rate of 3.152 Mbps. This is slightly more than twice the line speed for 24-channel D1, D2, or D3 channel banks. This is because with D4 channel banks, rather than transmit a single framing bit with each frame, a 10-bit frame synchronization pattern is used. Consequently, the total number of bits in a D4 (DS-1C) TDM frame is

$$\frac{8 \text{ bits}}{\text{channel}} \times \frac{48 \text{ channels}}{\text{frame}} = \frac{384 \text{ bits}}{\text{frame}} + \frac{10 \text{ syn bits}}{\text{frame}} = \frac{394 \text{ bits}}{\text{frame}}$$

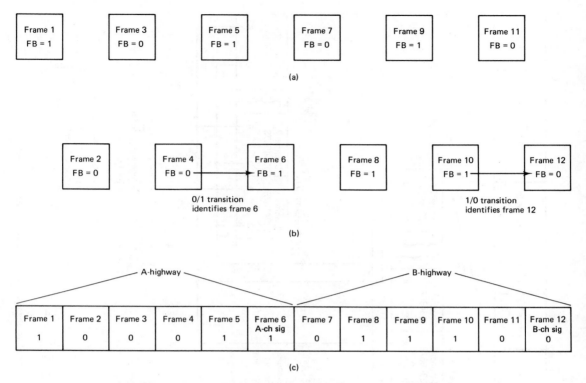

Figure 5-5 Framing bit sequence for the T1 superframe format: (a) frame synchronizing bits (odd-numbered frames); (b) signaling frame alignment bits (even-numbered frames); (c) composite frame alignment.

and the line speed is

$$\text{line speed} = \frac{394 \text{ bits}}{\text{frame}} \times \frac{8000 \text{ frames}}{\text{second}} = 3.152 \text{ Mbps}$$

The framing for the DS-1 (T1) system or the framing pattern for the DS-1C (T1C) time-division-multiplexed carrier systems are added to the multiplexed digital signal at the output of the multiplexer. Figure 5-7 shows the framing bit circuitry for the 24-channel T1 carrier system using either D1, D2, or D3 channel banks (DS-1). Note that the bit rate at the output of the TDM multiplexer is 1.536 Mbps and the bit rate at the output of the 193-bit shift register is 1.544 Mbps. The difference (8 kbps) is because of the added framing bit.

CCITT TIME-DIVISION-MULTIPLEXED CARRIER SYSTEM

Figure 5-8 shows the frame alignment for the CCITT (Comité Consultatif International Téléphonique et Télégraphique) European standard PCM-TDM system. With the

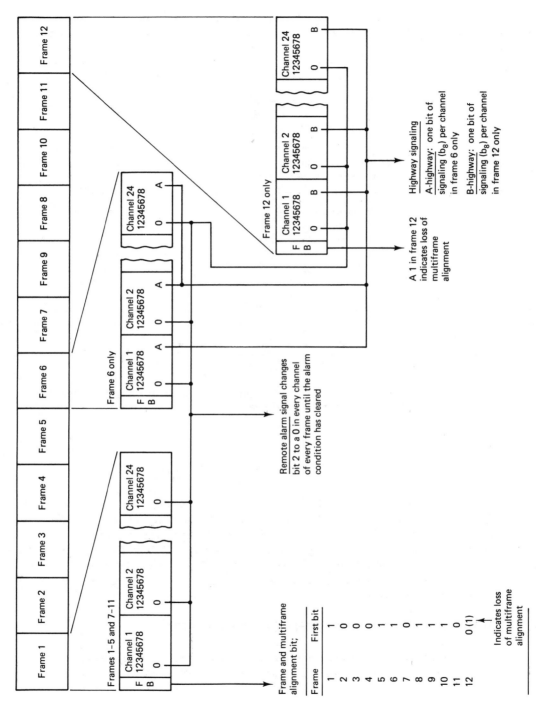

Figure 5-6 T1 carrier frame, sample, and signaling alignment for D2 and D3 channel banks.

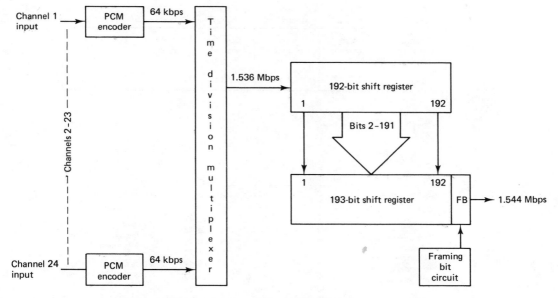

Figure 5-7 Framing bit circuitry for the DS-1 T1 carrier system.

CCITT system, an 125-μs frame is divided into 32 equal time slots. Time slot 0 is used for a frame alignment pattern and for an alarm channel. Time slot 17 is used for a common signaling channel. The signaling for all the voice band channels is accomplished on the common signaling channel. Consequently, there are 30 voice band channels time-division multiplexed into each CCITT frame.

With the CCITT standard, each time slot has 8 bits. Consequently, the total number of bits per frame is

$$\frac{8 \text{ bits}}{\text{time slot}} \times \frac{32 \text{ time slots}}{\text{frame}} = \frac{256 \text{ bits}}{\text{frame}}$$

and the line speed is

$$\text{line speed} = \frac{256 \text{ bits}}{\text{frame}} \times \frac{8000 \text{ frames}}{\text{second}} = 2.048 \text{ Mbps}$$

CODECS

A *codec* is a large-scale-integration (LSI) chip designed for use in the telecommunications industry for *p*rivate *b*ranch *ex*changes (PBXs), central office switches, digital handsets, voice store-and-forward systems, and digital echo suppressors. Essentially, the codec is applicable for any purpose that requires the digitizing of analog signals, such as in a PCM-TDM carrier system.

Time slot 0	Time slot 1	Time slots 2–16	Time slot 17	Time slots 18–30	Time slot 31
Framing and alarm channel	Voice channel 1	Voice channels 2–15	Common signaling channel	Voice channels 16–29	Voice channel 30
8 bits	8 bits	112 bits	8 bits	112 bits	8 bits

(a)

Time slot 17

	Bits	
Frame	1234	5678
0	0000	xyxx
1	ch 1	ch 16
2	ch 2	ch 17
3	ch 3	ch 18
4	ch 4	ch 19
5	ch 5	ch 20
6	ch 6	ch 21
7	ch 7	ch 22
8	ch 8	ch 23
9	ch 9	ch 24
10	ch 10	ch 25
11	ch 11	ch 26
12	ch 12	ch 27
13	ch 13	ch 28
14	ch 14	ch 29
15	ch 15	ch 30

16 frames equal one multiframe; 500 multiframes are transmitted each second

x = spare
y = loss of multiframe alignment if a 1

4 bits per channel are transmitted once every 16 frames, resulting in a 500-bps signaling rate for each channel

(b)

Figure 5-8 CCITT TDM frame alignment and common signaling channel alignment: (a) CCITT TDM frame (125 μs, 256 bits, 2.048 Mbps); (b) common signaling channel.

"Codec" is a generic term that refers to the *co*ding and *dec*oding functions performed by a device that converts analog signals to digital codes and digital codes to analog signals. Recently developed codecs are called *combo* chips because they combine codec and filter functions in the same LSI package. The input/output filter performs the following functions: bandlimiting, noise rejection, antialiasing, and reconstruction of analog audio waveforms after decoding. The codec performs the following functions: analog sampling, encoding/decoding (analog-to-digital and digital-to-analog conversions), and digital companding.

2913/14 COMBO CHIP

The 2913/14 is a combo chip that can provide the analog-to-digital and the digital-to-analog conversions and the transmit and receive filtering necessary to interface a full-duplex (four-wire) voice telephone circuit to the PCM highway of a TDM carrier system. Essentially, the 2913/14 combo chip replaces the older 2910A/11A codec and 2912A filter chip. The 2913 (20-pin package)/2914 (24-pin package) combo chip is manufactured with HMOS technology. There is a newer CHMOS version (29C13/14) that is functionally identical to the HMOS version except that it comes in a 28-pin package and has three low-power modes of operation. In addition, the 2916/17 and 29C16/17 are 16-pin limited-feature versions of the 2913/14 and 29C13/14. The following discussion is limited to the 2914 combo chip, although extrapolation to the other versions is quite simple.

Table 5-1 lists several of the combo chips available and their prominent features. Table 5-2 lists the pin names for the 2914 and gives a brief description of each of their functions. Figure 5-9 shows the block diagram of a 2914 combo chip.

General Operation

The following major functions are provided by the 2914 combo chip:

1. Bandpass filtering of the analog signals prior to encoding and after decoding
2. Encoding and decoding of voice and call progress signals
3. Encoding and decoding of signaling and supervision information
4. Digital companding

TABLE 5-1   FEATURES OF SEVERAL CODEC/FILTER COMBO CHIPS

2916 (16-pin)	2917 (16-pin)	2913 (20-pin)	2914 (24-pin)
μ-law companding only	A-law companding only	μ/A-law companding	μ/A-law companding
Master clock 2.048 MHz only	Master clock 2.048 MHz only	Master clock 1.536 MHz, 1.544 MHz, or 2.048 MHz	Master clock 1.536 MHz, 1.544 MHz, or 2.048 MHz
Fixed data rate	Fixed data rate	Fixed data rate	Fixed data rate
Variable data rate 64 kbps–2.048 Mbps	Variable data rate 64 kbps–4.048 Mbps	Variable data rate 64 kbps–4.096 Mbps	Variable data rate 64 kbps–4.096 Mbps
78-dB dynamic	78-dB dynamic range	78-dB dynamic range	78-dB dynamic range
ATT D3/4 compatible	ATT D3/4 compatible	ATT D3/4 compatible	ATT D3/4 compatible
Single-ended input Single-ended output	Single-ended input Single-ended output	Differential input Differential output	Differential input Differential output
Gain adjust transmit only	Gain adjust transmit only	Gain adjust transmit and receive	Gain adjust transmit and receive
Synchronous clocks	Synchronous clocks	Synchronous clocks	Synchronous clocks Asynchronous clocks
——	——	——	Analog loopback
——	——	——	Signaling

TABLE 5-2 2914 COMBO CHIP

Symbol	Name	Function
VBB	Power (−5 V)	Negative supply voltage.
PWRO+	Receive power amplifier output	Noninverting output of the receive power amplifier. This output can drive transformer hybrids or high-impedance loads directly in either a differential or single-ended mode.
PWRO−	Receive power amplifier output	Inverting output of the receive power amplifier. Functionally, PWRO− is identical and complementary to PWRO+.
GSR	Receive gain control	Input to the gain-setting network on the receive power amplifier. Transmission level can be adjusted over a 12-dB range depending on the voltage at GSR.
$\overline{\text{PDN}}$	Power-down select	When $\overline{\text{PDN}}$ is a TTL high, the 2914 is active. When $\overline{\text{PDN}}$ is low, the 2914 is powered down.
CLKSEL	Master clock frequency select	Input that must be pinstrapped to reflect the master clock frequency at CLKX, CLKR. CLKSEL = VBB .... 2.048 MHz CLKSEL = GRDD .... 1.544 MHz CLKSEL = VCC .... 1.536 MHz
LOOP	Analog loopback	When this pin is a TTL high, the analog output (PWRO+) is internally connected to the analog input (VFXI+), GSR is internally connected to PWRO−, and VFXI− is internally connected to GSX.
SIGR	Receive signaling bit output	Signaling bit output from the receiver. In the fixed data rate mode only, SIGR outputs the logic state of the eighth bit of the PCM word in the most recent signaling frame.
DCLKR	Receive variable data rate	Selects either the fixed or variable data rate mode of operation. When DCLKR is tied to VBB, the fixed data rate mode is selected. When DCLKR is not connected to VBB, the 2914 operates in the variable data rate mode and will accept TTL input levels from 64 kHz to 4096 MHz.
DR	Receive PCM highway input	PCM data are clocked in on this lead on eight consecutive negative transitions of the receive data rate clock; CLKR in the fixed data rate mode and DCLKR in the variable data rate mode.
FSR	Receive frame synchronization clock	8-kHz frame synchronization clock input/time slot enable for the receive channel. Also in the fixed data rate mode, this lead designates the signaling and nonsignaling frames. In the variable data rate mode this signal must remain active high for the entire length of the PCM word (8 PCM bits). The receive channel goes

TABLE 5-2 (continued)

Symbol	Name	Function
		into the standby mode whenever this input is TTL low for 300 ms.
GRDD	Digital ground	Digital ground for all internal logic circuits. This pin is not internally tied to GRDA.
CLKR	Receive master clock	Receive master clock and data rate clock in the fixed data rate mode, master clock only in the variable data rate mode.
CLKX	Transmit master clock	Transmit master clock and data rate clock in the fixed data rate mode, master clock only in the variable data rate mode.
FSX	Transmit frame synchronization clock	8-kHz frame synchronization clock input/time slot enable for the transmit channel. Operates independently but in an analogous manner to FSR.
DX	Transmit PCM output	PCM data are clocked out on this lead on eight consecutive positive transitions of the transmit data rate clock; CLKX in the fixed data rate mode and DCLKX in the variable data rate mode.
$\overline{\text{TSX}}$/ DCLKX	Times-slot strobe/ buffer enable Transmit variable data rate	Transmit channel timeslot strobe (output) or data rate clock (input) for the transmit channel. In the fixed data rate mode, this pin is an open drain output designed to be used as an enable signal for a three-state buffer. In variable data rate mode, this pin is the transmit data rate clock input which can operate at data rates between 64 kbps and 4096 kbps.
SIGX/ ASEL	Transmit signaling input μ- or A-law select	A dual-purpose pin. When connected to VBB, A-law companding is selected. When it is not connected to VBB, this pin is a TTL-level input for signaling bits. This input is substituted into the least significant bit position of the PCM word during every signaling frame.
GRDA	Analog ground	Analog ground return for all internal voice circuits. Not internally connected to GRDD.
VFXI+	Noninverting analog input	Noninverting analog input to uncommitted transmit operational amplifier.
VFXI−	Inverting analog input	Inverting analog input to uncommitted transmit operational amplifier.
GSX	Transmit gain control	Output terminal of on-chip uncommitted operational amplifier. Internally, this is the voice signal input to the transmit BPF.
VCC	Power (+5 V)	Positive supply voltage.

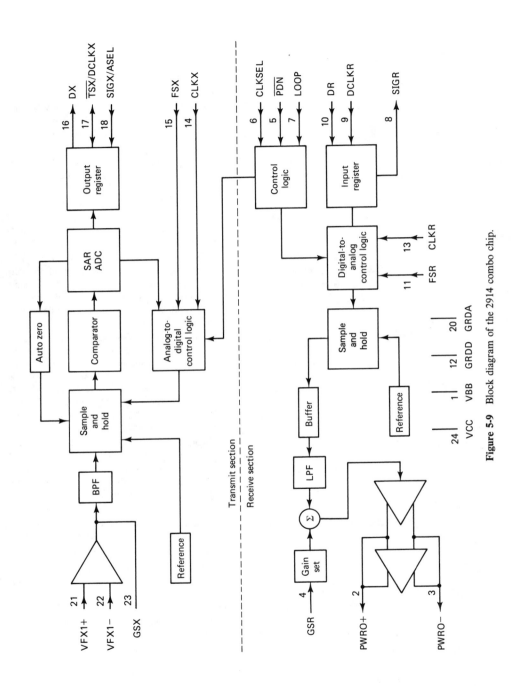

Figure 5-9 Block diagram of the 2914 combo chip.

184

System Reliability Features

The 2914 combo chip is powered up by pulsing the *transmit frame synchronization input* (FSX) and/or the *receive frame synchronization input* (FSR), while a TTL high (inactive condition) is applied to the *power down select pin* ($\overline{\text{PDN}}$) and all clocks and power supplies are connected. The 2914 has an internal reset on all power-ups (when VBB or VCC are applied or temporarily interrupted). This ensures the validity of the digital output and thereby maintains the integrity of the PCM highway.

On the transmit channel, PCM *data output* (DX) and *transmit timeslot strobe* ($\overline{\text{TSX}}$) are held in a high-impedance state for approximately four frames (500 μs) after power-up. After this delay DX, $\overline{\text{TSX}}$, and signaling are functional and will occur in the proper time slots. Due to the auto-zeroing circuit on the transmit channel, the analog circuit requires approximately 60 ms to reach equilibrium. Therefore, signaling information such as on/off hook detection is available almost immediately while analog input signals are not available until after the 60-ms delay.

On the receive channel, the *signaling bit output pin* SIGR is also held low (inactive) for approximately 500 μs after power-up and remains inactive until updated by reception of a signaling frame.

$\overline{\text{TSX}}$ and DX are placed in the high-impedance state and SIGR is held low for approximately 20 μs after an interruption of the *master clock* (CLKX). Such an interruption could be caused by some kind of fault condition.

Power-Down and Standby Modes

To minimize power consumption, two power-down modes are provided in which most 2914 functions are disabled. Only the power-down, clock, and frame synchronization buffers are enabled in these modes.

The power-down is enabled by placing an external TTL low signal on $\overline{\text{PDN}}$. In this mode power consumption is reduced to an average of 5 mW.

The standby mode for the transmit and receive channels is separately controlled by removing FSX and/or FSR.

Fixed-Data-Rate Mode

In the *fixed-data-rate mode*, the master *transmit* and *receive clocks* (CLKX and CLKR) perform the following functions:

1. Provide the master clock for the on-board switched capacitor filters
2. Provide the clock for the analog-to-digital and digital-to-analog converters
3. Determine the input and output data rates between the codec and the PCM highway

Therefore, in the fixed-data-rate mode, the transmit and receive data rates must be either 1.536, 1.544, or 2.048 Mbps, the same as the master clock rate.

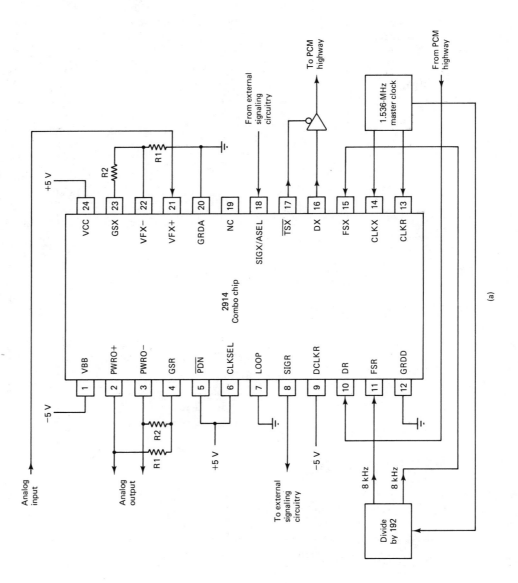

(a)

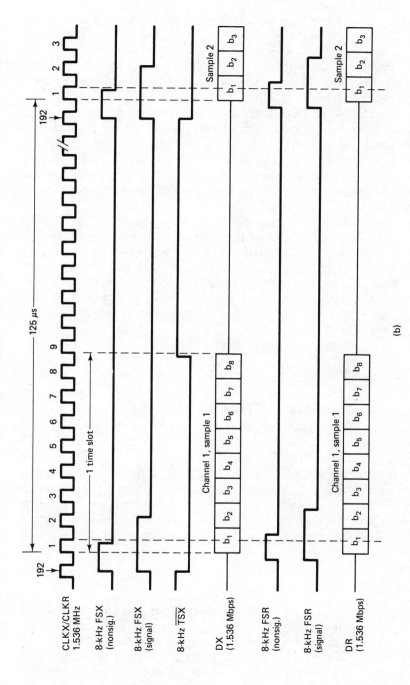

Figure 5-10 Single-channel PCM system using the 2914 combo chip in the fixed-data-rate mode: (a) block diagram; (b) timing sequence.

(b)

187

Transmit and receive frame synchronizing pulses (FSX and FSR) are 8-kHz inputs which set the transmit and receive sampling rates and distinguish between *signaling* and *nonsignaling* frames. $\overline{\text{TSX}}$ is a *time-slot strobe buffer enable* output which is used to gate the PCM word onto the PCM highway when an external buffer is used to drive the line. $\overline{\text{TSX}}$ is also used as an external gating pulse for a time-division multiplexer (see Figure 5-10).

Data are transmitted to the PCM highway from DX on the first eight positive transitions of CLKX following the rising edge of FSX. On the receive channel, data are received from the PCM highway from DR on the first eight falling edges of CLKR after the occurrence of FSR. Therefore, the occurrence of FSX and FSR must be synchronized between codecs in a multiple-channel system to ensure that only one codec is transmitting to or receiving from the PCM highway at any given time.

Figure 5-10 shows the block diagram and timing sequence for a single-channel PCM system using the 2914 combo chip in the fixed-data-rate mode and operating with a master clock frequency of 1.536 MHz. In the fixed-data-rate mode, data are inputted and outputted in short bursts. (This mode of operation is sometimes called the *burst mode*.) With only a single channel, the PCM highway is active only $\frac{1}{24}$ of the total frame time. Additional channels can be added to the system provided that their transmissions are synchronized so that they do not occur at the same time as transmissions from any other channel.

From Figure 5-10 the following observations can be made:

1. The input and output bit rates from the codec are equal to the master clock frequency, 1.536 Mbps.
2. The codec inputs and outputs 64,000 PCM bits per second.
3. The data output (DX) and data input (DR) are active only $\frac{1}{24}$ of the total frame time (125 μs).

To add channels to the system shown in Figure 5-10, the occurrence of the FSX, FSR, and $\overline{\text{TSX}}$ signals for each additional channel must be synchronized so that they follow a timely sequence and do not allow more than one codec to transmit or receive at the same time. Figure 5-11 shows the block diagram and timing sequence for a 24-channel PCM-TDM system operating with a master clock frequency of 1.536 MHz.

Variable-Data-Rate Mode

The *variable-data-rate mode* allows for a flexible data input and output clock frequency. It provides the ability to vary the frequency of the transmit and receive bit clocks. In the variable data rate mode, a master clock frequency of 1.536, 1.544, or 2.048 MHz is still required for proper operation of the on-board bandpass filters and the analog-to-digital and digital-to-analog converters. However, in the variable-data-rate mode, DCLKR and DCLKX become the data clocks for the receive and transmit PCM highways, respectively. When FSX is high, data are transmitted onto the PCM highway on the next eight consecutive positive transitions of DCLKX. Similarly, while FSR is high, data from the PCM highway are clocked into the codec on the next eight consecutive negative transitions of DCLKR. This mode of operation is sometimes called the *shift register mode*.

On the transmit channel, the last transmitted PCM word is repeated in all remaining time slots in the 125-μs frame as long as DCLKX is pulsed and FSX is held active high. This feature allows the PCM word to be transmitted to the PCM highway more than once per frame. Signaling is not allowed in the variable-data-rate mode because this mode provides no means to specify a signaling frame.

Figure 5-12 shows the block diagram and timing sequence for a two-channel PCM-TDM system using the 2914 combo chip in the variable-data-rate mode with a master clock frequency of 1.536 MHz, a sample rate of 8 kHz, and a transmit and receive data rate of 128 kbps.

With a sample rate of 8 kHz, the frame time is 125 μs. Therefore, one 8-bit PCM word from each channel is transmitted and/or received during each 125-μs frame. For 16 bits to occur in 125 μs, a 128-kHz transmit and receive data clock is required.

$$\frac{1 \text{ channel}}{8 \text{ bits}} \times \frac{1 \text{ frame}}{2 \text{ channels}} \times \frac{125 \ \mu s}{\text{frame}} = \frac{125 \ \mu s}{16 \text{ bits}} = \frac{7.8125 \ \mu s}{\text{bit}}$$

$$\text{bit rate} = \frac{1}{t_b} = \frac{1}{7.8125 \ \mu s} = 128 \text{ kbps}$$

The transmit and receive enable signals (FSX and FSR) for each codec are active for one-half of the total frame time. Consequently, 8-kHz, 50% duty cycle transmit and receive data enable signals (FXS and FXR) are fed directly to one codec and fed to the other codec 180° out of phase (inverted), thereby enabling only one codec at a time.

To expand to a four-channel system, simply increase the transmit and receive data clock rates to 256 kHz and change the enable signals to an 8-kHz, 25% duty cycle pulse.

Supervisory Signaling

With the 2914 combo chip, *supervisory signaling* can be used only in the fixed-data-rate mode. A transmit signaling frame is identified by making the FSX and FSR pulses twice their normal width. During a transmit signaling frame, the signal present on input SIGX is substituted into the least significant bit position (b_1) of the encoded PCM word. At the receive end, the signaling bit is extracted from the PCM word prior to decoding and placed on output SIGR until updated by reception of another signaling frame.

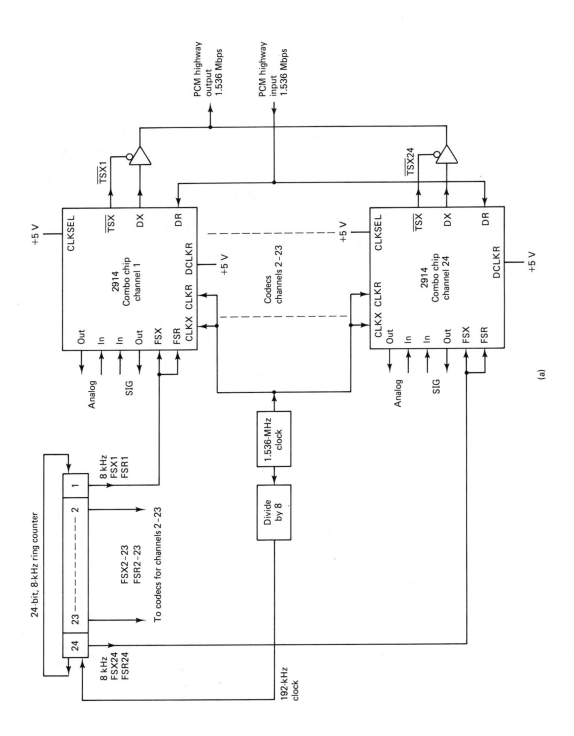

(a)

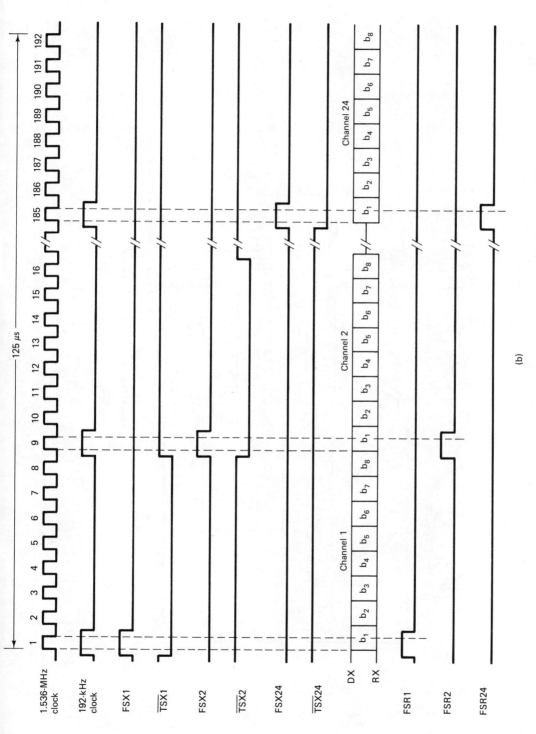

Figure 5-11 24-channel PCM-TDM system using the 2914 combo chip in the fixed-data-rate mode and operating with a master clock frequency of 1.536 MHz: (a) block diagram; (b) timing diagram.

(b)

191

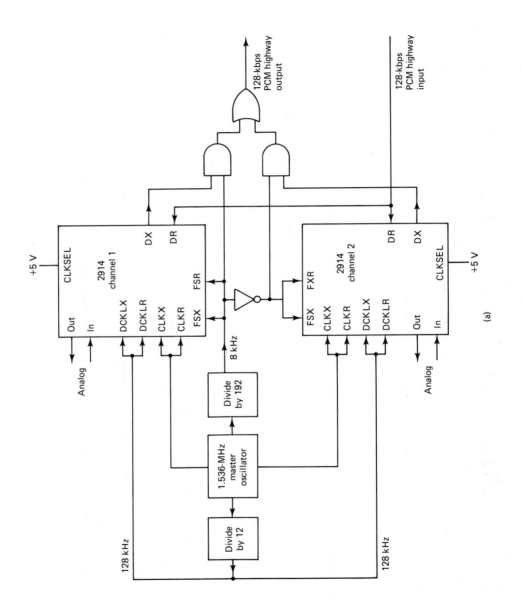

(a)

192

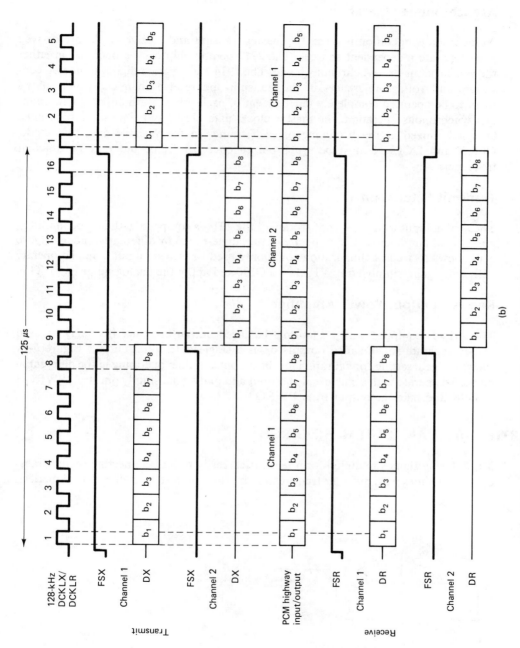

Figure 5-12 Two-channel PCM-TDM system using the 2914 combo chip in the variable-data-rate mode with a master clock frequency of 1.536 MHz: (a) block diagram; (b) timing diagram.

Asynchronous Operation

Asynchronous operation is when the master transmit and receive clocks are derived from separate independent sources. The 2914 combo chip can be operated in either the synchronous or asynchronous mode. The 2914 has separate digital-to-analog converters and voltage references in the transmit and receive channels, which allows them to be operated completely independent of each other. With either synchronous or asynchronous operation, the master clock, data clock, and time-slot strobe must be synchronized at the beginning of each frame. In the variable data rate mode, CLKX and DCLKX must be synchronized once per frame but may be different frequencies.

Transmit Filter Gain

The analog input to the transmit section of the 2914 is equipped with an uncommitted operational amplifier that can operate in the single-ended or differential mode. Figure 5-13 shows a circuit configuration commonly used to provide input gain. To operate with unity gain, simply strap VFXI− to GSX and apply the analog input to VFXI+.

Receive Output Power Amplifier

The 2914 is equipped with an internal balanced output amplifier that may be used as two separate single ended outputs or as a single differential output. Figure 5-14 shows the gain setting configuration for the output amplifier operating in the differential mode. To operate with a single ended output and unity gain, simply pin strap PWRO− to GSR and take the output from PWRO+.

NORTH AMERICAN DIGITAL HIERARCHY

Multiplexing signals in digital form lends itself easily to interconnecting digital transmission facilities with different transmission bit rates. Figure 5-15 shows the American

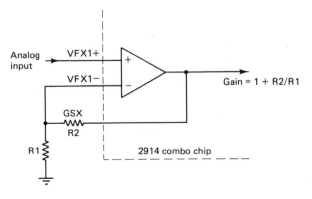

Figure 5-13 Transmit filter gain circuit.

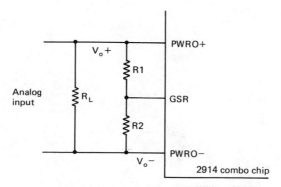

Figure 5-14 Receive output power amplifier. PWRO+ and PWRO− are low-impedance complementary outputs. The voltages at the nodes are V_o+ at PWRO+ and V_o- at PWRO−. R1 and R2 comprise a gain-setting resistor network with the center tap connected to the GSR input. A value greater than 10 kΩ for R1 and a value less than 100 kΩ for R2 is recommended because (1) the parallel combination of R1 + R2 and RL set the total load impedance to the analog sink, and (2) the total capacitance at the GSR input and the parallel combination of R1 and R2 define a time constant that has to be minimized to avoid inaccuracies. VA represents the maximum available digital miliwatt output response (VA = 3.006 V rms).

$$V_o = -\mathrm{A(VA)}$$

where

$$\mathrm{A} = \frac{1 + \mathrm{R1/R2}}{4 + \mathrm{R1/R2}}$$

For design purposes, a useful form is R1/R2 as a function of A.

$$\mathrm{R1/R2} = \frac{4\mathrm{A} - 1}{1 - \mathrm{A}}$$

Telephone and Telegraph Company's (AT&T) North American Digital Hierarchy for multiplexing digital signals with the same bit rates into a single pulse stream suitable for transmission on the next higher level of the hierarchy. To upgrade from one level in the hierarchy to the next higher level, special devices called *muldems* (*mul*tiplexers/*dem*ultiplexers) are used. Muldems can handle bit-rate conversions in both directions. The muldem designations (M12, M23, etc.) identify the input and output digital signals associated with that muldem. For instance, an M12 muldem is a multiplexer/demultiplexer that interfaces DS-1 and DS-2 *digital signals*. An M23 muldem interfaces DS-2 and DS-3 signals. DS-1 signals may be further multiplexed or line encoded and placed on specially conditioned lines called T1 lines. DS-2, DS-3, and DS-4 signals may be placed on T2, T3, and T4M lines, respectively.

Digital signals are routed at central locations called *digital cross-connects*. A digital cross-connect (DSX) provides a convenient place to make hardwire interconnects and to perform routine maintenance and troubleshooting. Each type of digital

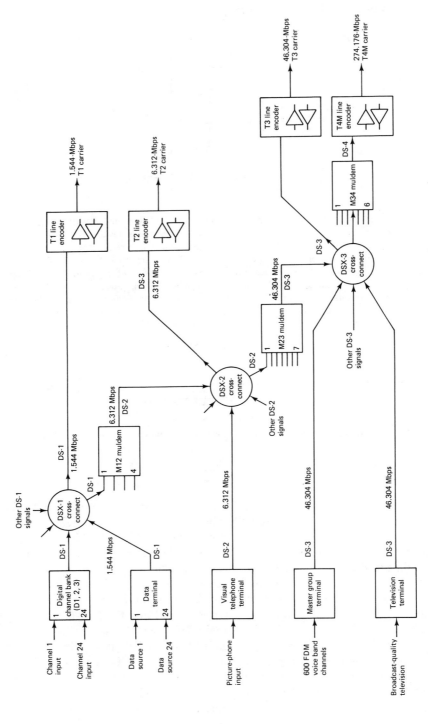

Figure 5-15 North American Digital Hierarchy.

signal (DS-1, DS-2, etc.) has its own digital switch (DSX-1, DSX-2, etc.). The output from a digital switch may be upgraded to the next higher level or line encoded and placed on their respective T lines (T1, T2, etc.).

Table 5-3 lists the digital signals, their bit rates, channel capacities, and services offered for the line types included in the North American Digital Hierarchy.

When the bandwidth of the signals to be transmitted is such that after digital conversion it occupies the entire capacity of a digital transmission line, a single-channel terminal is provided. Examples of such single-channel terminals are picturephone, mastergroup, and commercial television terminals.

Mastergroup and Commercial Television Terminals

Figure 5-16 shows the block diagram of a mastergroup and commercial television terminal. The mastergroup terminal receives voice band channels that have already been frequency-division multiplexed (a topic covered in Chapter 6) without requiring that each voice band channel be demultiplexed to voice frequencies. The signal processor provides frequency shifting for the mastergroup signal (shifts it from a 564- to 3084-kHz bandwidth to a 0- to 2520-kHz bandwidth) and dc restoration for the television signal. By shifting the mastergroup band, it is possible to sample at a 5.1-MHz rate. Sampling of the commercial television signal is at twice that rate or 10.2 MHz.

To meet the transmission requirements, a 9-bit PCM code is used to digitize each sample of the mastergroup or television signal. The digital output from the terminal is therefore approximately 46 Mbps for the mastergroup and twice that much (92 Mbps) for the television signal.

The digital terminal shown in Figure 5-16 has three specific functions: it converts the parallel data from the output of the encoder to serial data, it inserts frame synchronizing bits, and it converts the serial binary signal to a form more suitable for transmission. In addition, for the commercial television terminal, the 92-Mbps digital signal must be split into two 46-Mbps digital signals because there is no 92-Mbps line speed in the digital hierarchy.

TABLE 5-3 SUMMARY OF THE NORTH AMERICAN DIGITAL HIERARCHY

Line type	Digital signal	Bit rate (Mbps)	Channel capacities	Services offered
T1	DS-1	1.544	24	Voice band telephone
T1C	DS-1C	3.152	48	Voice band telephone
T2	DS-2	6.312	96	Voice band telephone and picture-phone
T3	DS-3	44.736	672	Voice band telephone, picturephone, and broadcast-quality television
T4M	DS-4	274.176	4032	Same as T3 except more capacity

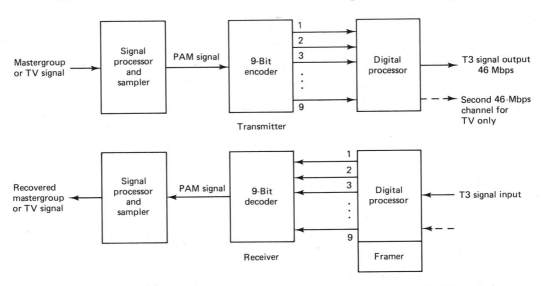

Figure 5-16 Block diagram of a mastergroup or commercial television digital terminal.

Picturephone Terminal

Essentially, *picturephone* is a low-quality video transmission for use between nondedicated subscribers. For economic reasons it is desirable to encode a picturephone signal into the T2 capacity of 6.312 Mbps, which is substantially less than that for commercial network broadcast signals. This substantially reduces the cost and makes the service affordable. At the same time, this permits the transmission of adequate detail and contrast resolution to satisfy the average picturephone subscriber. Picturephone service is ideally suited to a differential PCM code. Differential PCM is similar to conventional PCM except that the exact magnitude of a sample is not transmitted. Instead, only the difference between that sample and the previous sample is encoded and transmitted. To encode the difference between samples requires substantially fewer bits than encoding the actual sample.

Data Terminal

The portion of communications traffic that involves data (signals other than voice) is increasing exponentially. Also, in most cases, the data rates generated by each individual subscriber are substantially less than the data rate capacities of digital lines. Therefore, it seems only logical that terminals be designed that transmit data signals from several sources over the same digital line.

Data signals could be sampled directly; however, this would require excessively high sample rates resulting in excessively high transmission bit rates, especially for sequences of data with few or no transitions. A more efficient method is one that

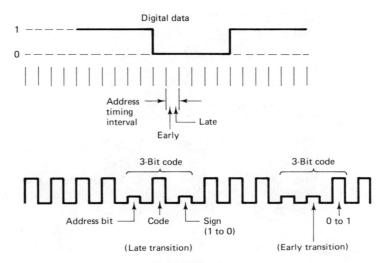

Figure 5-17 Data coding format.

codes the transition times. Such a method is shown in Figure 5-17. With the coding format shown, a 3-bit code is used to identify when transitions occur in the data and whether that transition is from a 1 to a 0, or vice versa. The first bit of the code is called the address bit. When this bit is a logic 1 this indicates that no transition occurred, a logic 0 indicates that a transition did occur. The second bit indicates whether the transition occurred during the first half (0) or during the second half (1) of the sample interval. The third bit indicates the sign or direction of the transition; a 1 for this bit indicates a 0-to-1 transition and a 0 indicates a 1-to-0 transition. Consequently, when there are no transitions in the data, a signal of all 1's is transmitted. Transmission of only the address bit would be sufficient; however, the sign bit provides a degree of error protection and limits error propagation (when one error leads to a second error, etc.). The efficiency of this format is approximately 33%; there are 3 code bits for each data bit. The advantage of using a coded format rather than the original data is that coded data are more efficiently substituted for voice in analog systems. To transmit a 250-kbps data signal, the same bandwidth is required to transmit 60 voice channels with analog multiplexing. With this coded format, a 50-kbps data signal displaces three 64-kbps PCM encoded channels, and a 250-kbps data stream displaces only 12 voice band channels.

LINE ENCODING

Line encoding involves converting standard logic levels (TTL, CMOS, etc.) to a form more suitable to telephone line transmission. Essentially, there are four primary factors that must be considered when selecting a line-encoding format:

1. Timing (clock) recovery
2. Transmission bandwidth
3. Ease of detection and decoding
4. Error detection

Transmission Voltages

Transmission voltages or levels can be categorized as either *unipolar* (UP) or *bipolar* (BP). Unipolar transmission of binary data involves the transmission of only a single nonzero voltage level (e.g., +V for a logic 1 and 0 V or ground for a logic 0). In bipolar transmission, two nonzero voltage levels are involved (e.g., +V for a logic 1 and −V for a logic 0).

Duty Cycle

The *duty cycle* of a binary pulse can also be used to categorize the type of transmission. If the binary pulse is maintained for the entire bit time, this is called *nonreturn-to-zero* (NRZ). If the active time of the binary pulse is less than 100% of the bit time, this is called *return-to-zero* (RZ).

Unipolar and bipolar transmission voltages and return-to-zero and nonreturn-to-zero encoding can be combined in several ways to achieve a particular line encoding scheme. Figure 5-18 shows five line-encoding possibilities.

In Figure 5-18a, there is only one nonzero voltage level (+V = logic 1); a zero voltage simply implies a binary 0. Also, each logic 1 maintains the positive voltage for the entire bit time (100% duty cycle). Consequently, Figure 5-18a represents a unipolar nonreturn-to-zero signal (UPNRZ). In Figure 5-18b, there are two nonzero voltages (+V = logic 1 and −V = logic 0) and a 100% duty cycle is used. Figure 5-18b represents a bipolar nonreturn-to-zero signal (BPNRZ). In Figure 5-18c, only one nonzero voltage is used but each pulse is active for only 50% of the bit time. Consequently, Figure 5-18c represents a unipolar return-to-zero signal (UPRZ). In Figure 5-18d, there are two nonzero voltages (+V = logic 1 and −V = logic 0). Also, each pulse is active only 50% of the total bit time. Consequently, Figure 5-18d represents a bipolar return-to-zero (BPRZ) signal. In Figure 5-18e, there are again two nonzero voltage levels (−V and +V), but here both polarities represent a logic 1 and 0 V represents a logic 0. This method of encoding is called *alternate mark inversion* (AMI). With AMI transmissions, each successive logic 1 is inverted in polarity from the previous logic 1. Because return-to-zero is used, this encoding technique is called *bipolar-return-to-zero alternate mark inversion* (BPRZ-AMI).

The method of line encoding used determines the minimum bandwidth required for transmission, how easily a clock may be extracted from it, how easily it may be decoded, and whether it offers a convenient means of detecting errors.

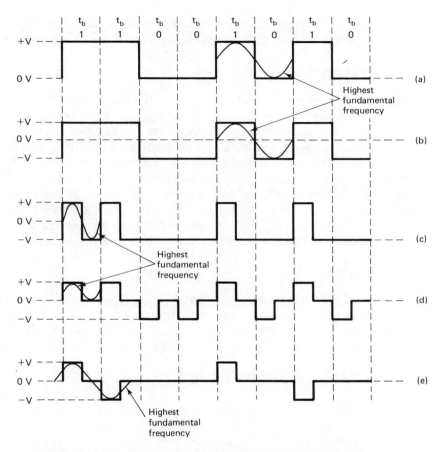

Figure 5-18 Line-encoding formats: (a) UPNRZ; (b) BPNRZ; (c) UPRZ; (d) BPRZ; (e) BPRZ-AMI.

Bandwidth Considerations

To determine the minimum bandwidth required to propagate a line-encoded signal, you must determine the highest fundamental frequency associated with it (see Figure 5-18). The highest fundamental frequency is determined from the worst-case (fastest transition) binary bit sequence. With UPNRZ, the worst-case condition is an alternating 1/0 sequence; the highest fundamental frequency takes the time of 2 bits and is therefore equal to one-half the bit rate. With BPNRZ, again the worst-case condition is an alternating 1/0 sequence and the highest fundamental frequency is one-half of the bit rate. With UPRZ, the worst-case condition is two successive 1's. The minimum bandwidth is therefore equal to the bit rate. With BPRZ, the worst-case condition is either successive 1's or 0's and the minimum bandwidth is again equal to the bit rate. With BPRZ-AMI, the worst-case condition is two or more consecutive 1's, and the minimum bandwidth is equal to one-half of the bit rate.

Clock Recovery

To recover and maintain clocking information from received data, there must be a sufficient number of transitions in the data signal. With UPNRZ and BPNRZ, a long string of consecutive 1's or 0's generates a data signal void of transitions and is therefore inadequate for clock synchronization. With UPRZ and BPRZ-AMI, a long string of 0's also generates a data signal void of transitions. With BPRZ, a transition occurs in each bit position regardless of whether the bit is a 1 or a 0. In the clock recovery circuit, the data are simply full-wave rectified to produce a data-independent clock equal to the receive bit rate. Therefore, BPRZ encoding is best suited for clock recovery. If long sequences of 0's are prevented from occurring, BPRZ-AMI encoding is sufficient to ensure clock synchronization.

Error Detection

With UPNRZ, BPNRZ, UPRZ, and BPRZ transmissions, there is no way to determine if the incoming data have errors. With BPRZ-AMI transmissions, an error in any bit will cause a bipolar violation (the reception of two or more consecutive 1's with the same polarity). Therefore, BPRZ-AMI has a built-in error detection mechanism.

Ease of Detection and Decoding

Because unipolar transmission involves the transmission of only one polarity voltage, there is a dc average voltage associated with the signal equal to $+V/2$. Assuming an equal probability of 1's and 0's occurring, bipolar transmissions have an average dc component of 0 V. A dc component is undesirable because it biases the input to a conventional threshold detector (a biased comparator) and could cause a misinterpretation of the logic condition of the received pulses. Therefore, bipolar transmission is better suited to data detection.

Table 5-4 summarizes the minimum bandwidth, average dc voltage, clock recovery, and error detection capabilities of the line-encoding formats shown in Figure 5-18. From Table 5-4 it can be seen that BPRZ-AMI encoding has the best overall characteristics and is therefore the most common method used.

T CARRIERS

T carriers involve the transmission of PCM-encoded time-division-multiplexed digital signals. In addition, T carriers utilize special line-encoded signals and metallic cables that have been conditioned to meet the relatively high bandwidths required for high-speed digital transmissions. Digital signals deteriorate as they are propagated along a cable due to power loss in the metallic conductors and the low-pass filtering inherent

TABLE 5-4   LINE-ENCODING SUMMARY

Encoding format	Minimum BW	Average DC	Clock recovery	Error detection
UPNRZ	$F_b/2$[a]	+V/2	Poor	No
BPNRZ	$F_b/2$[a]	0 V[a]	Poor	No
UPRZ	F_b	+V/2	Good	No
BPRZ	F_b	0 V[a]	Best[a]	No
BPRZ-AMI	$F_b/2$[a]	0 V[a]	Good	Yes[a]

[a] Denotes best performance or quality.

in parallel wire transmission lines. Consequently, *regenerative repeaters* must be placed at periodic intervals. The distance between repeaters is dependent on the transmission bit rate and the line-encoding technique used.

Figure 5-19 shows the block diagram of a regenerative repeater. Essentially, there are three functional blocks: an amplifier-equalizer, a timing circuit, and the regenerator. The amplifier-equalizer shapes the incoming digital signal and raises their power level so that a pulse/no pulse decision can be made by the regenerator circuit. The timing circuit recovers the clocking information from the received data and provides the proper timing information to the regenerator so that decisions can be made at the optimum time that minimizes the chance of an error occurring. Spacing of the repeaters is designed to maintain an adequate signal-to-noise ratio for error-free performance. The signal-to-noise ratio (S/N) at the output of a regenerator is exactly what it was at the output of the transmit terminal or at the output of the previous regenerator (i.e., the S/N does not deteriorate as a digital signal propagates through a regenerator; in fact, a regenerator reconstructs the original pulses and produces the original S/N ratio).

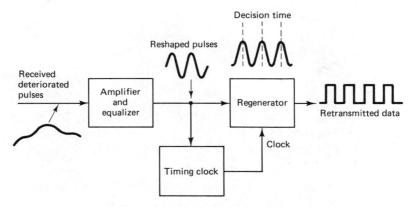

Figure 5-19   Regenerative repeater block diagram.

T1 and T1C Carrier Systems

The T1 carrier system utilizes PCM and TDM techniques to provide short-haul transmission of 24 voice band signals. The lengths of T1 carrier systems range from about 5 to 50 miles. T1 carriers use BPRZ-AMI encoding with regenerative repeaters placed every 6000 ft; 6000 ft was chosen because telephone company manholes are located at approximately 6000-ft intervals and these same manholes are used for placement of the repeaters, facilitating convenient installation, maintenance, and repair. The transmission medium for T1 carriers is either a 19- or 22-gauge wire pair.

Because T1 carriers use BPRZ-AMI encoding, they are susceptible to losing synchronization on a long string of consecutive 0's. With a folded binary PCM code, the possibility of generating a long string of consecutive 0's is high (whenever a channel is idle it generates a ±0-V code which is either seven or eight consecutive 0's). If two or more adjacent voice channels are idle, there is a high probability that a long string of consecutive 0's will be transmitted. To reduce this possibility, the PCM code is inverted prior to transmission and inverted again at the receiver prior to decoding. Consequently, the only time a long string of consecutive 0's is transmitted is when two or more adjacent voice band channels each encode the maximum positive sample voltage, which is unlikely to happen.

With T1 and T1C carrier systems, provisions are taken to prevent more than 14 consecutive 0's from occurring. The transmissions from each frame are monitored for the presence of either 15 consecutive 0's or any one PCM sample (8 bits) without at least one nonzero bit. If either of these conditions occurs, a 1 is substituted into the appropriate bit position. The worst-case conditions are as follows:

```
                 MSB      LSB MSB      LSB
Original        1000 0000  0000 0001    14 consecutive 0's
DS-1 signal                            (no substitution)


                 MSB      LSB MSB      LSB
Original        1000 0000  0000 0000    15 consecutive 0's
DS-1 signal

Substituted     1000 0000  0000 0010
DS-1 signal                       ↑
                            Substituted
                            bit
```

A 1 is substituted into the second least significant bit. This introduces an encoding error equal to twice the amplitude resolution. This bit is selected rather than the least significant bit because, with the superframe format, during every sixth frame the LSB is the signaling bit and to alter it would alter the signaling word.

```
                     MSB      LSB MSB      LSB MSB      LSB
         Original    1010 1000  0000 0000  0000 0001
         DS-1 signal
```

```
Substituted   1010  1000    0000 0010   0000 0001
DS-1 signal                    ↑
                        Substituted
                           bit
```

The process shown is used for T1 and T1C carrier systems. Also, if at any time 32 consecutive 0's are received, it is assumed that the system is not generating pulses and is therefore out of service; this is because the occurrence of 32 consecutive 0's is highly unlikely.

T2 Carrier System

The T2 carrier utilizes PCM to time-division multiplex 96 voice band channels into a single 6.312-Mbps data signal for transmission up to 500 miles over a special LOCAP cable. A T2 carrier is also used to carry a single picturephone signal. T2 carriers use BPRZ-AMI encoding. However, because of the higher transmission rate, clock synchronization becomes more critical. A sequence of six consecutive 0's could be sufficient to cause loss of clock synchronization. Therefore, T2 carrier systems use an alternative method of ensuring that ample transitions occur in the data. This method is called *binary six zero substitution* (B6ZS).

With B6ZS, whenever six consecutive 0's occur, one of the following codes is substituted in its place: $0-+0+-$ or $0+-0-+$. The $+$ and $-$ represent positive and negative logic 1s. A zero simply indicates a logic 0 condition. The 6-bit code substituted for the six 0's is selected to purposely cause a bipolar violation. If the violation is caught at the receiver and the B6ZS code is detected, the original six 0's can be substituted back into the data signal. The substituted patterns cause a bipolar violation in the second and fifth bits of the substituted pattern. If DS-2 signals are multiplexed to form DS-3 signals, the B6ZS code must be detected and stripped from the DS-2 signal prior to DS-3 multiplexing. An example of B6ZS is as follows:

```
                MSB      LSB MSB       LSB MSB
Original       +000  −0+0   000−  0000      000+· · ·
data signal                        ‿‿‿‿‿‿‿‿‿
                                     6 0's


                             Substituted
                               pattern
                             ⁀‿‿‿‿‿‿‿
Encoded        +000  −0+0  ╱000−  0−+0   ╱ +−0+
data                             ↗       ↗


                    Bipolar      ╱
                    violations ╱
```

T3 Carrier System

A T3 carrier time-division multiplexes 672, PCM-encoded voice channels for transmission over a single metallic cable. The transmission rate for T3 signals is 44.736 Mbps. The encoding technique used with T3 carriers is *binary three zero substitution* (B3ZS). Substitutions are made for any occurrence of three consecutive 0's. There are four substitution patterns used: 00−, −0−, 00+, and +0+. The pattern chosen should cause a bipolar error in the third substitute bit. An example is as follows:

```
              MSB      LSB MSB      LSB MSB      LSB
   Original   0+00 00−0    +000 0−00    +−00 00+0
   data         ⌣⌣          ⌣⌣            ⌣⌣
               3 0's        3 0's         3 0's

                    Substituted pattern

              ⌢⌢          ⌢⌢            ⌢⌢
   Encoded    0+00 +0−0    +−0− 0−00    +−00 −0+0
   data           ╲        ╱                 ╱
                  Bipolar
                  violations ╱
```

T4M Carrier System

A T4M carrier time-division multiplexes 4032 PCM-encoded voice band channels for transmission over a single coaxial cable up to 500 miles. The transmission rate is sufficiently high that substitute patterns are impractical. Instead, T4M-carriers transmit scrambled unipolar NRZ digital signals where the scrambling and descrambling functions are performed in the subscriber's terminal equipment.

FRAME SYNCHRONIZATION

With TDM systems it is imperative that a frame is identified and that individual time slots (samples) within the frame are also identified. To acquire frame synchronization, there is a certain amount of overhead that must be added to the transmission. There are five methods commonly used to establish frame synchronization: added digit framing, robbed digit framing, added channel framing, statistical framing, and unique line signal framing.

Added Digit Framing

T1 carriers using D1, D2, or D3 channel banks use *added digit framing*. There is a special *framing digit* (framing pulse) added to each frame. Consequently, for an 8-kHz sample rate (125-μs frame), there are 8000 digits added per second. With T1 carriers, an alternating 1/0 frame synchronizing pattern is used.

To acquire frame synchronization, the receive terminal searchs through the incoming data until it finds the alternating 1/0 sequence used for the framing bit pattern. This encompasses testing a bit, counting off 193 bits, then testing again for the opposite condition. This process continues until an alternating 1/0 sequence is found. Initial frame synchronization is dependent on the total frame time, the number of bits per frame, and the period of each bit. Searching through all possible bit positions requires N tests, where N is the number of bit positions in the frame. On the average, the receiving terminal dwells at a false framing position for two frame periods during a search; therefore, the maximum average synchronization time is

$$\text{synchronization time} = 2NT = 2N^2t$$

where

$T =$ frame period or Nt
$N =$ number of bits per frame
$t =$ bit time

For the T1 carrier, $N = 193$, $T = 125$ μs, and $t = 0.648$ μs; therefore, a maximum of 74,498 bits must be tested and the maximum average synchronization time is 48.25 μs.

Robbed Digit Framing

When a short frame time is used, added digit framing is very inefficient. This occurs in single-channel PCM systems such as those used in television terminals. An alternative solution is to replace the least significant bit of every nth frame with a framing bit. The parameter n is chosen as a compromise between reframe time and signal impairment. For $n = 10$, the SQR is impaired by only 1 dB. *Robbed digit framing* does not interrupt transmission, but instead, periodically replaces information bits with forced data errors to maintain clock synchronization.

Added Channel Framing

Essentially, *added channel framing* is the same as added digit framing except that digits are added in groups or words instead of as individual bits. The CCITT multiplexing scheme previously discussed uses added channel framing. One of the 32 time slots in each frame is dedicated to a unique synchronizing sequence. The average frame synchronization time for added channel framing is

$$\text{synchronization time (bits)} = \frac{N^2}{2(2^L - 1)}$$

where

$N =$ number of bits per frame
$L =$ number of bits in the frame code

For the CCITT 32-channel system, $N = 256$ and $L = 8$. Therefore, the average number of bits needed to acquire frame synchronization is 128.5. At 2.048 Mbps, the synchronization time is approximately 62.7 μs.

Statistical Framing

With *statistical framing*, it is not necessary to either rob or add digits. With the Gray code, the second bit is a 1 in the central half of the code range and 0 at the extremes. Therefore, a signal that has a centrally peaked amplitude distribution generates a high probability of a 1 in the second digit. A mastergroup signal has such a distribution. With a mastergroup encoder, the probability that the second bit will be a 1 is 95%. For any other bit it is less than 50%. Therefore, the second bit can be used for a framing bit.

Unique Line Code Framing

With *unique line code framing*, the framing bit is different from the information bits. It is either made higher or lower in amplitude or of a different time duration. The earliest PCM/TDM systems used unique line code framing. D1 channel banks used framing pulses that were twice the amplitude of normal data bits. With unique line code framing, added digit or added word framing can be used with it or data bits can be used to simultaneously convey information and carry synchronizing signals.

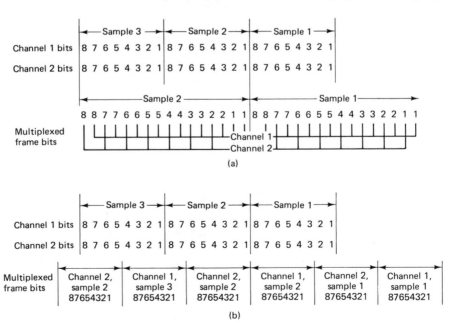

Figure 5-20   Interleaving: (a) bit; (b) word.

The advantage of unique line code framing is synchronization is immediate and automatic. The disadvantage is the additional processing requirements required to generate and recognize the unique framing bit.

BIT INTERLEAVING VERSUS WORD INTERLEAVING

When time-division multiplexing two or more PCM systems, it is necessary to interleave the transmissions from the various terminals in the time domain. Figure 5-20 shows two methods of interleaving PCM transmissions: *bit interleaving* and *word interleaving*.

T1 carrier systems use word interleaving; 8-bit samples from each channel are interleaved into a single 24-channel TDM frame. Higher-speed TDM systems and delta modulation systems use bit interleaving. The decision as to which type of interleaving to use is usually determined by the nature of the signals to be multiplexed.

QUESTIONS

5-1. Define *multiplexing*.

5-2. Describe time-division multiplexing.

5-3. Describe the Bell System T1 carrier system.

5-4. What is the purpose of the signaling bit?

5-5. What is frame synchronization? How is it achieved in a PCM/TDM system?

5-6. Describe the superframe format. Why is it used?

5-7. What is a codec? A combo chip?

5-8. What is a fixed-data-rate mode?

5-9. What is a variable-data-rate mode?

5-10. What is a DSX? What is it used for?

5-11. Explain *line encoding*.

5-12. Briefly explain unipolar and bipolar transmission.

5-13. Briefly explain return-to-zero and nonreturn-to-zero transmission.

5-14. Contrast the bandwidth considerations of return-to-zero and nonreturn-to-zero transmission.

5-15. Contrast the clock recovery capabilities with return-to-zero and nonreturn-to-zero transmission.

5-16. Contrast the error detection and decoding capabilities of return-to-zero and nonreturn-to-zero transmission.

5-17. What is a regenerative repeater?

5-18. Explain B6ZS and B3ZS. When or why would you use one rather than the other?

5-19. Briefly explain the following framing techniques: added digit framing, robbed digit framing, added channel framing, statistical framing, and unique line code framing.

5-20. Contrast bit and word interleaving.

PROBLEMS

5-1. A PCM/TDM system multiplexes 24 voice band channels. Each sample is encoded into 7 bits and a framing bit is added to each frame. The sampling rate is 9000 samples/second. BPRZ-AMI encoding is the line format. Determine:
(a) Line speed in bits per second.
(b) Minimum Nyquist bandwidth.

5-2. A PCM/TDM system multiplexes 32 voice band channels each with a bandwidth of 0 to 4 kHz. Each sample is encoded with an 8-bit PCM code. UPNRZ encoding is used. Determine:
(a) Minimum sample rate.
(b) Line speed in bits per second.
(c) Minimum Nyquist bandwidth.

5-3. For the following bit sequence, draw the timing diagram for UPRZ, UPNRZ, BPRZ, BPNRZ, and BPRZ-AMI encoding:

    bit stream: 1 1 1 0 0 1 0 1 0 1 1 0 0

5-4. Encode the following BPRZ-AMI data stream with B6ZS and B3ZS.

    + − 0 0 0 0 + − + 0 − 0 0 0 0 0 + − 0 0 + − + 0

Chapter 6

FREQUENCY-DIVISION MULTIPLEXING

INTRODUCTION

In *frequency-division multiplexing* (FDM), multiple sources that originally occupied the same frequency spectrum are each converted to a different frequency band and transmitted simultaneously over a single transmission medium. Thus many relatively narrowband channels can be transmitted over a single wideband transmission system.

FDM is an analog multiplexing scheme; the information entering an FDM system is analog and it remains analog throughout transmission. An example of FDM is the AM commercial broadcast band, which occupies a frequency spectrum from 535 to 1605 kHz. Each station carries an intelligence signal with a bandwidth of 0 to 5 kHz. If the audio from each station were transmitted with their original frequency spectrum, it would be impossible to separate one station from another. Instead, each station amplitude modulates a different carrier frequency and produces a 10-kHz double-sideband signal. Because adjacent stations' carrier frequencies are separated by 10 kHz, the total commercial AM band is divided into 107 10-kHz frequency slots stacked next to each other in the frequency domain. To receive a particular station, a receiver is simply tuned to the frequency band associated with that station's transmissions. Figure 6-1 shows how commercial AM broadcast station signals are frequency-division multiplexed and transmitted over a single transmission medium (free space).

There are many other applications for FDM such as commercial FM and television broadcasting and high-volume telecommunications systems. Within any of the

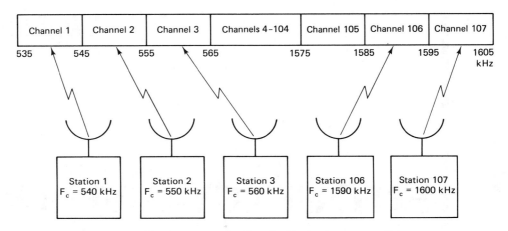

Figure 6-1 Frequency-division-multiplexing commercial AM broadcast band stations.

commercial broadcast bands, each station's transmissions are independent of all the other stations' transmissions. Consequently, the multiplexing (stacking) process is accomplished without any synchronization between stations. With a high-volume telephone communication system, many voice band telephone channels may originate from a common source and terminate in a common destination. The source and destination terminal equipment is most likely a high-capacity *electronic switching system* (ESS). Because of the possibility of a large number of narrowband channels originating and terminating at the same location, all multiplexing and demultiplexing operations must be synchronized.

AT&T's FDM HIERARCHY

Although AT&T is no longer the only long-distance common carrier in the United States, they still provide a vast majority of the long-distance services and if for no other reason than their overwhelming size, have essentially become the standards organization for the telephone industry in North America.

AT&T's nationwide communications network is subdivided into two classifications: *short haul* (short distance) and *long haul* (long distance). The T1 carrier explained in Chapter 5 is an example of a short-haul communications system.

Long-Haul Communications with FDM

Figure 6-2 shows AT&T's North American FDM Hierarchy for long-haul communications. Only a transmit terminal is shown, although a complete set of inverse functions must be performed at the receiving terminal.

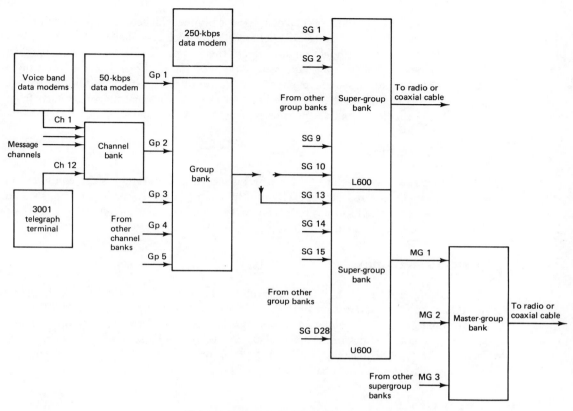

Figure 6-2 AT&T's long-haul FDM hierarchy.

Message Channel

The *message channel* is the basic building block of the FDM hierarchy. The basic message channel was originally intended for voice transmission, although it now includes any transmissions that utilize voice band frequencies (0 to 4 kHz) such as voice band data circuits. The basic voice band (VB) circuit is called a 3002 channel and is actually bandlimited to a 300- to 3000-Hz band, although for practical considerations, it is considered a 4-kHz channel. The basic 3002 channel can be subdivided into 24 narrower 3001 (telegraph) channels that have been frequency-division multiplexed to form a single 3002 channel.

Basic Group

A *group* is the next higher level in the FDM hierarchy above the basic message channel and is, consequently, the first multiplexing step for the message channels.

A basic group is comprised of 12 voice band channels stacked on top of each other in the frequency domain. The 12-channel modulating block is called an *A-type* (analog) channel bank. The 12-channel *group* output of the A-type channel bank is the standard building block for most long-haul *broadband* communications systems. Additions and deletions in total system capacity are accomplished with a minimum of one group (12 VB channels). The A-type channel bank has generically progressed from the early A1 channel bank to the most recent A6 channel bank.

Basic Supergroup

The next higher level in the FDM hierarchy shown in Figure 6-2 is the combination of five groups into a *supergroup*. The multiplexing of five groups is accomplished in a group bank. A single supergroup can carry information from 60 VB channels or handle high-speed data up to 250 kbps.

Basic Mastergroup

The next higher level in the FDM hierarchy is the basic *mastergroup*. A mastergroup is comprised of 10 supergroups (10 supergroups of five groups each = 600 VB channels). Supergroups are combined in supergroup banks to form mastergroups. There are two categories of mastergroups (U600 and L600) which occupy different frequency bands. The type of mastergroup used depends on the system capacity and whether the transmission medium is a coaxial cable or a microwave radio.

Larger Groupings

Master groups can be further multiplexed in mastergroup banks to form *jumbogroups*, *multi-jumbogroups*, and *superjumbogroups*. A basic FDM/FM microwave radio channel carries three mastergroups (1800 VB channels), a jumbogroup has 3600 VB channels, and a superjumbogroup has three jumbogroups (10,800 VB channels).

COMPOSITE BASEBAND SIGNAL

Baseband describes the modulating signal (intelligence) in a communications system. A single message channel is baseband. A group, supergroup, or mastergroup is also baseband. The composite baseband signal is the total intelligence signal prior to modulation of the final carrier. In Figure 6-2 the output of a channel bank is baseband. Also, the output of a group or supergroup bank is baseband. The final output of the FDM multiplexer is the *composite* (total) baseband. The formation of the composite baseband signal can include channel, group, supergroup, and mastergroup banks, depending on the capacity of the system.

Formation of a Group

Figure 6-3a shows how a group is formed with an A-type channel bank. Each voice band channel is bandlimited with an antialiasing filter prior to modulating the channel carrier. FDM uses single-sideband suppressed carrier (SSBSC) modulation. The combination of the balanced modulator and the bandpass filter make up the SSBSC modulator. A balanced modulator is a double-sideband suppressed carrier modulator and the bandpass filter is tuned to the difference between the carrier and the input voice band frequencies (LSB). The ideal input frequency range for a single voice band channel is 0 to 4 kHz. The carrier frequencies for the channel banks are determined from the following expression:

$$F_c = 112 - 4n \text{ kHz}$$

where n is the channel number. Table 6-1 lists the carrier frequencies for channels 1 through 12. Therefore, for channel 1, a 0- to 4-kHz band of frequencies modulates a 108-kHz carrier. Mathematically, the output of the channel 1 bandpass filter is

$$F_{\text{out}} = F_c - F_i$$

where

F_c = channel carrier frequency ($112 - 4n$ kHz)
F_i = channel frequency spectrum (0 to 4 kHz)

For channel 1:

$$F_{\text{out}} = 108 \text{ kHz} - (0 \text{ to } 4 \text{ kHz}) = 104 \text{ to } 108 \text{ kHz}$$

For channel 2:

$$F_{\text{out}} = 104 \text{ kHz} - (0 \text{ to } 4 \text{ kHz}) = 100 \text{ to } 104 \text{ kHz}$$

**TABLE 6-1   CHANNEL
CARRIER FREQUENCIES**

Channel	Carrier frequency (kHz)
1	108
2	104
3	100
4	96
5	92
6	88
7	84
8	80
9	76
10	72
11	68
12	64

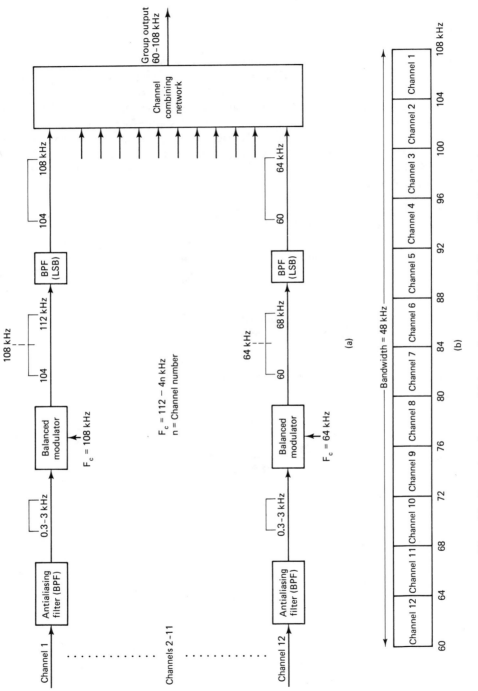

Figure 6-3 Formation of a group: (a) A-type channel bank block diagram; (b) output spectrum.

For channel 12:

$$F_{out} = 64 \text{ kHz} - (0 \text{ to } 4 \text{ kHz}) = 60 \text{ to } 64 \text{ kHz}$$

The outputs from the 12 A-type channel modulators are summed in the *linear* combiner to produce the total group spectrum shown in Figure 6-3b (60 to 108 kHz). Note that the total group bandwidth is equal to 48 kHz (12 channels $\times$ 4 kHz).

Formation of a Supergroup

Figure 6-4a shows how a supergroup is formed with a group bank and combining a network. Five groups are combined to form a supergroup. The frequency spectrum for each group is 60 to 108 kHz. Each group is mixed with a different group carrier frequency in a balanced modulator then bandlimited with a bandpass filter tuned to the difference frequency band (LSB) to produce a SSBSC signal. The group carrier frequencies are derived from the following expression:

$$F_c = 372 + 48n \text{ kHz}$$

where n is the group number. Table 6-2 lists the carrier frequencies for groups 1 through 5. For group 1, a 60- to 108-kHz group signal modulates a 420-kHz group carrier frequency. Mathematically, the output of the group 1 bandpass filter is

$$F_{out} = F_c - F_i$$

where

F_c = group carrier frequency ($372 + 48n$ kHz)
F_i = group frequency spectrum (60 to 108 kHz)

TABLE 6-2 GROUP CARRIER FREQUENCIES

Group	Carrier frequency (kHz)
1	420
2	468
3	516
4	564
5	612

For group 1:

$$F_{out} = 420 \text{ kHz} - (60 \text{ to } 108 \text{ kHz}) = 312 \text{ to } 360 \text{ kHz}$$

For group 2:

$$F_{out} = 468 \text{ kHz} - (60 \text{ to } 108 \text{ kHz}) = 360 \text{ to } 408 \text{ kHz}$$

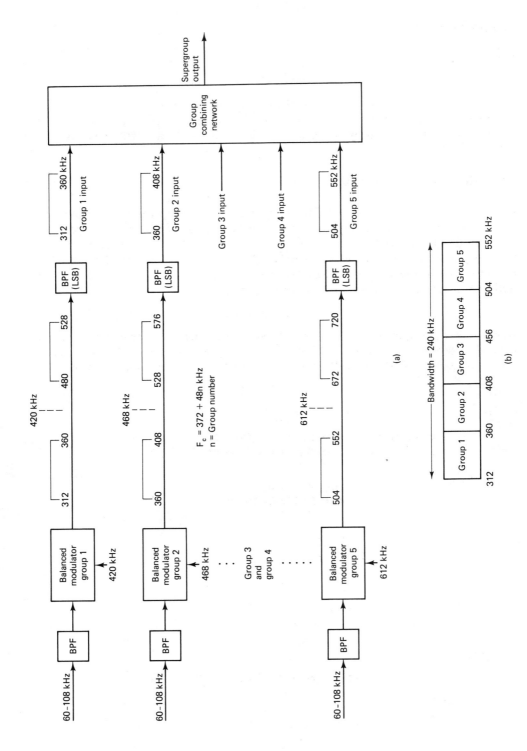

Figure 6-4 Formation of a supergroup: (a) group bank and combining network block diagram; (b) output spectrum.

For group 5:

$$F_{out} = 612 \text{ kHz} - (60 \text{ to } 108 \text{ kHz}) = 504 \text{ to } 552 \text{ kHz}$$

The outputs from the five group modulators are summed in the linear combiner to produce the total supergroup spectrum shown in Figure 6-4b (312 to 552 kHz). Note that the total supergroup bandwidth is equal to 240 kHz (60 channels $\times$ 4 kHz).

Formation of a Mastergroup

There are two types of mastergroups: L600 and U600 type. The L600 mastergroup is used for low-capacity microwave systems, while the U600 mastergroup may be further multiplexed and used for higher-capacity microwave radio systems.

U600 Mastergroup. Figure 6-5a shows how a U600 mastergroup is formed with a supergroup bank and combining network. Ten supergroups are combined to form a mastergroup. The frequency spectrum for each supergroup is 312 to 552 kHz. Each supergroup is mixed with a different supergroup carrier frequency in a balanced modulator. The output is then bandlimited to the difference frequency band (LSB) to form a SSBSC signal. The 10 supergroup carrier frequencies are listed in Table 6-3. For supergroup 13, a 312- to 552-kHz supergroup band of frequencies modulates a 1116-kHz carrier frequency. Mathematically, the output from the supergroup 13 bandpass filter is

$$F_{out} = F_c - F_i$$

where

F_c = supergroup carrier frequency
F_i = supergroup frequency
    spectrum (312 to 552 kHz)

For supergroup 13:

$$F_{out} = 1116 \text{ kHz} - (312 \text{ to } 552 \text{ kHz}) = 564 \text{ to } 804 \text{ kHz}$$

For supergroup 14:

$$F_{out} = 1364 \text{ kHz} - (312 \text{ to } 552 \text{ kHz}) = 812 \text{ to } 1052 \text{ kHz}$$

For supergroup D28:

$$F_{out} = 3396 \text{ kHz} - (312 \text{ to } 552 \text{ kHz}) = 2844 \text{ to } 3084 \text{ kHz}$$

The outputs from the 10 supergroup modulators are summed in the linear summer to produce the total mastergroup spectrum shown in Figure 6-4b (564 to 3084 kHz). Note that between any two adjacent supergroups there is a void band of frequencies that is not included within any supergroup band. These voids are called *guard bands*. The guard bands are necessary because the demultiplexing process

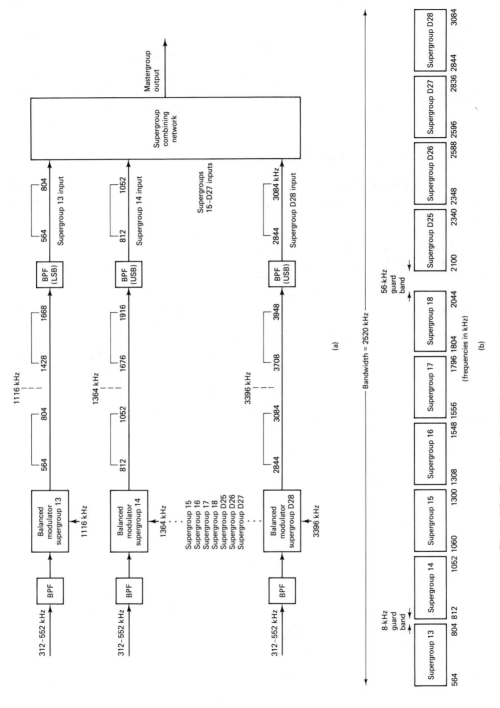

Figure 6-5 Formation of a U600 mastergroup: (a) supergroup bank and combining network block diagram; (b) output spectrum.

TABLE 6-3 SUPERGROUP CARRIER FREQUENCIES FOR A U600 MASTERGROUP

Supergroup	Carrier frequency (kHz)
13	1116
14	1364
15	1612
16	1860
17	2108
18	2356
D25	2652
D26	2900
D27	3148
D28	3396

is accomplished through filtering and down-converting. Without the guard bands, it would be difficult to separate one supergroup from an adjacent supergroup. The guard bands reduce the *quality factor* (*Q*) required to perform the necessary filtering. The guard band is 8 kHz between all supergroups except 18 and D25, where it is 56 kHz. Consequently, the bandwidth of a U600 mastergroup is 2520 kHz (564 to 3084 kHz), which is greater than is necessary to stack 600 voice band channels (600 × 4 kHz = 2400 kHz).

Guard bands were not necessary between adjacent groups because the group frequencies are sufficiently low and it is relatively easy to build bandpass filters to separate one group from another.

In the channel bank, the antialiasing filter at the channel input passes a 0.3-to 3-kHz band. The separation between adjacent channel carrier frequencies is 4 kHz. Therefore, there is a 1300-Hz guard band between adjacent channels. This is shown in Figure 6-6.

L600 Mastergroup. With an L600 mastergroup, 10 supergroups are combined as with the U600 mastergroup except that the supergroup carrier frequencies are lower. Table 6-4 lists the supergroup carrier frequencies for a L600 mastergroup. With an L600 mastergroup, the composite baseband spectrum occupies a lower-frequency band than the U-type mastergroup (Figure 6-7). An L600 mastergroup is not further multiplexed. Therefore, the maximum channel capacity for a microwave or coaxial cable system using a single L600 mastergroup is 600 voice band channels.

Formation of a Radio Channel

A *radio channel* comprise either a single L600 mastergroup or up to three U600 mastergroups (1800 voice band channels). Figure 6-8 shows how an 1800-channel composite FDM baseband signal is formed for transmission over a single microwave

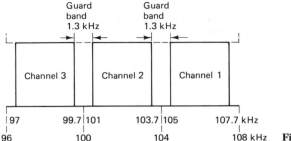

Figure 6-6   Channel guard bands.

radio channel. Mastergroup 1 is transmitted directly as is, while mastergroups 2 and 3 undergo an additional multiplexing step. The three mastergroups are summed in a mastergroup combining network to produce the output spectrum shown in Figure 6-8b. Note the 80-kHz guard band between adjacent mastergroups.

The system shown in Figure 6-8 can be increased from 1800 voice band channels to 1860 by adding an additional supergroup (supergroup 12) directly to mastergroup 1. The addition 312- to 552-kHz supergroup band extends the output spectrum to 312 to 8284 kHz.

Frequency Translation in FDM

Essentially, FDM is the process of transposing or translating a given input frequency band to some higher frequency band, where it is combined with other translated signals. With the L1860 FDM system shown in Figure 6-2, a single voice band channel may undergo as many as four frequency translations before it is transmitted. When troubleshooting an FDM system, it is necessary that the frequency band of a given channel be known at the various levels of multiplexing.

**TABLE 6-4   SUPERGROUP
CARRIER FREQUENCIES FOR A
L600 MASTERGROUP**

Supergroup	Carrier frequency (kHz)
1	612
2	Direct
3	1116
4	1364
5	1612
6	1860
7	2108
8	2356
9	1860
10	3100

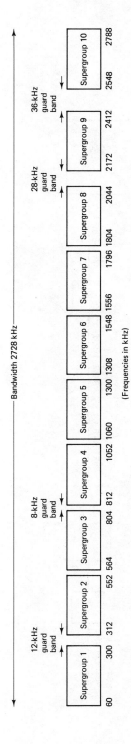

Figure 6-7 L600 mastergroup.

223

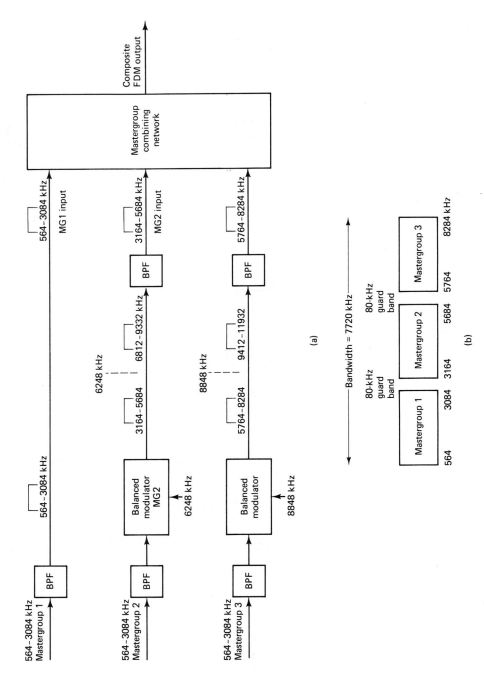

Figure 6-8 Three-mastergroup radio channel: (a) block diagram; (b) output spectrum.

EXAMPLE 6-1

For a single-voice-band channel:

     (a) Determine its frequency band at the output of the channel, group, supergroup, and mastergroup combiners when it is assigned to channel 4, group 2, supergroup 16, and mastergroup 2.

     (b) Determine the frequency that a 1-kHz tone on the same channel would translate to.

Solution   (a) For an ideal bandwidth of 0 to 4 kHz, the frequency band at the channel, group, supergroup, and master group combiners is determined as follows:

$$\text{channel bank out} = 96 \text{ kHz} - (0 \text{ to } 4 \text{ kHz}) = 92 \text{ to } 96 \text{ kHz}$$

$$\text{GP bank out} = 468 \text{ kHz} - (92 \text{ to } 96 \text{ kHz}) = 372 \text{ to } 376 \text{ kHz}$$

$$\text{SG bank out} = 1860 \text{ kHz} - (372 \text{ to } 376 \text{ kHz}) = 1484 \text{ to } 1488 \text{ kHz}$$

$$\text{MG bank out} = 6248 \text{ kHz} - (1484 \text{ to } 1488 \text{ kHz}) = 4760 \text{ to } 4764 \text{ kHz}$$

(b) For a 1-kHz test tone,

$$\text{channel bank out} = 96 \text{ kHz} - 1 \text{ kHz} = 95 \text{ kHz}$$

$$\text{GP bank out} = 468 \text{ kHz} - 95 \text{ kHz} = 373 \text{ kHz}$$

$$\text{SG bank out} = 1860 \text{ kHz} - 373 \text{ kHz} = 1487 \text{ kHz}$$

$$\text{MG bank out} = 6248 \text{ kHz} - 1487 \text{ kH} = 4761 \text{ kHz}$$

L CARRIERS

L carrier systems transmit frequency-division-multiplexed voice band signals over a coaxial cable for distances up to 4000 miles. L carriers have generically progressed from the early L1 and L3 systems to the high-capacity L4, L5, and L6 systems. L carrier systems combine many coaxial cables into a single tube and carry dozens of mastergroups and literally thousands of two-way simultaneous voice transmissions. In the near future, L cables are destined to be replaced by even higher-capacity fiber-optic systems.

Carrier Synchronization

With FDM, the receive channel, group, supergroup, and mastergroup carrier frequencies must be synchronized to the transmit carrier frequencies. If they are not synchronized, the recovered voice band signals will be offset in frequency from their original spectrum by the difference in the two carrier frequencies. FDM uses single-sideband suppressed carrier transmission. The carriers are suppressed in the balanced modulators at the transmit terminal and therefore cannot be recovered in the receive terminal directly from the composite baseband signal. Consequently, a carrier pilot frequency is transmitted together with the baseband signal for the purpose of carrier synchronization.

The channel, group, supergroup, and mastergroup carrier frequencies are all integral multiples of 4 kHz. Therefore, if the carrier frequencies at the transmit and receive terminals are derived from a single 4-kHz master oscillator, all of the transmit and receive carriers will be synchronized.

In an FDM communications system, one station is designated as the *master station*. Every other station in the system is a slave. That is, there is a single 4-kHz master oscillator from which all carrier frequencies in the system are derived. The 4-kHz master oscillator is multiplied to either a 64-, 312-, or 552-kHz pilot frequency, combined with the composite baseband signal, and transmitted to each slave station. The slave stations detect the pilot, divide it down to a 4-kHz base frequency, and synchronize their 4-kHz slave oscillators to it. Each slave station then multiplies the synchronous 4-kHz signal to produce synchronous channel, group, supergroup, and mastergroup carrier frequencies. If the master 4-kHz oscillator drifts in frequency, each slave station's 4-kHz oscillator tracks the frequency shift and the system remains synchronous.

D-Type Supergroups

With the U600 mastergroup, supergroups 25 through 28 are preceded by the letter "D." These supergroup carrier frequencies are derived in a slightly different fashion than the other carrier frequencies. Except for the D-supergroups, all carrier frequencies are generated through integer multiplication of the 4-kHz base frequency. The D-supergroups are generated through both multiplication and heterodyning. The supergroup 15- to 18-carrier frequencies are mixed with a 1040-kHz harmonic and the sum frequencies become the D25 to D28 carrier frequencies.

This method of deriving the higher supergroup carrier frequencies has no effect on the frequency offset introduced by frequency drift of the 4-kHz base oscillator. However, this technique reduces the amount of *phase jitter* (incidental phase modulation generally caused by ac ripple present in dc power supplies) in the higher supergroup carrier frequencies. When two frequencies with phase jitter are combined in a mixer, the total phase jitter in the sum frequency is less than the algebraic sum of the phase jitter of the two signals. Consequently, using D-type supergroup carriers has no effect on the frequency shift but reduces the magnitude of the total phase jitter in the higher supergroups. Before this technique was introduced, the voice band channels located in the higher supergroups could not be used for voice band data transmission when PSK modulation was used because the phase jitter caused excessive transmission errors.

Amplitude Regulation

Figure 6-9a shows the gain characteristics for an ideal transmission medium. For the ideal situation (Figure 6-9a), the gain for all baseband frequencies is the same. In a more practical situation, the gain is not the same for all frequencies (Figure 6-9b). Therefore, the demultiplexed voice band channels do not have the same ampli-

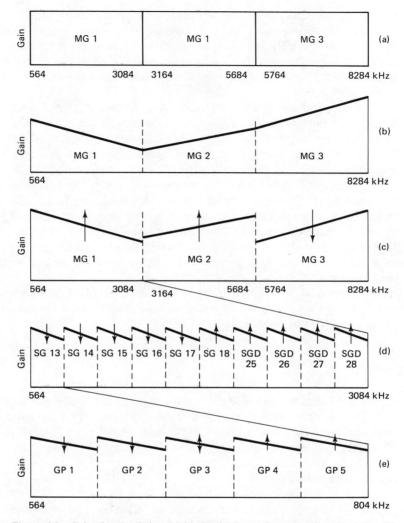

Figure 6-9 Gain characteristics: (a) ideal gain versus frequency characteristics; (b) amplitude distortion; (c) mastergroup regulation; (d) supergroup regulation; (e) group regulation.

tude characteristics as the original voice band signals had. This is called *amplitude distortion*. To reduce amplitude distortion, filters with the opposite characteristics as those introduced in the transmission medium can be added to the system, thus canceling the distortion. This is impractical because every transmission system has different characteristics and a special filter would have to be designed and built for each system.

Automatic gain devices (regulators) are used in the receiver demultiplexing equipment to compensate for amplitude distortion introduced in the transmission medium. Amplitude regulation is accomplished in several stages. First, the amplitude of each mastergroup is adjusted or regulated (mastergroup regulation; Figure 6-9c), then each supergroup within each mastergroup is regulated (supergroup regulation; Figure 6-9d). The last stage of regulation is performed at a group level (group regulation; Figure 6-9e).

Regulation is performed by monitoring the power level of a mastergroup, supergroup, or group *pilot*, then regulating the entire frequency band associated with it, depending on the pilot level. Pilots are monitored rather than the actual signal level because the signal levels vary depending on how many channels are in use at a given time. A pilot is a continuous signal with a constant power level.

Figure 6-10 shows how the regulation pilots are nested within the composite baseband signal. Each group has a 104.08-kHz pilot added to it in the channel combining network. Consequently, each supergroup has five group pilots. The group 1 pilot is also the supergroup pilot. Thus each mastergroup has 50 group pilots, of which 10 are also supergroup pilots. A separate 2840-kHz mastergroup pilot is added to each mastergroup in the supergroup combining network, making a total of 51 pilots per mastergroup.

Figure 6-11 is a partial block diagram for an FDM demultiplexer that shows how the pilots are monitored and used to separately regulate the mastergroups, supergroups, and groups automatically.

HYBRID DATA

With *hybrid* data it is possible to combine digitally encoded signals with FDM signals and transmit them as one composite baseband signal. There are four primary types of hybrid data: data under voice (DUV), data above voice (DAV), data above video (DAVID), and data in voice (DIV).

Data under Voice

Figure 6-12a shows the block diagram of AT&T's 1.544-Mbps *data under FDM voice* system. With the L1800 FDM system explained earlier in this chapter, the 0 to 564-kHz frequency spectrum is void of baseband signals. With FM transmission, the lower baseband frequencies realize the highest signal-to-noise ratios. Consequently, the best portion of the baseband spectrum was unused. DUV is a means of utilizing this spectrum for the transmission of digitally encoded signals. A T1 carrier system can be converted to a quasi-analog signal and then frequency-division multiplexed onto the lower portion of the FDM spectrum.

In Figure 6-12a, the *elastic store* removes timing jitter from the incoming data stream. The data are then *scrambled* to suppress the discrete high-power spectral components. The advantage of scrambling is the randomized data output spectrum is continuous and has a predictable effect on the FDM radio system. In other words,

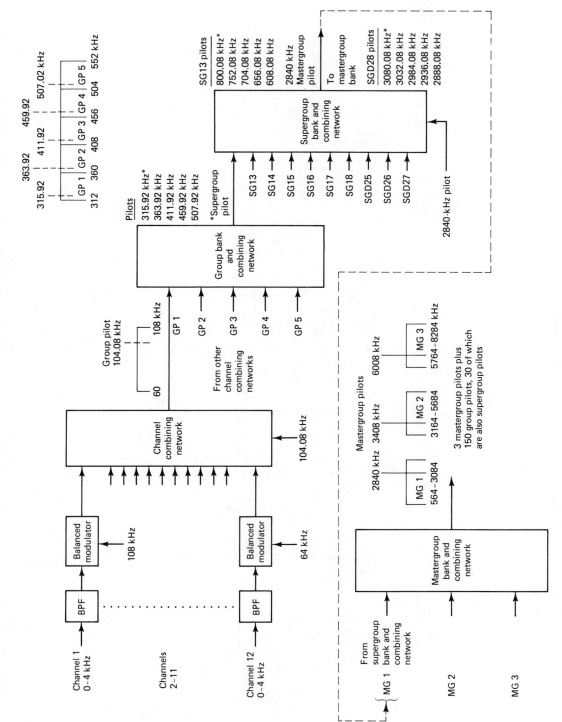

Figure 6-10 Pilot insertion.

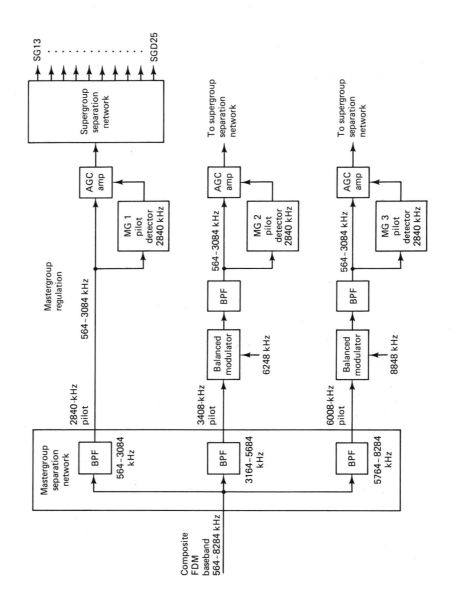

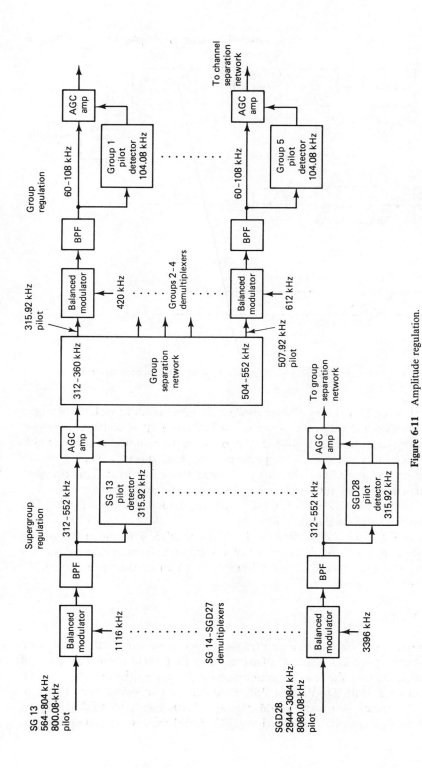

Figure 6-11 Amplitude regulation.

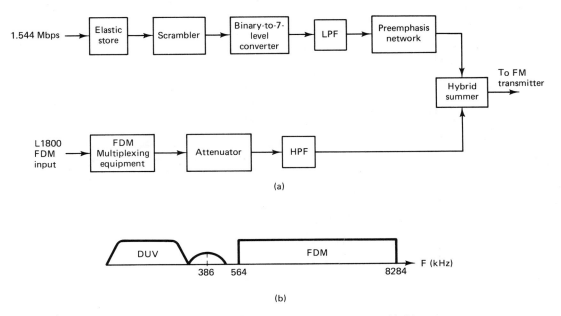

Figure 6-12   Data under voice (DUV): (a) block diagram; (b) frequency spectrum.

the data present a load to the system equivalent to adding additional FDM voice channels. The serial seven-level *partial response* encoder (correlative coder) compresses the data bandwidth and allows a 1.544-Mbps signal to be transmitted in a bandwidth less than 400 kHz. The low-pass filter performs the final spectral shaping of the digital information and suppresses the spectral power above 386 kHz. This prevents the DUV information from interfering with the 386-kHz pilot control tone. The DUV signal is preemphasized and combined with the L1800 baseband signal. The output spectrum is shown in Figure 6-12b.

AT&T uses DUV for *digital data service* (DDS). DDS is intended to provide a communications medium for the transfer of digital data from station to station without the use of a data modem. DDS circuits are guaranteed to average 99.5% error-free seconds at 56 kbps.

Data above Voice

Figure 6-13 shows the block diagram and frequency spectrum for a *data above voice* system. The advantage of DAV is for FDM systems that extend into the low end of the baseband spectrum; the low-frequency baseband does not have to be vacated for data transmission. With DAV, data PSK modulates a carrier which is then up-converted to a frequency above the FDM message. With DAV, up to 3.152 Mbps can be cost-effectively transmitted using existing FDM/FM microwave systems (Chapter 7).

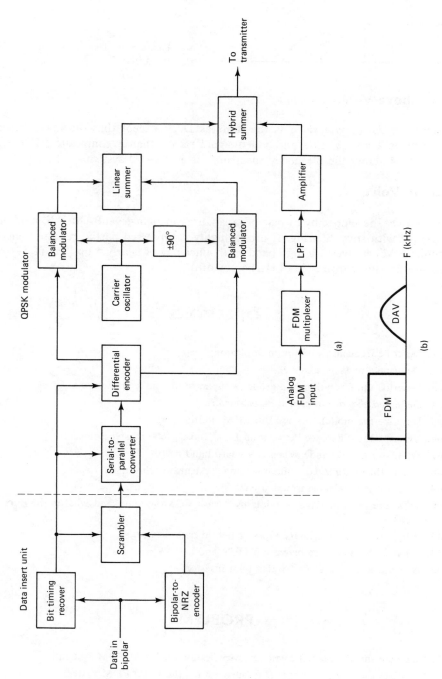

Figure 6-13 Data above voice (DAV): (a) block diagram; (b) frequency spectrum.

Figure 6-14 Data above video (DAVID).

Data above Video

Essentially, *data above video* is the same as DAV except that the lower baseband spectrum is a *vestigal sideband* video signal rather than a composite FDM signal. Figure 6-14 shows the frequency spectrum for a DAVID system.

Data in Voice

Data in voice, developed by Fujitsu of Japan, uses an eight-level PAM-VSB modulation technique with steep filtering. It uses a highly compressed partial response encoding technique which gives it a high bandwidth efficiency of nearly 5 bps/Hz (1.544-Mbps data are transmitted in a 344-kHz bandwidth).

QUESTIONS

6-1. Describe frequency-division multiplexing.

6-2. Describe a message channel.

6-3. Describe the formation of a group, a supergroup, and a mastergroup.

6-4. Define *baseband* and *composite baseband*.

6-5. Describe the modulators used in FDM multiplexers.

6-6. Describe the difference between an L600 and a U600 mastergroup.

6-7. What is a guard band? When is a guard band used?

6-8. Are FDM-multiplexed communications systems synchronous? Explain.

6-9. Why are D-type supergroups used?

6-10. What are the two types of pilots used with FDM systems, and what is the purpose of each?

6-11. What are the four primary types of hybrid data networks?

6-12. What is the difference between a DUV and a DAV network?

6-13. At what level is the 104.08-kHz pilot inserted?

PROBLEMS

6-1. Calculate the 12 channel carrier frequencies for the U600 FDM system.

6-2. Calculate the five group carrier frequencies for the U600 FDM system.

6-3. Calculate the frequency range for a single channel at the output of the channel, group, supergroup, and mastergroup combining networks for channel 3, group 4, supergroup 15, mastergroup 2.

6-4. Determine the frequency that a 1-kHz test tone will translate to at the output of the channel, group, supergroup, and mastergroup combining networks for channel 5, group 5, supergroup 27, mastergroup 3.

6-5. Determine the frequency at the output of the mastergroup combining network for a group pilot of 104.08 kHz on group 2, supergroup 13, mastergroup 2.

6-6. Calculate the frequency range for group 4, supergroup 18, mastergroup 1 at the output of the mastergroup combining network.

6-7. Calculate the frequency range for supergroup 15, mastergroup 2 at the output of the mastergroup combining network.

Chapter 7

MICROWAVE COMMUNICATIONS AND SYSTEM GAIN

INTRODUCTION

Presently, terrestrial (earth) *microwave radio relay systems* provide less than half of the total message circuit mileage in the United States. However, at one time microwave systems carried the bulk of long-distance communications for the public telephone network, military and governmental agencies, and specialized private communications networks. There are many different types of microwave systems that operate over distances varing from 15 to 4000 miles in length. *Intrastate* or *feeder service* systems are generally categorized as *short haul* because they are used for relatively short distances. *Long-haul* radio systems are those used for relatively long distances, such as interstate and backbone route applications. Microwave system capacities range from less than 12 voice band channels to more than 22,000. Early microwave systems carried frequency-division-multiplexed voice band circuits and used conventional, noncoherent frequency modulation techniques. More recently developed microwave systems carry pulse-code-modulated time-division-multiplexed voice band circuits and use more modern digital modulation techniques, such as phase shift keying and quadrature amplitude modulation. This chapter deals primarily with conventional FDM/FM microwave systems, and Chapter 8 deals with the more modern PCM/PSK techniques.

SIMPLIFIED MICROWAVE SYSTEM

A simplified block diagram of a microwave radio system is shown in Figure 7-1. The *baseband* is the composite signal that modulates the FM carrier and may comprise one or more of the following:

236

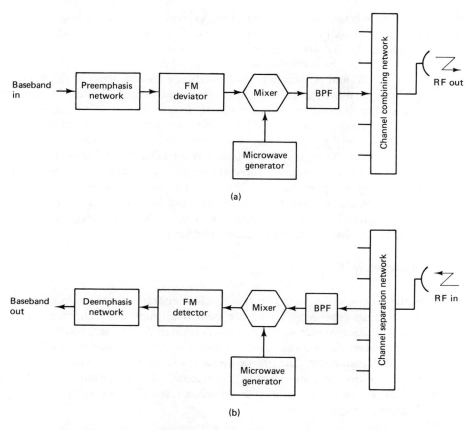

Figure 7-1 Simplified block diagram of a microwave system: (a) transmitter; (b) receiver.

1. Frequency-division-multiplexed voice band channels
2. Time-division-multiplexed voice band channels
3. Broadcast-quality composite video or picturephone

Microwave Transmitter

In the *microwave transmitter* (Figure 7-1a), a *preemphasis* network precedes the FM deviator. The preemphasis network provides an artificial boost in amplitude to the higher baseband frequencies. This allows the lower baseband frequencies to frequency modulate the IF carrier and the higher baseband frequencies to phase modulate it. This scheme assures a more uniform signal-to-noise ratio throughout the entire baseband spectrum. An FM deviator provides the modulation of the IF carrier which eventually becomes the main microwave carrier. Typically, IF carrier frequencies are between 60 and 80 MHz, with 70 MHz the most common. *Low-index* frequency modulation is used in the FM deviator. Typically, modulation indices are kept between 0.5 and 1. This produces a *narrowband* FM signal at the output of the deviator. Consequently, the IF bandwidth resembles conventional AM and is approximately equal to twice the highest baseband frequency.

The IF and its associated sidebands are up-converted to the microwave region by the AM mixer, microwave oscillator, and bandpass filter. Mixing, rather than multiplying, is used to translate the IF frequencies to RF frequencies because the modulation index is unchanged by the heterodyning process. Multiplying the IF carrier would also multiply the frequency deviation and the modulation index, thus increasing the bandwidth. Typically, frequencies above 1000 MHz (1 GHz) are considered microwave frequencies. Presently, there are microwave systems operating with carrier frequencies up to approximately 18 GHz. The most common microwave frequencies currently being used are the 2-, 4-, 6-, 12-, and 14-GHz bands. The channel combining network provides a means of connecting more than one microwave transmitter to a single transmission line feeding the antenna.

Microwave Receiver

In the receiver (Figure 7-1b), the channel separation network provides the isolation and filtering necessary to separate individual microwave channels and direct them to their respective receivers. The bandpass filter, AM mixer, and microwave oscillator down-convert the RF microwave frequencies to IF frequencies and pass them on to the FM demodulator. The FM demodulator is a conventional, *noncoherent* FM detector (i.e., a discriminator or a ratio detector). At the output of the FM detector, a deemphasis network restores the baseband signal to its original amplitude versus frequency characteristics.

MICROWAVE REPEATERS

The permissible distance between a microwave transmitter and its associated microwave receiver depends on several system variables, such as transmitter output power, receiver noise threshold, terrain, atmospheric conditions, system capacity, reliability objectives, and performance expectations. Typically, this distance is between 15 and 40 miles. Longhaul microwave systems span distances considerably longer than this. Consequently, a single-hop microwave system, such as the one shown in Figure 7-1, is inadequate for most practical system applications. With systems that are longer than 40 miles or when geographical obstructions, such as a mountain, block the transmission path, *repeaters* are needed. A microwave repeater is a receiver and a transmitter placed back to back or in tandem with the system. A block diagram of a microwave repeater is shown in Figure 7-2. The repeater station receives a signal, amplifies and reshapes it, then retransmits the signal to the next repeater or terminal station downline from it.

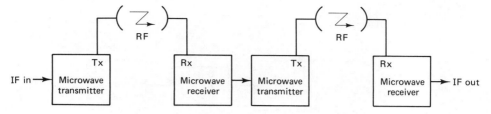

Figure 7-2 Microwave repeater.

Basically, there are two types of microwave repeaters: *baseband* and *IF* (Figure 7-3). IF repeaters are also called *heterodyne* repeaters. With an IF repeater (Figure 7-3a), the received RF carrier is down-converted to an IF frequency, amplified, re-shaped, up-converted to an RF frequency, and then retransmitted. The signal is never demodulated beyond IF. Consequently, the baseband intelligence is unmodified by the repeater. With a baseband repeater (Figure 7-3b), the received RF carrier is down-converted to an IF freqency, amplified, filtered, and then further demodulated to baseband. The baseband signal, which is typically frequency-division-multiplexed voice band channels, is further demodulated to a mastergroup, supergroup, group, or even channel level. This allows the baseband signal to be reconfigured to meet the routing needs of the overall communications network. Once the baseband signal has been reconfigured, it FM modulates an IF carrier which is up-converted to an RF carrier and then retransmitted.

Figure 7-3c shows another baseband repeater configuration. The repeater de-modulates the RF to baseband, amplifies and reshapes it, then modulates the FM carrier. With this technique, the baseband is not reconfigured. Essentially, this configu-ration accomplishes the same thing that an IF repeater accomplishes. The difference is that in a baseband configuration, the amplifier and equalizer act on baseband frequen-cies rather than IF frequencies. The baseband frequencies are generally less than 9 MHz, whereas the IF frequencies are in the range 60 to 80 MHz. Consequently, the filters and amplifiers necessary for baseband repeaters are simpler to design and less expensive than the ones required for IF repeaters. The disadvantage of a baseband configuration is the addition of the FM terminal equipment.

DIVERSITY

Microwave systems use *line-of-sight* transmission. There must be a direct, line-of-sight signal path between the transmit and the receive antennas. Consequently, if that signal path undergoes a severe degradation, a service interruption will occur. *Diversity* suggests that there is more than one transmission path or method of transmis-sion available between a transmitter and a receiver. In a microwave system, the purpose of using diversity is to increase the reliability of the system by increasing its availability. When there is more than one transmission path or method of transmission available, the system can select the path or method that produces the highest-quality received

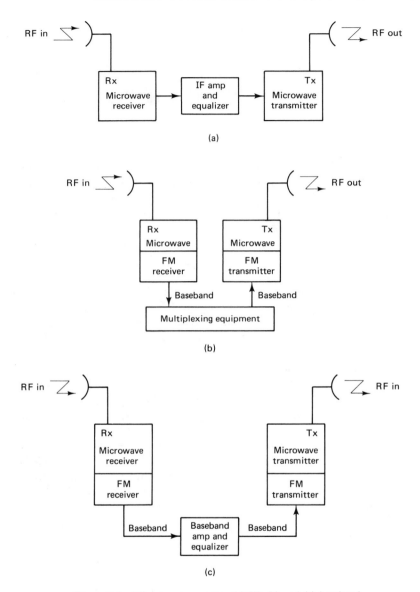

Figure 7-3   Microwave repeaters: (a) IF; (b) and (c) baseband.

signal. Generally, the highest quality is determined by evaluating the carrier-to-noise (C/N) ratio at the receiver input or by simply measuring the received carrier power. Although there are many ways of achieving diversity, the most common methods used are *frequency*, *space*, and *polarization*.

Frequency Diversity

Frequency diversity is simply modulating two different RF carrier frequencies with the same IF intelligence, then transmitting both RF signals to a given destination. At the destination, both carriers are demodulated, and the one that yields the better-quality IF signal is selected. Figure 7-4 shows a single-channel frequency-diversity microwave system.

In Figure 7-4a, the IF input signal is fed to a power splitter, which directs it to microwave transmitters A and B. The RF outputs from the two transmitters are combined in the channel-combining network and fed to the transmit antenna. At the receive end (Figure 7-4b), the channel separator directs the A and B RF carriers to their respective microwave receivers, where they are down-converted to IF. The quality detector circuit determines which channel, A or B, is the higher quality and directs that channel through the IF switch to be further demodulated to baseband. Many of the temporary, adverse atmospheric conditions that degrade an RF signal are frequency selective; they may degrade one frequency more than another. Therefore, over a given period of time, the IF switch may switch back and forth from receiver A to receiver B, and vice versa many times.

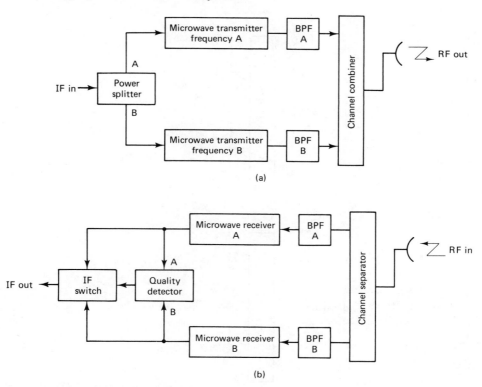

Figure 7-4 Frequency diversity microwave system: (a) transmitter; (b) receiver.

Space Diversity

With space diversity, the output of a transmitter is fed to two or more antennas that are physically separated by an appreciable number of wavelengths. Similarly, at the receiving end, there may be more than one antenna providing the input signal to the receiver. If multiple receiving antennas are used, they must also be separated by an appreciable number of wavelengths. Figure 7-5 shows a single-channel space-diversity microwave system.

When space diversity is used, it is important that the electrical distance from a transmitter to each of its antennas and to a receiver from each of its antennas is an equal multiple of wavelengths long. This is to ensure that when two or more signals of the same frequency arrive at the input to a receiver, they are in phase and additive. If received out of phase, they will cancel and, consequently, result in less received signal power than if simply one antenna system were used. Adverse atmospheric conditions are often isolated to a very small geographical area. With space diversity, there is more than one transmission path between a transmitter and a receiver. When adverse atmospheric conditions exist in one of the paths, it is unlikely that the alternate path is experiencing the same degradation. Consequently, the proba-

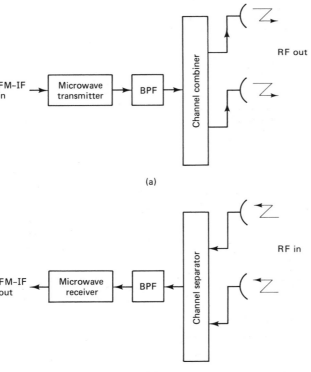

Figure 7-5 Space-diversity microwave system: (a) transmitter; (b) receiver.

bility of receiving an acceptable signal is higher when space diversity is used than when no diversity is used. An alternate method of space diversity uses a single transmitting antenna and two receiving antennas separated vertically. Depending on the atmospheric conditions at a particular time, one of the receiving antennas should be receiving an adequate signal. Again, there are two transmission paths that are unlikely to be affected simultaneously by fading.

Polarization Diversity

With polarization diversity, a single RF carrier is propagated with two different electromagnetic polarizations (vertical and horizontal). Electromagnetic waves of different polarizations do not necessarily experience the same transmission impairments. Polarization diversity is generally used in conjunction with space diversity. One transmit/receive antenna pair is vertically polarized and the other is horizontally polarized. It is also possible to use frequency, space, and polarization diversity simultaneously.

PROTECTION SWITCHING

Radio path losses vary with atmospheric conditions. Over a period of time, the atmospheric conditions between transmitting and receiving antenna can vary significantly, causing a corresponding reduction in the received signal strength of 20, 30, 40, or more dB. This reduction in signal strength is referred to as a *radio fade*. *Automatic gain control circuits*, built into radio receivers, can compensate for fades of 25 to 40 dB, depending on the system design. However, when fades in excess of 40 dB occur, this is equivalent to a total loss of the received signal. When this happens, service continuity is lost. To avoid a service interruption during periods of deep fades or equipment failures, alternate facilities are temporarily made available in what is called a *protection switching* arrangement. Essentially, there are two types of protection switching arrangements: *hot standby* and *diversity*. With hot standby protection, each working radio channel has a dedicated backup or spare channel. With diversity protection, a single backup channel is made available to as many as 11 working channels. Hot standby systems offer 100% protection for each working radio channel. A diversity system offers 100% protection only to the first working channel that fails. If two radio channels fail at the same time, a service interruption will occur.

Hot Standby

Figure 7-6a shows a single-channel hot standby protection switching arrangement. At the transmitting end, the IF goes into a *head-end bridge*, which splits the signal power and directs it to the working and the spare (standby) microwave channels simultaneously. Consequently, both the working and standby channels are carrying the same baseband information. At the receiving end, the IF switch passes the IF signal from the working channel to the FM terminal equipment. The IF switch continuously monitors the received signal power on the working channel and if it fails, switches to the standby channel. When the IF signal on the working channel is restored, the IF switch resumes its normal position.

Diversity

Figure 7-6b shows a diversity protection switching arrangement. This system has two working channels (channel 1 and channel 2), one spare channel, and an *auxiliary* channel. The IF switch at the receive end continuously monitors the receive signal strength of both working channels. If either one should fail, the IF switch detects a loss of carrier and sends back to the transmitting station IF switch a VF (*voice frequency*) tone-encoded signal that directs it to switch the IF signal from the failed channel onto the space microwave channel. When the failed channel is restored, the IF switches resume their normal positions. The auxiliary channel simply provides a transmission path between the two IF switches. Typically, the auxiliary channel is a low-capacity low-power microwave radio that is designed to be used for maintenance channels only.

Reliability

The number of repeater stations between protection switches depends on the *reliability objectives* of the system. Typically, there are between two and six repeaters between switching stations.

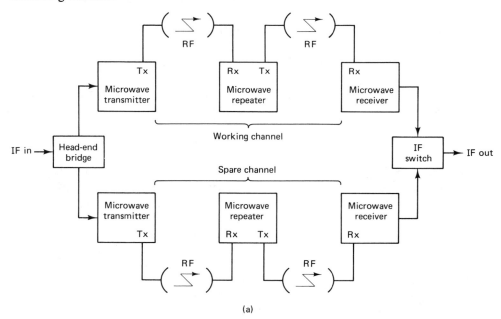

(a)

Figure 7-6   Microwave protection switching arrangements: (a) hot standby; (b) diversity.

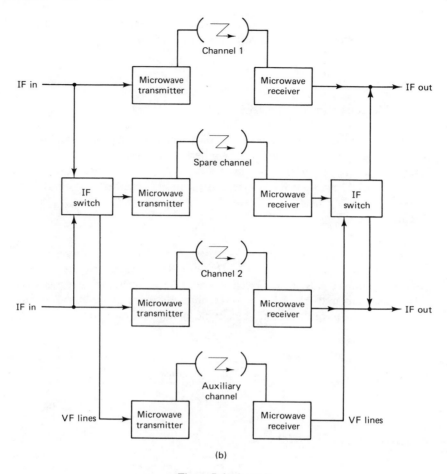

(b)

Figure 7-6 (cont'd)

As you can see, diversity systems and protection switching arrangements are quite similar. The primary difference between the two is that diversity systems are permanent arrangements and are intended only to compensate for temporary, abnormal atmospheric conditions between only two selected stations in a system. Protection switching arrangements, on the other hand, compensate for both radio fades and equipment failures and may include from six to eight repeater stations between switches. Protection channels may also be used as temporary communication facilities, while routine maintenance is performed on a regular working channel. With a protection switching arrangement, all signal paths and radio equipment are protected. Diversity is used selectively, that is, only between stations that historically experience severe fading a high percentage of the time.

A statistical study of outage time (i.e., service interruptions) caused by radio fades, equipment failures, and maintenance is important in the design of a microwave radio system. From such a study, engineering decisions can be made on the type of diversity system and protection switching arrangement best suited for a particular application.

MICROWAVE RADIO STATIONS

Basically, there are two types of microwave stations: terminals and repeaters. *Terminal stations* are points in the system where baseband signals are either originated or terminated. *Repeater stations* are points in a system where baseband signals may be reconfigured or where RF carriers are simply "repeated" or amplified.

Terminal Station

Essentially, a terminal station consists of four major sections: the baseband, wire line entrance link (WLEL), FM-IF, and RF sections. Figure 7-7 shows the block diagram of the baseband, WLEL, and FM-IF sections. As mentioned previously, the baseband may be one of several different types of signals. For our example, frequency-division-multiplexed voice band channels are used.

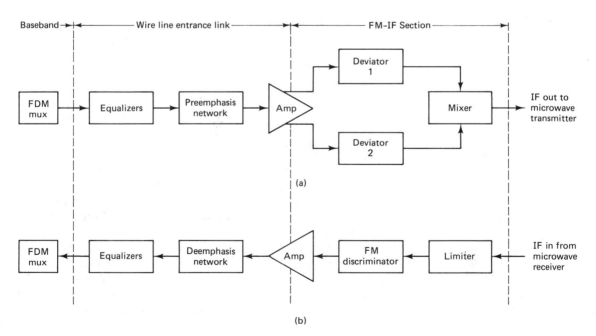

Figure 7-7 Microwave terminal station, baseband, wire line entrance link, and FM-IF: (a) transmitter; (b) receiver.

Wire line entrance link (WLEL).   Very often in large communications networks such as the American Telephone and Telegraph Company (AT&T), the building that houses the radio station is quite large. Consequently, it is desirable that similar equipment be physically placed at a common location (i.e., all FDM equipment in the same room). This simplifies alarm systems, providing dc power to the equipment, maintenance, and other general cabling requirements. Dissimilar equipment may be separated by a considerable distance. For example, the distance between the FDM multiplexing equipment and the FM-IF section is typically several hundred feet and in some cases several miles. For this reason a WLEL is required. A WLEL serves as the interface between the multiplex terminal equipment and the FM-IF equipment. A WLEL generally consists of an amplifier and an equalizer (which together compensate for cable transmission losses) and a level-shaping device commonly called pre- and deemphasis networks.

IF section.   The FM terminal equipment shown in Figure 7-7 generates a frequency-modulated IF carrier. This is accomplished by mixing the outputs of two deviated oscillators that differ in frequency by the desired IF carrier. The oscillators are deviated in phase opposition, which reduces the magnitude of phase deviation required of a single deviator by a factor of 2. This technique also reduces deviation linearity requirements for the oscillators and provides for the partial cancellation of unwanted modulation products. Again, the receiver is a conventional noncoherent FM detector.

RF section.   A block diagram of the RF section of a microwave terminal station is shown in Figure 7-8. The IF signal enters the transmitter (Figure 7-8a) through a protection switch. The IF and compression amplifiers help keep the IF signal power constant and at approximately the required input level to the transmit modulator (*transmod*). A transmod is a balanced modulator that when used in conjunction with a microwave generator, power amplifier, and bandpass filter, up-converts the IF carrier to an RF carrier and amplifies the RF to the desired output power. Power amplifiers for microwave radios must be capable of amplifying very high frequencies and pass a very wide bandwidth signal. *Klystron tubes*, *traveling-wave tubes* (TWTs), and *IMPATT* (*imp*act/*a*valanche and *T*ransit *T*ime) diodes are several of the devices currently being used in microwave power amplifiers. Because high-gain antennas are used and the distance between microwave stations is relatively short, it is not necessary to develop a high output power from the transmitter output amplifiers. Typical gains for microwave antennas range from 40 to 80 dB, and typical transmitter output powers are between 0.5 and 10 W.

A *microwave generator* provides the RF carrier input to the up-converter. It is called a microwave generator rather than an oscillator because it is difficult to construct a stable circuit that will oscillate in the gigahertz range. Instead, a crystal-controlled oscillator operating in the range 5 to 25 MHz is used to provide a base frequency that is multiplied up to the desired RF carrier frequency.

An *isolator* is a unidirectional device often made from a ferrite material. The

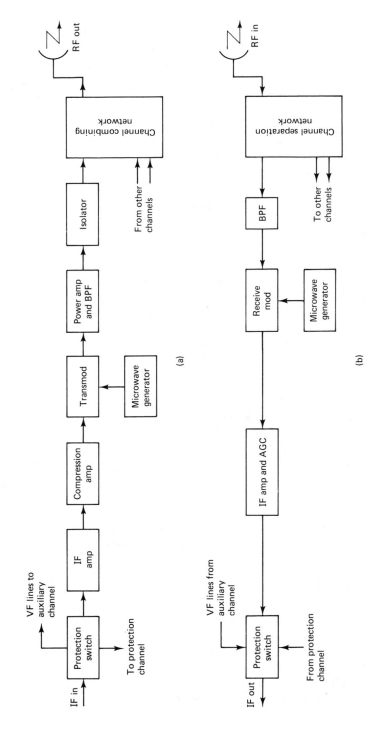

Figure 7-8 Microwave terminal station: (a) transmitter; (b) receiver.

248

isolator is used in conjunction with a channel-combining network to prevent the output of one transmitter from interfering with the output of another transmitter.

The RF receiver (Figure 7-8b) is essentially the same as the transmitter except that it works in the opposite direction. However, one difference is the presence of an IF amplifier in the receiver. This IF amplifier has an *automatic gain control* (AGC) circuit. Also, very often, there are no RF amplifiers in the receiver. Typically, a very sensitive, low-noise-balanced demodulator is used for the receive demodulator (receive mod). This eliminates the need for an RF amplifier and improves the overall signal-to-noise ratio. When RF amplifiers are required, high-quality, *low-noise amplifiers* (LNAs) are used. Examples of commonly used LNAs are tunnel diodes and parametric amplifiers.

Repeater Station

Figure 7-9 shows the block diagram of a microwave IF repeater station. The received RF signal enters the receiver through the channel separation network and bandpass filter. The receive mod down-converts the RF carrier to IF. The IF AMP/AGC and equalizer circuits amplify and reshape the IF. The equalizer compensates for *gain versus frequency nonlinearities* and *envelope delay distortion* introduced in the system. Again, the transmod up-converts the IF to RF for retransmission. However, in a repeater station, the method used to generate the RF microwave carrier frequencies is slightly different from the method used in a terminal station. In the IF repeater, only one microwave generator is required to supply both the transmod and the receive mod with an RF carrier signal. The microwave generator, shift oscillator, and shift modulator allow the repeater to receive one RF carrier frequency, down-convert it to IF, and then up-convert the IF to a different RF carrier frequency (Figure 7-10a). It is possible for station C to receive the transmissions from both station A and station B simultaneously (this is called *multihop interference*). This can occur only when three stations are placed in a geographical straight line in the system. To prevent this from occurring, the allocated bandwidth for the system is divided in half, creating a low-frequency and a high-frequency band. Each station, in turn, alternates from a low-band to a high-band transmit carrier frequency (Figure 7-10b). If a transmission from station A is received by station C, it will be rejected in the channel separation network and cause no interference. This is called a high/low microwave repeater system. The rules are simple: If a repeater station receives a low-band RF carrier, it retransmits a high-band RF carrier, and vice versa. The only time that multiple carriers of the same frequency can be received is when a transmission from one station is received from another station that is three hops away. This is unlikely to happen.

Another reason for using a high/low-frequency scheme is to prevent the power that "leaks" out the back and sides of a transmit antenna from interferring with the signal entering the input of a neaby receive antenna. This is called *ringaround*. All antennas, no matter how high their gain or how directive their radiation pattern, radiate a small percentage of their power out the back and sides; giving a finite

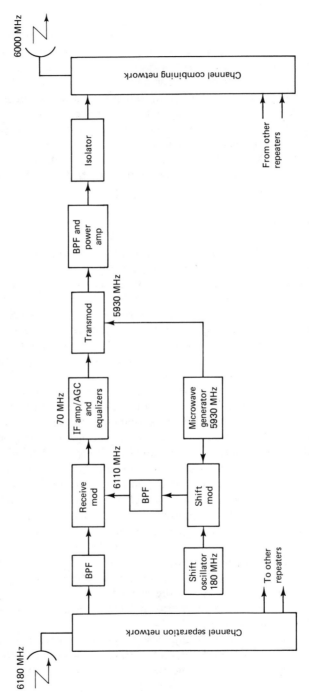

Figure 7-9 Microwave IF repeater station.

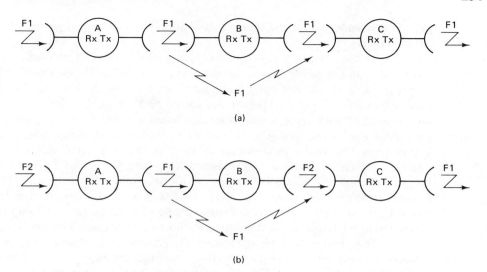

(a)

(b)

Figure 7-10 (a) Multihop interference and (b) high/low microwave system.

front-to-back ratio for the antenna. Although the front-to-back ratio of a typical microwave antenna is quite high, the relatively small amount of power that is radiated out the back of the antenna may be quite substantial compared to a normal received carrier power in the system. If the transmit and receive carrier frequencies are different, filters in the receiver separation network will prevent ringaround from occurring.

A high/low microwave repeater station (Figure 7-10b) needs two microwave carrier supplies for the down- and up-converting process. Rather than use two microwave generators, a single generator together with a shift oscillator, a shift modulator, and a bandpass filter can generate the two required signals. One output from the microwave generator is fed directly into the transmod and another output (from the same microwave generator) is mixed with the shift oscillator signal in the shift modulator to produce a second microwave carrier frequency. The second microwave carrier frequency is offset from the first by the shift oscillator frequency. The second microwave carrier frequency is fed into the receive modulator.

EXAMPLE 7-1

In Figure 7-9 the received RF carrier frequency is 6180 MHz, and the transmitted RF carrier frequency is 6000 MHz. With a 70-MHz IF frequency, a 5930-MHz microwave generator frequency, and a 180-MHz shift oscillator frequency, the output filter of the shift mod must be tuned to 6110 MHz. This is the sum of the microwave generator and the shift oscillator frequencies (5930 MHz + 180 MHz = 6110 MHz).

This process does not reduce the number of oscillators required, but it is simpler and cheaper to build one microwave generator and one relatively low-frequency shift oscillator than to build two microwave generators. The obvious disadvantage of the high/low scheme is that the number of channels available in a given bandwidth is cut in half.

Figure 7-11 shows a high/low-frequency plan with eight channels (four high-band and four low-band). Each channel occupies a 29.7-MHz bandwidth. The west terminal transmits the low-band frequencies and receives the high-band frequencies. Channel 1 and 3 (Figure 7-11a) are designated as *V channels*. This means that they are propagated with vertical polarization. Channels 2 and 4 are designated as H or horizontally polarized channels. This is not a polarization diversity system. Channels 1 through 4 are totally independent of each other; they carry different baseband information. The transmission of *orthogonally* polarized carriers (90° out of phase) further enhances the isolation between the transmit and receive signals. In the west-to-east direction, the repeater receives the low-band and transmits the high-band frequencies. After channel 1 is received and down-converted to IF, it is up-converted to a different RF frequency and polarization for retransmission. The low-band channel 1 corresponds to the high-band channel 11, channel 2 to channel 12, and so on. The east-to-west direction (Figure 7-11b) propagates the high- and low-band carriers in the sequence opposite to the west-to-east system. The polarizations are also reversed. If some of the power from channel 1 of the west terminal were to propagate directly to the east terminal receiver, it has a different frequency and polarization than channel 11's transmissions. Consequently, it would not interfere with the reception of channel 11 (no multihop interference). Also, note that none of the transmit or receive channels at the repeater station has both the same frequency and polarization. Consequently, the interference from the transmitters to the receivers due to ringaround is insignificant.

SYSTEM GAIN

In its simplest form, *system gain* is the difference between the nominal output power of a transmitter and the minimum input power to a receiver. System gain must be greater than or equal to the sum of all the gains and losses incurred by a signal as it propagates from a transmitter to a receiver. In essence, it represents the net loss of a radio system. System gain is used to predict the reliability of a system for given system parameters. Mathematically, system gain is

$$G_s = P_t - C_{\min}$$

where

$$G_s = \text{system gain (dB)}$$
$$P_t = \text{transmitter output power (dBm)}$$
$$C_{\min} = \text{minimum receiver input power for a given quality objective (dBm)}$$

and where

$$P_t - C_{\min} \geq \text{losses} + \text{gains}$$

Gains:

A_t = transmit antenna gain (dB) relative to an isotropic radiator

A_r = receive antenna gain (dB) relative to an isotropic radiator

Losses:

L_p = free-space path loss between antennas (dB)

L_f = waveguide feeder loss (dB) between the distribution network (channel combining network or channel separation network) and its respective antenna (see Table 7-1)

L_b = total coupling or branching loss (dB) in the circulators, filters, and distribution network between the output of a transmitter or the input to a receiver and its respective waveguide feed (see Table 7-1)

FM = fade margin for a given reliability objective

Mathematically, system gain is

$$G_s = P_t - C_{\min} \geq \text{FM} + L_p + L_f + L_b - A_t - A_r \qquad (7\text{-}1)$$

where all values are expressed in dB or dBm. Because system gain is indicative of a net loss, the losses are represented with positive dB values and the gains are represented with negative dB values. Figure 7-12 shows an overall microwave system diagram and indicates where the respective losses and gains are incurred.

TABLE 7-1   SYSTEM GAIN PARAMETERS

Frequency (GHz)	Feeder loss, L_f		Branching loss (dB) Diversity:		Antenna gain, A_t or A_r	
	Type	Loss (dB/100 m)	Frequency	Space	Size (m)	Gain (dB)
1.8	Air-filled coaxial cable	5.4	5	2	1.2	25.2
					2.4	31.2
					3.0	33.2
					3.7	34.7
7.4	EWP 64 eliptical waveguide	4.7	3	2	1.5	38.8
					2.4	43.1
					3.0	44.8
					3.7	46.5
8.0	EWP 69 eliptical waveguide	6.5	3	2	2.4	43.8
					3.0	45.6
					3.7	47.3
					4.8	49.8

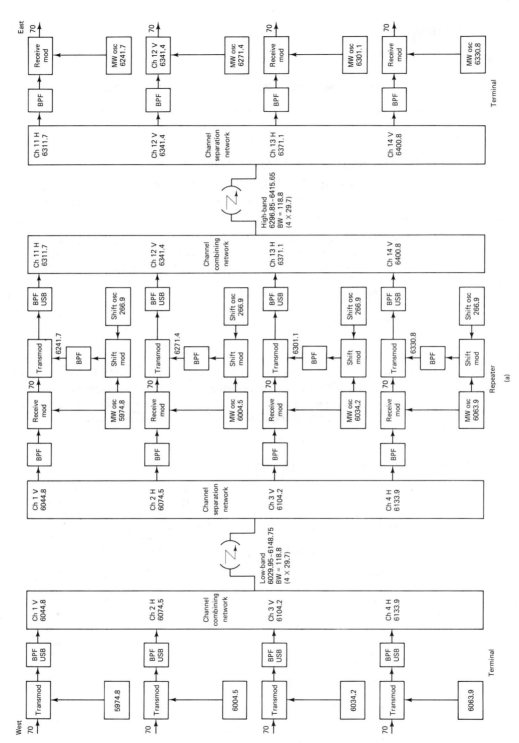

Figure 7-11 Eight-channel high/low frequency plan: (a) west to east; (b) east to west. All frequencies in megahertz.

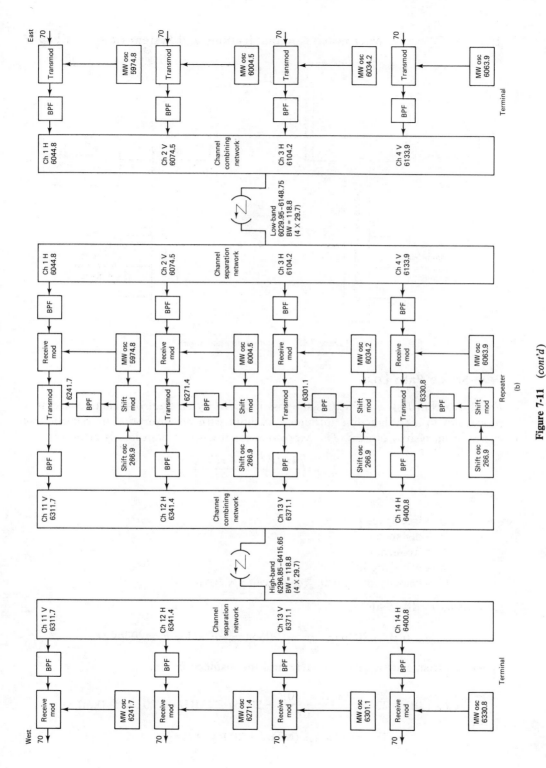

Figure 7-11 *(cont'd)*

255

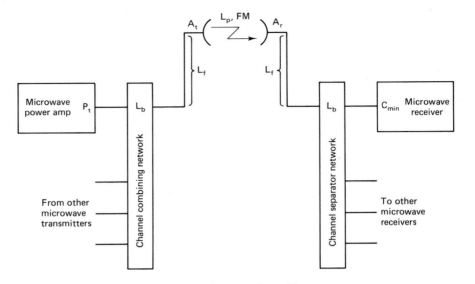

Figure 7-12   System gains and losses.

Free-Space Path Loss

Free-space path loss is defined as the loss incurred by an electromagnetic wave as it propagates in a straight line through a vacuum with no absorption or reflection of energy from nearby objects. The expression for free-space path loss is given as

$$L_p = \left(\frac{4\pi D}{\lambda}\right)^2 = \left(\frac{4\pi FD}{C}\right)^2$$

where

$$L_p = \text{free-space path loss}$$
$$D = \text{distance}$$
$$F = \text{frequency}$$
$$\lambda = \text{wavelength}$$
$$C = \text{velocity of light in free space } (3 \times 10^8 \text{ m/s})$$

Converting to dB yields

$$L_p \text{ (dB)} = 20 \log \frac{4\pi FD}{C} = 20 \log \frac{4\pi}{C} + 20 \log F + 20 \log D$$

When the frequency is given in MHz and the distance in km,

$$L_p \text{ (dB)} = 20 \log \frac{4\pi (10)^6 (10)^3}{3 \times 10^8} + 20 \log F \text{ (MHz)} + 20 \log D \text{ (km)}$$

$$= 32.4 + 20 \log F \text{ (MHz)} + 20 \log D \text{ (km)} \tag{7-2}$$

When the frequency is given in GHz and the distance in km,

$$L_p \text{ (dB)} = 92.4 + 20 \log F \text{ (GHz)} + 20 \log D \text{ (km)} \qquad (7\text{-}3)$$

Similar conversions can be made using distance in miles, frequency in kHz, and so on.

EXAMPLE 7-2

For a carrier frequency of 6 GHz and a distance of 50 km, determine the free-space path loss.

Solution

$$L_p \text{ (dB)} = 32.4 + 20 \log 6000 + 20 \log 50$$

$$= 32.4 + 75.6 + 34$$

$$= 142 \text{ dB}$$

or

$$L_p \text{ (dB)} = 92.4 + 20 \log 6 + 20 \log 50$$

$$= 92.4 + 15.6 + 34$$

$$= 142 \text{ dB}$$

Fade Margin

Essentially, *fade margin* is a "fudge factor" included in the system gain equation that considers the nonideal and less predictable characteristics of radio-wave propagation, such as *multipath propagation* (*multipath loss*) and *terrain sensitivity*. These characteristics cause temporary, abnormal atmospheric conditions that alter the free-space path loss and are usually detrimental to the overall system performance. Fade margin also considers system reliability objectives. Thus fade margin is included in the system gain equation as a loss.

Solving the Barnett–Vignant reliability equations for a specified annual system availability for an unprotected, nondiversity system yields the following expression:

$$\text{FM} = \underbrace{30 \log D}_{\substack{\text{multipath} \\ \text{effect}}} + \underbrace{10 \log (6ABF)}_{\substack{\text{terrain} \\ \text{sensitivity}}} - \underbrace{10 \log (1 - R)}_{\substack{\text{reliability} \\ \text{objectives}}} - \underbrace{70}_{\text{constant}} \qquad (7\text{-}4)$$

where

$$\text{FM} = \text{fade margin (dB)}$$
$$D = \text{distance (km)}$$
$$F = \text{frequency (GHz)}$$
$$R = \text{reliability expressed as a decimal (i.e., } 99.99\% = 0.9999 \text{ reliability)}$$
$$1 - R = \text{reliability objective for a one-way 400-km route}$$
$$A = \text{roughness factor}$$

$= 4$ over water or a very smooth terrain

$= 1$ over an average terrain

$= 0.25$ over a very rough, mountainous terrain

$B =$ factor to convert a worst-month probability to an annual probability

$= 1$ to convert an annual availability to a worst-month basis

$= 0.5$ for hot humid areas

$= 0.25$ for average inland areas

$= 0.125$ for very dry or mountainous areas

EXAMPLE 7-3

Consider a space-diversity microwave radio system operating at an RF carrier frequency of 1.8 GHz. Each station has a 2.4-m-diameter parabolic antenna that is fed by 100 m of air-filled coaxial cable. The terrain is smooth and the area has a humid climate. The distance between stations is 40 km. A reliability objective of 99.99% is desired. Determine the system gain.

Solution   Substituting into Equation 7-4, we find that the fade margin is

$$FM = 30 \log 40 + 10 \log (6)\,(4)\,(0.5)\,(1.8) - 10 \log (1 - 0.9999) - 70$$

$$= 48.06 + 13.34 - (-40) - 70$$

$$= 48.06 + 13.34 + 40 - 70$$

$$= 31.4 \text{ dB}$$

Substituting into Equation 7-3, we obtain path loss

$$L_p = 92.4 + 20 \log 1.8 + 20 \log 40$$

$$= 92.4 + 5.11 + 32.04$$

$$= 129.55 \text{ dB}$$

From Table 7-1,

$$L_b = 4 \text{ dB } (2 + 2 = 4)$$

$$L_f = 10.8 \text{ dB } (100 \text{ m} + 100 \text{ m} = 200 \text{ m})$$

$$A_t = A_r = 31.2 \text{ dB}$$

Substituting into Equation 7-1 gives us system gain

$$G_s = 31.4 + 129.55 + 10.8 + 4 - 31.2 - 31.2 = 113.35 \text{ dB}$$

The results indicate that for this sytem to perform at 99.99% reliability with the given terrain, distribution networks, transmission lines, and antennas, the transmitter output power must be at least 113.35 dB more than the minimum receive signal level.

Receiver Threshold

Carrier-to-noise (C/N) is probably the most important parameter considered when evaluating the performance of a microwave communications system. The minimum wideband carrier power (C_{min}) at the input to a receiver that will produce a usable baseband output is called the receiver *threshold* or, sometimes, receiver *sensitivity*. The receiver threshold is dependent on the wideband noise power present at the input of a receiver, the noise introduced within the receiver, and the noise sensitivity of the baseband detector. Before C_{min} can be calculated, the input noise power must be determined. The input noise power is expressed mathematically as

$$N = KTB$$

where

N = noise power
K = Boltzmann's constant (1.38×10^{-23})
T = equivalent noise temperature of the receiver (K) (room temperature = 290 K)
B = noise bandwidth (Hz)

Expressed in dBm,

$$N \text{ (dBm)} = 10 \log \frac{KTB}{0.001} = 10 \log \frac{KT}{0.001} + 20 \log B$$

For a 1-Hz bandwidth at room temperature,

$$N = 10 \log \frac{(1.38 \times 10^{-23})\,(290)}{0.001} + 10 \log 1$$

$$= -174 \text{ dBm}$$

Thus

$$N \text{ (dBm)} = -174 \text{ dBm} + 10 \log B \tag{7-5}$$

EXAMPLE 7-4

For an equivalent noise bandwidth of 10 MHz, determine the noise power.

Solution Substituting into Equation 7-5 yields

$$N = -174 \text{ dBm} + 10 \log (10 \times 10^6)$$

$$= -174 \text{ dBm} + 70 \text{ dB} = -104 \text{ dBm}$$

If the minimum C/N requirement for a receiver with a 10-MHz noise bandwidth is 24 dB, the minimum receive carrier power is

$$C_{min} = \frac{C}{N} \text{ (dB)} + N \text{ (dB)}$$

$$= 24 \text{ dB} + (-104 \text{ dBm}) = -80 \text{ dBm}$$

For a system gain of 113.35 dB, it would require a minimum transmit carrier power (P_t) of

$$P_t = G_s + C_{min}$$

$$= 113.35 \text{ dB} + (-80 \text{ dBm}) = 33.35 \text{ dBm}$$

This indicates that a minimum transmit power of 33.35 dBm (2.16 W) is required to achieve a carrier-to-noise ratio of 24 dB with a system gain of 113.35 dB and a bandwidth of 10 MHz.

Carrier-to-Noise versus Signal-to-Noise

Carrier-to-noise (C/N) is the ratio of the wideband "carrier" (actually, not just the carrier, but rather the carrier and its associated sidebands) to the wideband noise power (the noise bandwidth of the receiver). C/N can be determined at an RF or an IF point in the receiver. Essentially, C/N is a *predetection* (before the FM demodulator) signal-to-noise ratio. Signal-to-noise (S/N) is a *postdetection* (after the FM demodulator) ratio. At a baseband point in the receiver, a single voice band channel can be separated from the rest of the baseband and measured independently. At an RF or IF point in the receiver, it is impossible to separate a single voice band channel from the composite FM signal. For example, a typical bandwidth for a single microwave channel is 30 MHz. The bandwidth of a voice band channel is 4 kHz. C/N is the ratio of the power of the composite RF signal to the total noise power in the 30-MHz bandwidth. S/N is the ratio of the power of a single voice band channel to the noise power in a 4-kHz bandwidth.

Noise Figure

In its simplest form, *noise figure* (F) is the signal-to-noise ratio of an ideal noiseless device divided by the S/N ratio at the output of an amplifier or a receiver. In a more practical sense, noise figure is defined as the ratio of the S/N ratio at the input to a device divided by the S/N ratio at the output. Mathematically, noise figure is

$$F = \frac{(S/N)_{in}}{(S/N)_{out}} \quad \text{and} \quad F \text{ (dB)} = 10 \log \frac{(S/N)_{in}}{(S/N)_{out}}$$

Thus noise figure is a ratio of ratios. The noise figure of a totally noiseless device is 1 or 0 dB. Remember, the noise present at the input to an amplifier is amplified by the same gain as the signal. Consequently, only noise added within the amplifier can decrease the signal-to-noise ratio at the output and increase the noise figure. (Keep in mind, the higher the noise figure, the worse the S/N ratio at the output.)

Essentially, noise figure indicates the relative increase of the noise power to the increase in signal power. A noise figure of 10 means that the device added sufficient noise to reduce the S/N ratio by a factor of 10, or the noise power increased tenfold in respect to the increase in signal power.

When two or more amplifiers or devices are cascaded together (Figure 7-13), the total noise figure (NF) is an accumulation of the individual noise figures. Mathematically, the total noise figure is

$$NF = F_1 + \frac{F_2 - 1}{A_1} + \frac{F_3 - 1}{A_1 A_2} + \frac{F_4 - 1}{A_1 A_2 A_3} \quad \text{etc.} \quad (7\text{-}6)$$

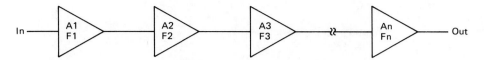

Figure 7-13   Total noise figure.

where

NF = total noise figure
F_1 = noise figure of amplifier 1
F_2 = noise figure of amplifier 2
F_3 = noise figure of amplifier 3
A_1 = gain of amplifier 1
A_2 = gain of amplifier 2

Note: Noise figures and gains are expressed as absolute values rather than dB.

It can be seen that the noise figure of the first amplifier (F1) contributes the most toward the overall noise figure. The noise introduced in the first stage is amplified by each of the succeeding amplifiers. Therefore, when compared to the noise introduced in the first stage, the noise added by each succeeding amplifier is effectively reduced by a factor equal to the product of the gains of the preceding amplifiers.

When precise noise calculations (0.1 dB or less) are necessary, it is generally more convenient to express noise figure in terms of noise temperature or equivalent noise temperature rather than as an absolute power (Chapter 8). Because noise power (N) is proportional to temperature, the noise present at the input to a device can be expressed as a function of the device's environmental temperature (T) and its equivalent noise temperature (T_e). To convert noise figure to a term dependent on temperature only, refer to Figure 7-14.

Let

N_d = noise power added by a single amplifier

Then

$$N_d = KT_eB$$

where T_e is the equivalent noise temperature. Let

N_o = total output noise power of an amplifier
N_i = total input noise power of an amplifier
A = gain of an amplifier

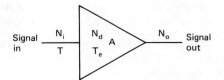

Figure 7-14   Noise figure as a function of temperature.

Therefore,

N_o may be expressed as
$$N_o = AN_i + AN_d$$

and

$$N_o = AKTB + AKT_eB$$

Simplifying yields

$$N_o = AKB \, (T + T_e)$$

and the overall noise figure (NF) equals

$$NF = \frac{(S/N)_{in}}{(S/N)_{out}} = \frac{S/N_i}{AS/N_o} = \frac{N_o}{AN_i} = \frac{AKB \, (T + T_e)}{AKTB} \tag{7-7}$$

$$= \frac{T + T_e}{T} = 1 + \frac{T_e}{T}$$

EXAMPLE 7-5

In Figure 7-13, let $F_i = F_2 = F_3 = 3$ dB and $A_1 = A_2 = A_3 = 10$ dB. Solve for the total noise figure.

Solution   Substituting into Equation 7-6 (*note*: All gains and noise figures have been converted to absolute values) yields

$$NF = F_1 + \frac{F_2 - 1}{A_1} + \frac{F_3 - 1}{A_1 A_2}$$

$$= 2 + \frac{2 - 1}{10} + \frac{2 - 1}{100}$$

$$= 2.11 \text{ or } 10 \log 2.11 = 3.24 \text{ dB}$$

An overall noise figure of 3.24 dB indicates that the S/N ratio at the output of A3 is 3.24 dB less than the S/N ratio at the input to A1.

The noise figure of a receiver must be considered when determining C_{min}. The noise figure is included in the system gain equation as an equivalent loss. (Essentially, a gain in the total noise power is equivalent to a corresponding loss in the signal power.)

EXAMPLE 7-6

Refer to Figure 7-15. For a system gain of 112 dB, a total noise figure of 6.5 dB, an input noise power of −104 dBm, and a minimum $(S/N)_{out}$ of the FM demodulator of 32 dB, determine the minimum receive carrier power and the minimum transmit power.

Solution   To achieve a S/N ratio of 32 dB out of the FM demodulator, an input C/N of 15 dB is required (17 dB of improvement due to FM quieting). Solving for the receiver input carrier-to-noise ratio gives

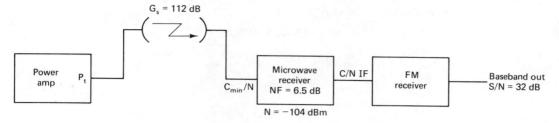

Figure 7-15 System gain example.

$$\frac{C_{min}}{N} = \frac{C}{N} + NF = 15 \text{ dB} + 6.5 \text{ dB} = 21.5 \text{ dB}$$

Thus

$$C_{min} = \frac{C_{min}}{N} + N$$

$$= 21.5 \text{ dB} + (-104 \text{ dBm}) = -82.5 \text{ dBm}$$

$$P_t = G_s + C_{min}$$

$$= 112 \text{ dB} + (-82.5 \text{ dBm}) = 29.5 \text{ dBm}$$

EXAMPLE 7-7

For the system shown in Figure 7-16, determine the following: G_s, C_{min}/N, C_{min}, N, G_s, and P_t.

Solution The minimum C/N at the input to the FMR is 23 dB.

$$\frac{C_{min}}{N} = \frac{C}{N} + NF$$

$$= 23 \text{ dB} + 4.24 \text{ dB} = 27.24 \text{ dB}$$

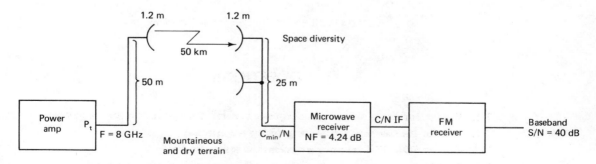

Reliability objective = 99.999%
Bandwidth = 6.3 MHz

Figure 7-16 System gain example.

Substituting into Equation 7-5 yields

$$N = -174 \text{ dBm} + 10 \log B$$

$$= -174 \text{ dBm} + 68 \text{ dB} = -106 \text{ dBm}$$

$$C_{min} = \frac{C_{min}}{N} + N$$

$$= 27.24 \text{ dB} + (-106 \text{ dBm}) = -78.76 \text{ dBm}$$

Substituting into Equation 7-4 gives us

$$FM = 30 \log 50 + 10 \log [(6) (0.25) (0.125) (8)]$$

$$- 10 \log (1 - 0.99999) - 70$$

$$= 32.76 \text{ dB}$$

Substituting into Equation 7-3, we have

$$L_p = 92.4 \text{ dB} + 20 \log 8 + 20 \log 50$$

$$= 92.4 \text{ dB} + 18.06 \text{ dB} + 33.98 \text{ dB} = 144.44 \text{ dB}$$

From Table 7-1,

$$L_b = 4 \text{ dB}$$

$$L_f = 0.75 (6.5 \text{ dB}) = 4.875 \text{ dB}$$

$$A_t = A_r = 37.8 \text{ dB}$$

Note: The gain of an antenna increases or decreases proportional to the square of its diameter (i.e., if its diameter changes by a factor of 2, its gain changes by a factor of 4 or 6 dB).

Substituting into Equation 7-1 yields

$$G_s = 32.76 + 144.44 + 4.875 + 4 - 37.8 - 37.8 = 110.475 \text{ dB}$$

$$P_t = G_s + C_{min}$$

$$= 110.475 \text{ dB} + (-78.76 \text{ dBm}) = 31.715 \text{ dBm}$$

QUESTIONS

7-1. What constitutes a short-haul microwave system? A long-haul microwave system?

7-2. Describe the baseband signal for a microwave system.

7-3. Why do FDM/FM microwave systems use low-index FM?

7-4. Describe a microwave repeater. Contrast baseband and IF repeaters.

7-5. Define *diversity*. Describe the three most commonly used diversity schemes.

7-6. Describe a protection switching arrangement. Contrast the two types of protection switching arrangements.

7-7. Briefly describe the four major sections of a microwave terminal station.

7-8. Define *ringaround*.

7-9. Briefly describe a high/low microwave system.

7-10. Define *system gain*.

7-11. Define the following terms: free-space path loss, branching loss, and feeder loss.

7-12. Define *fade margin*. Describe multipath losses, terrain, sensitivity, and reliability objectives and how they effect fade margin.

7-13. Define *receiver threshold*.

7-14. Contrast carrier-to-noise ratio and signal-to-noise ratio.

7-15. Define *noise figure*.

PROBLEMS

7-1. Calculate the noise power at the input to a receiver that has a radio carrier frequency of 4 GHz and a bandwidth of 30 MHz (assume room temperature).

7-2. Determine the path loss for a 3.4-GHz signal propagating 20,000 m.

7-3. Determine the fade margin for a 60-km microwave hop. The RF carrier frequency is 6 GHz, the terrain is very smooth and dry, and the reliability objective is 99.95%.

7-4. Determine the noise power for a 20-MHz bandwidth at the input to a receiver with an input noise temperature of 290°C.

7-5. For a system gain of 120 dB, a minimum input C/N of 30 dB, and an input noise power of -115 dBm, determine the minimum transmit power (P_t).

7-6. Determine the amount of loss contributed to a reliability objective of 99.98%.

7-7. Determine the terrain sensitivity loss for a 4-GHz carrier that is propagating over a very dry, mountainous area.

7-8. A frequency-diversity microwave system operates at an RF carrier frequency of 7.4 GHz. The IF is a low-index frequency-modulated subcarrier. The baseband signal is the 1800-channel FDM system described in Chapter 6 (564 to 8284 kHz). The antennas are 4.8-m-diameter parabolic dishes. The feeder lengths are 150 m at one station and 50 m at the other station. The reliability objective is 99.999%. The system propagates over an average terrain that has a very dry climate. The distance between stations is 50 km. The minimum carrier-to-noise ratio at the receiver input is 30 dB. Determine the following: fade margin, antenna gain, free-space path loss, total branching and feeder losses, receiver input noise power, C_{min}, minimum transmit power, and system gain.

7-9. Determine the overall noise figure for a receiver that has two RF amplifiers each with a noise figure of 6 dB and a gain of 10 dB, a mixer down-converter with a noise figure of 10 dB, and a conversion gain of -6 dB, and 40 dB of IF gain with a noise figure of 6 dB.

7-10. A microwave receiver has a total input noise power of -102 dBm and an overall noise figure of 4 dB. For a minimum C/N ratio of 20 dB at the input to the FM detector, determine the minimum receive carrier power.

Chapter 8

SATELLITE COMMUNICATIONS

INTRODUCTION

In the early 1960s, the American Telephone and Telegraph Company (AT&T) released studies indicating that a few powerful satellites of advanced design could handle more traffic than the entire AT&T long-distance communications network. The cost of these satellites was estimated to be only a fraction of the cost of equivalent terrestrial microwave facilities. Unfortunately, because AT&T was a utility, government regulations prevented them from developing the satellite systems. Smaller and much less lucrative corporations were left to develop the satellite systems, and AT&T continued to invest billions of dollars each year in conventional terrestrial microwave systems. Because of this, early developments in satellite technology were slow in coming.

Throughout the years the prices of most goods and services have increased substantially; however, satellite communications services have become more affordable each year. In most instances, satellite systems offer more flexibility than submarine cables, buried underground cables, line-of-sight microwave radio, tropospheric scatter radio, or optical fiber systems.

Essentially, a satellite is a radio repeater in the sky (*transponder*). A satellite system consists of a transponder, a ground-based station to control its operation, and a user network of earth stations that provide the facilities for transmission and reception of communications traffic through the satellite system. Satellite transmissions are categorized as either *bus* or *payload*. The bus includes control mechanisms that support the payload operation. The payload is the actual user information that is conveyed through the system. Although in recent years new data services and television broadcasting are more and more in demand, the transmission of conventional speech telephone signals (in analog or digital form) is still the bulk of the satellite payload.

266

HISTORY OF SATELLITES

The simplest type of satellite is a *passive reflector*, a device that simply "bounces" a signal from one place to another. The moon is a natural satellite of the earth and, consequently, in the late 1940s and early 1950s, became the first satellite transponder. In 1954, the U.S. Navy successfully transmitted the first messages over this earth-to-moon-to-earth relay. In 1956, a relay service was established between Washington, D.C., and Hawaii and, until 1962, offered reliable long-distance communications. Service was limited only by the availability of the moon.

In 1957, Russia launched *Sputnik I*, the first *active* earth satellite. An active satellite is capable of receiving, amplifying, and retransmitting information to and from earth stations. *Sputnik I* transmitted telemetry information for 21 days. Later in the same year, the United States launched *Explorer I*, which transmitted telemetry information for nearly 5 months.

In 1958, NASA launched *Score*, a 150-pound conical-shaped projectory. With an on-board tape recording, *Score* rebroadcasted President Eisenhower's 1958 Christmas message. *Score* was the first artificial satellite used for relaying terrestrial communications. *Score* was a *delayed repeater satellite*; it received transmissions from earth stations, stored them on magnetic tape, and rebroadcasted them to ground stations farther along its orbit.

In 1960, NASA in conjunction with Bell Telephone Laboratories and the Jet Propulsion Laboratory launched *Echo*, a 100-ft-diameter plastic balloon with an aluminum coating. *Echo* passively reflected radio signals from a large earth antenna. *Echo* was simple and reliable but required extremely high power transmitters at the earth stations. The first transatlantic transmission using a satellite transponder was accomplished using *Echo*. Also in 1960, the Department of Defense launched *Courier*. *Courier* transmitted 3 W of power and lasted only 17 days.

In 1962, AT&T launched *Telstar I*, the first satellite to receive and transmit simultaneously. The electronic equipment in *Telstar I* was damaged by radiation from the newly discovered Van Allen belts and, consequently, lasted only a few weeks. *Telstar II* was electronically identical to *Telstar I*, but it was made more radiation resistant. *Telstar II* was successfully launched in 1963. It was used for telephone, television, facsimile, and data transmissions. The first successful transatlantic transmission of video was accomplished with *Telstar II*.

Early satellites were both of the passive and active type. Again, a passive satellite is one that simply reflects a signal back to earth; there are no gain devices on board to amplify or repeat the signal. An active satellite is one that electronically repeats a signal back to earth (i.e., receives, amplifies, and retransmits the signal). An advantage of passive satellites is that they do not require sophisticated electronic equipment on board, although they are not necessarily void of power. Some passive satellites require a *radio beacon transmitter* for tracking and ranging purposes. A beacon is a continuously transmitted unmodulated carrier that an earth station can lock onto and use to align its antennas or to determine the exact location of the satellite. A disadvantage of passive satellites is their inefficient use of transmitted power. With

Echo, for example, only 1 part in every 10^{18} of the earth station transmitted power was actually returned to the earth station receiving antenna.

ORBITAL SATELLITES

The satellites mentioned thus far are of the *orbital* or *nonsynchronous* type. That is, they rotate around the earth in a low-altitude elliptical or circular pattern with an angular velocity greater than (*prograde*) or less than (*retrograde*) that of Earth. Consequently, they are continuously gaining or falling back on Earth and do not remain stationary to any particular point on Earth. Thus they have to be used when available, which may be as short a period of time as 15 minutes per orbit. Another disadvantage of orbital satellites is the need for complicated and expensive tracking equipment at the earth stations. Each Earth station must locate the satellite as it comes into view on each orbit and then lock its antenna onto the satellite and track it as it passes overhead. A major advantage of orbital satellites is that propulsion rockets are not required on board the satellites to keep them in their respective orbits.

One of the more interesting orbital satellite systems is the Soviet *Molniya* system. It is presently the only nonsynchronous-orbit commercial satellite system in use. *Molniya* uses a highly elliptical orbit with *apogee* at about 40,000 km and *perigee* at about 1000 km (see Figure 8-1). The apogee is the maximum distance from earth a satellite orbit reaches; the perigee is the minimum distance. With the *Molniya* system, the apogee is reached while over the northern hemisphere and the perigee while over the southern hemisphere. The size of the ellipse was chosen to make its period exactly one-half of a sidereal day (the time it takes the earth to rotate back to the same constellation). Because of its unique orbital pattern, the *Molniya* satellite is synchronous with the rotation of the earth. During its 12-h orbit, it spends about 11 h over the north hemisphere.

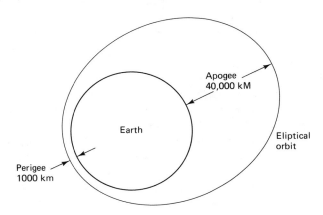

Figure 8-1 Soviet *Molniya* satellite orbit.

GEOSTATIONARY SATELLITES

Geostationary or *geosynchronous* satellites are satellites that orbit in a circular pattern with an angular velocity equal to that of Earth. Consequently, they remain in a fixed position in respect to a given point on Earth. An obvious advantage is they are available to all the earth stations within their *shadow* 100% of the time. The shadow of a satellite includes all earth stations that have a line-of-sight path to it and lie within the radiation pattern of the satellite's antennas. An obvious disadvantage is they require sophisticated and heavy propulsion devices on board to keep them

TABLE 8-1   CURRENT SATELLITE COMMUNICATIONS SYSTEMS

	Characteristic System				
	Westar	*Intelsat V*	*SBS*	*Fleet-satcom*	*Anik D*
Operator	Western Union Telegraph	Intelsat	Satellite Business Systems	U.S. Dept. of Defense	Telsat Canada
Frequency band	C	C and Ku	Ku	UHF, X	C, Ku
Coverage	Consus	Global, zonal, spot	Consus	Global	Canada, northern U.S.
Number of transponders	12	21	10	12	24
Transponder BW (MHz)	36	36–77	43	0.005–0.5	36
EIRP (dBw)	33	23.5–29	40–43.7	26–28	36
Multiple Access	FDMA, TDMA	FDMA, TDMA, reuse	TDMA	FDMA	FMDA
Modulation	FM, QPSK	FDM/FM, QPSK	QPSK	FM, QPSK	FDM, FM, FM/TVD, SCPC
Service	Fixed tele, TTY	Fixed tele, TVD	Fixed tele, TVD	Mobile military	Fixed tele

C-band: 3.4–6.425 GHz
Ku-band: 10.95–14.5 GHz
X-band: 7.25–8.4 GHz

TTY   teletype
TVD   TV distribution
FDMA frequency-division multiple access
TDMA time-division multiple access
Consus continental United States

in a fixed orbit. The orbital time of a geosynchronous satellite is 24 h, the same as Earth.

Syncom I, launched in February 1963, was the first attempt to place a geosynchronous satellite into orbit. Syncom I was lost during orbit injection. Syncom II and Syncom III were successfully launched in February 1963 and August 1964, respectively. The Syncom III satellite was used to broadcast the 1964 Olympic Games from Tokyo. The Syncom projects demonstrated the feasibility of using geosynchronous satellites.

Since the Syncom projects, a number of nations and private corporations have successfully launched satellites that are currently being used to provide national as well as regional and international global communications. There are more than 80 satellite communications systems operating in the world today. They provide worldwide fixed common-carrier telephone and data circuits; point-to-point cable television (CATV); network television distribution; music broadcasting; mobile telephone service; and private networks for corporations, governmental agencies, and military applications. A commercial global satellite network known as Intelsat (International Telecommunications Satellite Organization) is owned and operated by a consortium of more than 100 countries. Intelsat is managed by the designated communications entities in their respective countries. The Intelsat network provides high-quality, reliable service to its member countries. Table 8-1 is a partial list of current international and domestic satellite systems and their primary payload.

ORBITAL PATTERNS

Once projected, a satellite remains in orbit because the centrifugal force caused by its rotation around the earth is counterbalanced by the earth's gravitational pull. The closer to earth the satellite rotates, the greater the gravitational pull and the greater the velocity required to keep it from being pulled to earth. Low-altitude satellites that orbit close to Earth (100 to 300 miles in height) travel at approximately 17,500 miles per hour. At this speed, it takes approximately $1\frac{1}{2}$ h to rotate around the entire earth. Consequently, the time that the satellite is in line of sight of a particular earth station is only $\frac{1}{4}$ h or less per orbit. Medium-altitude satellites (6000 to 12,000 miles in height) have a rotation period of 5 to 12 h and remain in line of sight of a particular earth station for 2 to 4 h per orbit. High-altitude, geosynchronous satellites (19,000 to 25,000 miles in height) travel at approximately 6879 miles per hour and have a rotation period of 24 h, exactly the same as the earth. Consequently, they remain in a *fixed* position in respect to a given earth station and have a 24-h availability time. Figure 8-2 shows a low-, medium-, and high-altitude satellite orbit. It can be seen that three equally spaced, high-altitude geosynchronous satellites rotating around the earth above the equator can cover the entire earth except for the unpopulated areas of the north and south poles.

Figure 8-3 shows the three paths that a satellite may take as it rotates around the earth. When the satellite rotates in an orbit above the equator, it is called an

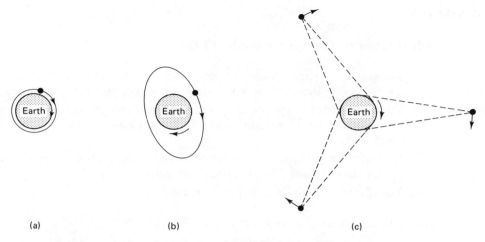

Figure 8-2   Satellite orbits: (a) low altitude (circular orbit, 100–300 mi); (b) medium altitude (elliptical orbit, 6000–12,000 mi); (c) high altitude (geosynchronous orbit, 19,000–25,000 mi).

equatorial orbit. When the satellite rotates in an orbit that takes it over the north and south poles, it is called a *polar orbit*. Any other orbital path is called an *inclined orbit*.

It is interesting to note that 100% of the earth's surface can be covered with a single satellite in a polar orbit. The satellite is rotating around the earth in a longitudinal orbit while the earth is rotating on a latitudinal axis. Consequently, the satellite's radiation pattern is a diagonal spiral around the earth which somewhat resembles a barber pole. As a result, every location on earth lies within the radiation pattern of the satellite twice each day.

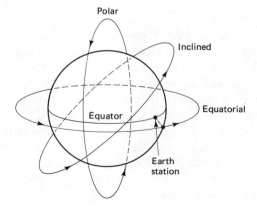

Figure 8-3   Satellite orbits.

SUMMARY

Advantages of Geosynchronous Orbits

1. The satellite remains almost stationary in respect to a given earth station. Consequently, expensive tracking equipment is not required at the earth stations.
2. There is no need to switch from one satellite to another as they orbit overhead. Consequently, there are no breaks in transmission because of the switching times.
3. High-altitude geosynchronous satellites can cover a much larger area of the earth than their low-altitude orbital counterparts.
4. The effects of Doppler shift are negligible.

Disadvantages of Geosynchronous Orbits

1. The higher altitudes of geosynchronous satellites introduce much longer propagation times. The round-trip propagation delay between two earth stations through a geosynchronous satellite is 500 to 600 ms.
2. Geosynchronous satellites require higher transmit powers and more sensitive receivers because of the longer distances and greater path losses.
3. High-precision spacemanship is required to place a geosynchronous satellite into orbit and to keep it there. Also, propulsion engines are required on board the satellites to keep them in their respective orbits.

LOOK ANGLES

To orient an antenna toward a satellite, it is necessary to know the *elevation angle* and *azimuth* (Figure 8-4). These are called the *look angles*.

Angle of Elevation

The angle of elevation is the angle formed between the plane of a wave radiated from an earth station antenna and the horizon, or the angle subtended at the earth station antenna between the satellite and the earth's horizon. The smaller the angle of elevation, the greater the distance a propagated wave must pass through the earth's atmosphere. As with any wave propagated through the earth's atmosphere, it suffers absorption and may also be severely contaminated by noise. Consequently, if the angle of elevation is too small and the distance the wave is within the earth's atmosphere is too long, the wave may deteriorate to a degree that it provides inadequate transmission. Generally, 5° is considered as the minimum acceptable angle of elevation. Figure 8-5 shows how the angle of elevation affects the signal strength of a propagated wave due to normal atmospheric absorption, absorption due to thick fog, and absorp-

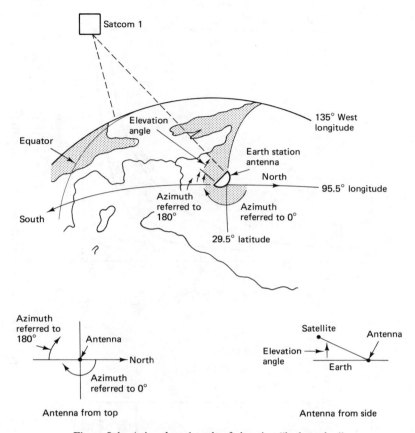

Figure 8-4 Azimuth and angle of elevation "look angles."

tion due to a heavy rain. It can be seen that the 14/12-GHz band (Figure 8-5b) is more severely affected than the 6/4-GHz band (Figure 8-5a). This is due to the smaller wavelengths associated with the higher frequencies. Also, at elevation angles less than 5°, the attenuation increases rapidly.

Azimuth

Azimuth is defined as the horizontal pointing angle of an antenna. It is measured in a clockwise direction in degrees from true north. The angle of elevation and the azimuth both depend on the latitude of the earth station and the longitude of both the earth station and the orbiting satellite. For a geosynchronous satellite in an equatorial orbit, the procedure is as follows: From a good map, determine the longitude and latitude of the earth station. From Table 8-2, determine the longitude of the satellite of interest. Calculate the difference, in degrees (ΔL), between the longitude

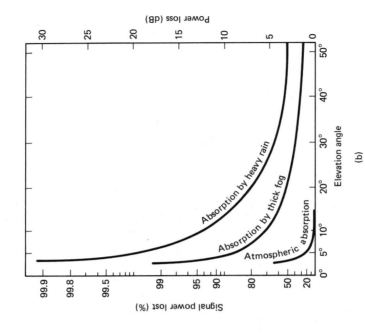

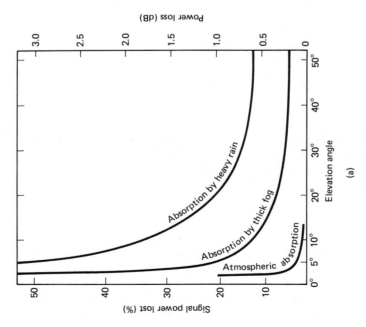

Figure 8-5 Attenuation due to atmospheric absorption: (a) 6/4-GHz band; (b) 14/12-GHz band.

**TABLE 8-2   LONGITUDINAL
POSITION OF SEVERAL CURRENT
SYNCHRONOUS SATELLITES
PARKED IN AN EQUATORIAL ARC[a]**

Satellite	Longitude (°W)
Anik 1	104
Anik 2	109
Anik 3	114
Westar I	99
Westar II	123.5
Westar III	91
Satcom 1	135
Satcom 2	119
Comstar D2	95
Palapa 1	277
Palapa 2	283

[a] 0° latitude.

of the satellite and the longitude of the earth station. Then, from Figure 8-6, determine the azimuth and elevation angle for the antenna. Figure 8-6 is for a geosynchronous satellite in an equatorial orbit.

EXAMPLE 8-1

An earth station is located at Houston, Texas, which has a longitude of 95.5°W and a latitude of 29.5°N. The satellite of interest is RCA's *Satcom 1*, which has a longitude of 135°W. Determine the azimuth and elevation angle for the earth station antenna.

Solution   First determine the difference between the longitude of the earth station and the satellite.

$$\Delta L = 135° - 95.5° = 39.5°$$

Locate the intersection of ΔL and the latitude of the earth station on Figure 8-6. From the figure the angle of elevation is approximately 35°, and the azimuth is approximately 59° west of south.

ORBITAL SPACING AND FREQUENCY ALLOCATION

Geosynchronous satellites must share a limited space and frequency spectrum within a given arc of a geostationary orbit. Satellites operating at or near the same frequency must be sufficiently separated in space to avoid interfering with each other (Figure 8-7). There is a realistic limit to the number of satellite structures that can be stationed (*parked*) within a given area in space. The required *spatial separation* is dependent on the following variables:

Azimuth angle referenced to 180° (degrees)

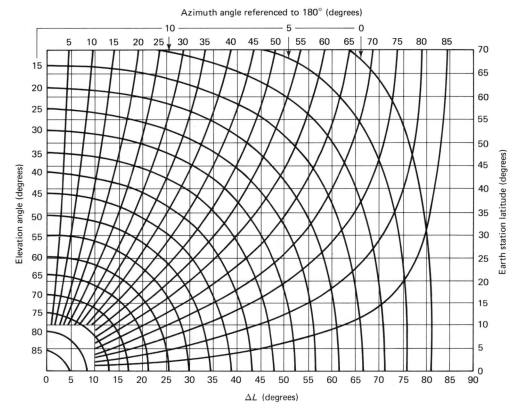

Figure 8-6 Azimuth and elevation angle for earth stations located in the northern hemisphere (referred to 180°).

1. Beamwidths and sidelobe radiation of both the earth station and satellite antennas
2. RF carrier frequency
3. Encoding or modulation technique used
4. Acceptable limits of interference
5. Transmit carrier power

Generally, 3 to 6° of spatial separation is required depending on the variables stated above.

The most common carrier frequencies used for satellite communications are the 6/4- and 14/12-GHz bands. The first number is the up-link (earth station-to-transponder) frequency, and the second number is the down-link (transponder-to-earth station) frequency. Different up-link and down-link frequencies are used to prevent ringaround from occurring (Chapter 7). The higher the carrier frequency, the smaller the diameter required of an antenna for a given gain. Most domestic

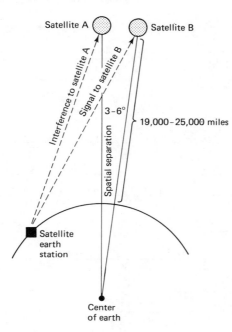

Figure 8-7 Spatial separation of satellites in geosynchronous orbit.

satellites use the 6/4-GHz band. Unfortunately, this band is also used extensively for terrestrial microwave systems. Care must be taken when designing a satellite network to avoid interference from or interference with established microwave links.

Certain positions in the geosynchronous orbit are in higher demand than the others. For example, the mid-Atlantic position which is used to interconnect North America and Europe is in exceptionally high demand. The mid-Pacific position is another.

The frequencies allocated by WARC (World Administrative Radio Conference) are summarized in Figure 8-8. Table 8-3 shows the bandwidths available for various services in the United States. These services include *fixed-point* (between earth stations located at fixed geographical points on earth), *broadcast* (wide-area coverage), *mobile* (ground-to-aircraft, ships, or land vehicles), and *intersatellite* (satellite-to-satellite cross-links).

RADIATION PATTERNS: FOOTPRINTS

The area of the earth covered by a satellite depends on the location of the satellite in its geosynchronous orbit, its carrier frequency, and the gain of its antennas. Satellite engineers select the antenna and carrier frequency for a particular spacecraft to concentrate the limited transmitted power on a specific area of the earth's surface. The geographical representation of a satellite antenna's radiation pattern is called a *footprint* (Figure 8-9). The contour lines represent limits of equal receive power density.

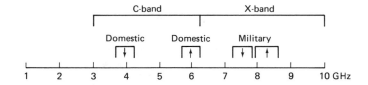

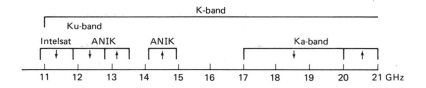

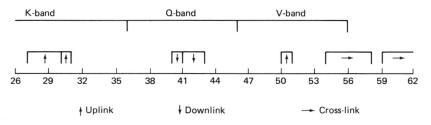

↑ Uplink   ↓ Downlink   → Cross-link

Figure 8-8   WARC satellite frequency assignments.

TABLE 8-3   RF SATELLITE BANDWIDTHS AVAILABLE IN THE UNITED STATES

| Band | Frequency band (GHz) | | Bandwidth (MHz) |
	Up-link	Down-link	
C	5.9–6.4	3.7–4.2	500
X	7.9–8.4	7.25–7.75	500
Ku	14–14.5	11.7–12.2	500
Ka	27–30	17–20	—
	30–31	20–21	—
V	50–51	40–41	1000
Q	—	41–43	2000
V	54–58		3900
(ISL)	59–64		5000

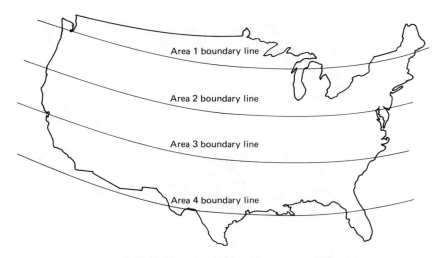

Figure 8-9   Satellite antenna radiation patterns ("footprints").

The radiation pattern from a satellite antenna may be categorized as either *spot*, *zonal*, or *earth* (Figure 8-10). The radiation patterns of earth coverage antennas have a beamwidth of approximately 17° and include coverage of approximately one-third of the earth's surface. Zonal coverage includes an area less than one-third of the earth's surface. Spot beams concentrate the radiated power in a very small geographic area.

Reuse

When an allocated frequency band is filled, additional capacity can be achieved by *reuse* of the frequency spectrum. By increasing the size of an antenna (i.e., increasing the antenna gain) the beamwidth of the antenna is also reduced. Thus different beams of the same frequency can be directed to different geographical areas of the earth. This is called frequency reuse. Another method of frequency reuse is to use dual polarization. Different information signals can be transmitted to different earth station receivers using the same band of frequencies simply by orienting their electromagnetic polarizations in an orthogonal manner (90° out of phase). Dual polarization is less effective because the earth's atmosphere has a tendency to reorient or repolarize an electromagnetic wave as it passes through. Reuse is simply another way to increase the capacity of a limited bandwidth.

SATELLITE SYSTEM LINK MODELS

Essentially, a satellite system consists of three basic sections: the uplink, the satellite transponder, and the downlink.

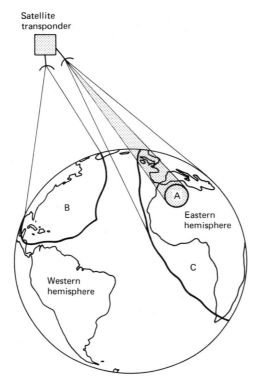

Figure 8-10   Beams: A, spot; B, zonal; C, earth.

Uplink Model

The primary component within the *uplink* section of a satellite system is the earth station transmitter. A typical earth station transmitter consists of an IF modulator, an IF-to-RF microwave up-converter, a high-power amplifier (HPA), and some means of bandlimiting the final output spectrum (i.e., an output bandpass filter). Figure 8-11 shows the block diagram of a satellite earth station transmitter. The IF modulator converts the input baseband signals to either an FM, a PSK, or a QAM modulated intermediate frequency. The up-converter (mixer and bandpass filter) converts the IF to an appropriate RF carrier frequency. The HPA provides adequate input sensitivity and output power to propagate the signal to the satellite transponder. HPAs commonly used are klystons and traveling-wave tubes.

Transponder

A typical *satellite transponder* consists of an input bandlimiting device (BPF), an input *low-noise amplifier* (LNA), a *frequency translator*, a low-level power amplifier, and an output bandpass filter. Figure 8-12 shows a simplified block diagram of a

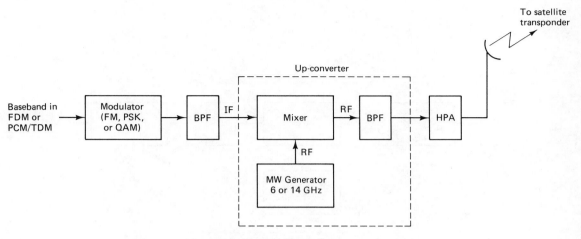

Figure 8-11 Satellite uplink model.

satellite transponder. This transponder is an RF-to-RF repeater. Other transponder configurations are IF and baseband repeaters similar to those used in microwave repeaters. In Figure 8-12, the input BPF limits the total noise applied to the input of the LNA. (A common device used as an LNA is a tunnel diode.) The output of the LNA is fed to a frequency translator (a shift oscillator and a BPF) which converts the high-band uplink frequency to the low-band downlink frequency. The low-level power amplifier, which is commonly a traveling-wave tube, amplifies the RF signal for transmission through the downlink to the earth station receivers. Each RF satellite channel requires a separate transponder.

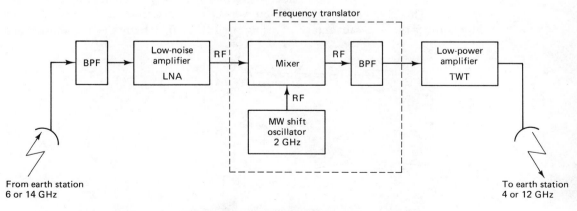

Figure 8-12 Satellite transponder.

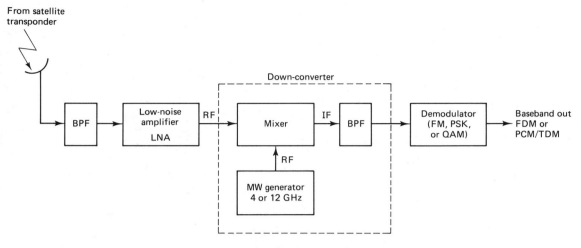

Figure 8-13   Satellite downlink model.

Downlink Model

An earth station receiver includes an input BPF, an LNA, and an RF-to-IF down-converter. Figure 8-13 shows a block diagram of a typical earth station receiver. Again, the BPF limits the input noise power to the LNA. The LNA is a highly sensitive, low-noise device such as a tunnel diode amplifier or a parametric amplifier. The RF-to-IF down-converter is a mixer/bandpass filter combination which converts the received RF signal to an IF frequency.

Cross-Links

Occasionally, there is an application where it is necessary to communicate between satellites. This is done using *satellite cross-links* or *intersatellite links* (ISLs), shown in Figure 8-14. A disadvantage of using an ISL is that both the transmitter and

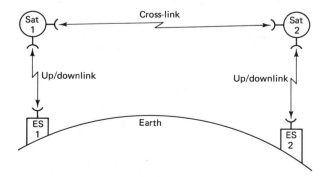

Figure 8-14   Intersatellite link.

receiver are *spacebound*. Consequently, both the transmitter's output power and the receiver's input sensitivity are limited.

SATELLITE SYSTEM PARAMETERS

Transmit Power and Bit Energy

High-power amplifiers used in earth station transmitters and the traveling-wave tubes typically used in satellite transponders are *nonlinear devices*; their gain (output power versus input power) is dependent on input signal level. A typical input/output power characteristic curve is shown in Figure 8-15. It can be seen that as the input power is reduced by 5 dB, the output power is reduced by only 2 dB. There is an obvious *power compression*. To reduce the amount of intermodulation distortion caused by the nonlinear amplification of the HPA, the input power must be reduced (*backed off*) by several dB. This allows the HPA to operate in a more *linear* region. The amount the input level is backed off is equivalent to a loss and is appropriately called *back-off loss* (L_{bo}).

To operate as efficiently as possible, a power amplifier should be operated as close as possible to saturation. The *saturated output power* is designated P_o (sat) or simply P_t. The output power of a typical satellite earth station transmitter is much higher than the output power from a terrestrial microwave power amplifier. Consequently, when dealing with satellite systems, P_t is generally expressed in dBW (decibels in respect to 1 W) rather than in dBm (decibels in respect to 1 mW).

Most modern satellite systems use either phase shift keying (PSK) or quadrature amplitude modulation (QAM) rather than conventional frequency modulation (FM).

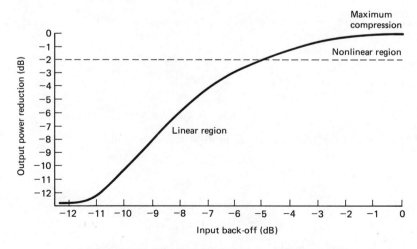

Figure 8-15 HPA input/output characteristic curve.

With PSK and QAM, the input baseband is generally a PCM-encoded, time-division-multiplexed signal which is digital in nature. Also, with PSK and QAM, several bits may be encoded in a single transmit signaling element (baud). Consequently, a parameter more meaningful than carrier power is *energy per bit* (E_b). Mathematically, E_b is

$$E_b = P_t T_b \tag{8-1a}$$

where

E_b = energy of a single bit (J/bit)
P_t = total carrier power (W)
T_b = time of a single bit (s)

or because $T_b = 1/F_b$, where F_b is the bit rate in bits per second.

$$E_b = \frac{P_t}{F_b} \tag{8-1b}$$

EXAMPLE 8-2

For a total transmit power (P_t) of 1000 W, determine the energy per bit (E_b) for a transmission rate of 50 Mbps.

Solution

$$T_b = \frac{1}{F_b} = \frac{1}{50 \times 10^6 \text{ bps}} = 0.02 \times 10^{-6} \text{ s}$$

(It appears that the units for T_b should be s/bit but the per bit is implied in the definition of T_b, time of bit.)

Substituting into Equation 8-1a yields

$$E_b = 1000 \text{ J/s} \, (0.02 \times 10^{-6} \text{ s/bit}) = 20 \, \mu\text{J}$$

(Again the units appear to be J/bit, but the per bit is implied in the definition of E_b, energy per bit.)

$$E_b = \frac{1000 \text{ J/s}}{50 \times 10^6 \text{ bps}} = 20 \, \mu\text{J}$$

Expressed as a log,

$$E_b = 10 \log(20 \times 10^{-6}) = -47 \text{ dBJ}$$

It is common to express P_t in dBW and E_b in dBW/bps. Thus

$$P_t = 10 \log 1000 = 30 \text{ dBW}$$

$$E_b = P_t - 10 \log F_b$$

$$= P_t - 10 \log(50 \times 10^6)$$

$$= 30 \text{ dBW} - 77 \text{ dB} = -47 \text{ dBW/bps}$$

or simply -47 dBW.

Effective Isotropic Radiated Power

Effective isotropic radiated power (EIRP) is defined as an equivalent transmit power and is expressed mathematically as

$$EIRP = P_r A_t$$

where

$EIRP$ = effective isotropic radiated power (W)
P_r = total power radiated from an antenna (W)
A_t = transmit antenna gain (W/W or a unitless ratio)

Expressed as a log,

$$EIRP \ (dBW) = P_r \ (dBW) + A_t \ (dB)$$

In respect to the transmitter output,

$$P_r = P_t - L_{bo} - L_{bf}$$

Thus

$$EIRP = P_t - L_{bo} - L_{bf} + A_t \qquad (8\text{-}2)$$

where

P_t = actual power output of the transmitter (dBW)
L_{bo} = back-off losses of HPA (dB)
L_{bf} = total branching and feeder loss (dB)
A_t = transmit antenna gain (dB)

EXAMPLE 8-3

For an earth station transmitter with an output power of 40 dBW (10,000 W), a back-off loss of 3 dB, a total branching and feeder loss of 3 dB, and a transmit antenna gain of 40 dB, determine the EIRP.

Solution   Substituting into Equation 8-2 yields

$$EIRP = P_t - L_{bo} - L_{bf} + A_t$$

$$= 40 \ dBW - 3 \ dB - 3dB + 40 \ dB = 74 \ dBW$$

Equivalent Noise Temperature

With terrestrial microwave systems, the noise introduced in a receiver or a component within a receiver was commonly specified by the parameter noise figure. In satellite communications systems, it is often necessary to differentiate or measure noise in increments as small as a tenth or a hundredth of a decibel. Noise figure, in its standard form, is inadequate for such precise calculations. Consequently, it is common to use *environmental temperature* (T) and *equivalent noise temperature* (T_e) when evaluating the performance of a satellite system. In Chapter 7 total noise power was expressed mathematically as

$$N = KTB$$

Rearranging and solving for T gives us

$$T = \frac{N}{KB}$$

where

> N = total noise power (W)
> K = Boltzmann's constant (J/K)
> B = bandwidth (Hz)
> T = temperature of the environment (K)

Again from Chapter 7 (Equation 7-7),

$$NF = 1 + \frac{T_e}{T}$$

where

> T_e = equivalent noise temperature (K)
> NF = noise figure (absolute value)
> T = temperature of the environment (K)

Rearranging Equation 7-7, we have

$$T_e = T(NF - 1)$$

Typically, equivalent noise temperatures of the receivers used in satellite transponders are about 1000 K. For earth station receivers T_e values are between 20 and 1000 K. Equivalent noise temperature is generally more useful when expressed logarithmically with the unit of dBK, as follows:

$$T_e \text{ (dBK)} = 10 \log T_e$$

For an equivalent noise temperature of 100 K, T_e (dBK) is

$$T_e \text{ (dBK)} = 10 \log 100 \text{ or } 20 \text{ dBK}$$

Equivalent noise temperature is a hypothetical value that can be calculated but cannot be measured. Equivalent noise temperature is often used rather than noise figure because it is a more accurate method of expressing the noise contributed by a device or a receiver when evaluating its performance. Essentially, equivalent noise temperature (T_e) is the noise present at the input to a device or amplifier plus the noise added internally by that device. This allows us to analyze the noise characteristics of a device by simply evaluating an equivalent input noise temperature. As you will see in subsequent discussions, T_e is a very useful parameter when evaluating the performance of a satellite system.

EXAMPLE 8-4

Convert noise figures of 4 and 4.01 to equivalent noise temperatures. Use 300 K for the environmental temperature.

Solution   Substituting into Equation 7-7 yields

$$T_e = T(\text{NF} - 1)$$

For NF = 4:

$$T_e = 300(4 - 1) = 900 \text{ K}$$

For NF = 4.01:

$$T_e = 300(4.01 - 1) = 903 \text{ K}$$

It can be seen that the 3° difference in the equivalent temperatures is 300 times as large as the difference between the two noise figures. Consequently, equivalent noise temperature is a more accurate way of comparing the noise performances of two receivers or devices.

Noise Density

Simply stated, *noise density* (N_o) is the total noise power normalized to a 1-Hz bandwidth, or the noise power present in a 1-Hz bandwidth. Mathematically, noise density is

$$N_o = \frac{N}{B} \quad \text{or} \quad KT_e \tag{8-3a}$$

where

N_o = noise density (W/Hz) (N_o is generally expressed as simply watts; the per hertz is implied in the definition of N_o)
N = total noise power (W)
B = bandwidth (Hz)
K = Boltzmann's constant (J/K)
T_e = equivalent noise temperature (K)

Expressed as a log,

$$N_o \text{ (dBW/Hz)} = 10 \log N - 10 \log B \tag{8-3b}$$

$$= 10 \log K + 10 \log T_e \tag{8-3c}$$

EXAMPLE 8-5

For an equivalent noise bandwidth of 10 MHz and a total noise power of 0.0276 pW, determine the noise density and equivalent noise temperature.

Solution   Substituting into Equation 8-3a, we have

$$N_o = \frac{N}{B} = \frac{276 \times 10^{-16}\,\text{W}}{10 \times 10^6\,\text{Hz}} = 276 \times 10^{-23}\,\frac{\text{W}}{\text{Hz}}$$

or simply, 276×10^{-23} W.

$$N_o = 10 \log (276 \times 10^{-23}) = -205.6\,\text{dBW/Hz}$$

or simply -205.6 dBW. Substituting into Equation 8-3b gives us

$$N_o = N\,(\text{dBW}) - B\,(\text{dB/Hz})$$

$$= -135.6\,\text{dBW} - 70\,(\text{dB/Hz}) = -205.6\,\text{dBW}$$

Rearranging Equation 8-3a and solving for equivalent noise temperature yields

$$T_e = \frac{N_o}{K}$$

$$= \frac{276 \times 10^{-23}\,\text{J/cycle}}{1.38 \times 10^{-23}\,\text{J/K}} = 200°\,\text{K/cycle}$$

$$= 10 \log 200 = 23\,\text{dBK}$$

$$= N_o\,(\text{dBW}) - 10 \log K$$

$$= -205.6\,\text{dBW} - (-228.6\,\text{dBWK}) = 23\,\text{dBK}$$

Carrier-to-Noise Density Ratio

C/N_o is the average wideband carrier power-to-noise density ratio. The *wideband carrier power* is the combined power of the carrier and its associated sidebands. The noise is the thermal noise present in a normalized 1-Hz bandwidth. The carrier-to-noise density ratio may also be written as a function of noise temperature. Mathematically, C/N_o is

$$\frac{C}{N_o} = \frac{C}{KT_e} \tag{8-4a}$$

Expressed as a log,

$$\frac{C}{N_o}\,(\text{dB}) = C\,(\text{dBW}) - N_o\,(\text{dBW}) \tag{8-4b}$$

Energy of Bit-to-Noise Density Ratio

E_b/N_o is one of the most important and most often used parameters when evaluating a digital radio system. The E_b/N_o ratio is a convenient way to compare digital systems that use different transmission rates, modulation schemes, or encoding techniques. Mathematically, E_b/N_o is

$$\frac{E_b}{N_o} = \frac{C/F_b}{N/B} = \frac{CB}{NF_b} \tag{8-5}$$

E_b/N_o is a convenient term used for digital system calculations and performance comparisons, but in the real world, it is more convenient to measure the wideband carrier power-to-noise density ratio and convert it to E_b/N_o. Rearranging Equation 8-5 yields the following expression:

$$\frac{E_b}{N_o} = \frac{C}{N} \times \frac{B}{F_b}$$

The E_b/N_o ratio is the product of the carrier-to-noise ratio (C/N) and the noise bandwidth-to-bit ratio (B/F_b). Expressed as a log,

$$\frac{E_b}{N_o}\,(\text{dB}) = \frac{C}{N}\,(\text{dB}) + \frac{B}{F_b}\,(\text{dB}) \tag{8-6}$$

The energy per bit (E_b) will remain constant as long as the total wideband carrier power (C) and the transmission rate (bps) remain unchanged. Also, the noise density (N_o) will remain constant as long as the noise temperature remains constant. The following conclusion can be made: For a given carrier power, bit rate, and noise temperature, the E_b/N_o ratio will remain constant regardless of the encoding technique, modulation scheme, or bandwidth used.

Figure 8-16 graphically illustrates the relationship between an expected probability of error $P(e)$ and the minimum C/N ratio required to achieve the $P(e)$. The C/N specified is for the minimum double-sided Nyquist bandwidth. Figure 8-17 graphically illustrates the relationship between an expected $P(e)$ and the minimum E_b/N_o ratio required to achieve that $P(e)$.

A $P(e)$ of 10^{-5} ($1/10^5$) indicates a probability that 1 bit will be in error for every 100,000 bits transmitted. $P(e)$ is analogous to the bit error rate (BER).

EXAMPLE 8-6

A coherent binary phase-shift-keyed (BPSK) transmitter operates at a bit rate of 20 Mbps. For a probability of error $P(e)$ of 10^{-4}:

(a) Determine the minimum theoretical C/N and E_b/N_o ratios for a receiver bandwidth equal to the minimum double-sided Nyquist bandwidth.

(b) Determine the C/N if the noise is measured at a point prior to the bandpass filter, where the bandwidth is equal to twice the Nyquist bandwidth.

(c) Determine the C/N if the noise is measured at a point prior to the bandpass filter where the bandwidth is equal to three times the Nyquist bandwidth.

Solution (a) With BPSK, the minimum bandwidth is equal to the bit rate, 20 MHz. From Figure 8-16, the minimum C/N is 8.8 dB. Substituting into Equation 8-6 gives us

$$\frac{E_b}{N_o}\,(\text{dB}) = \frac{C}{N}\,(\text{dB}) + \frac{B}{F_b}\,(\text{dB})$$

$$= 8.8\ \text{dB} + 10\log\frac{20 \times 10^6}{20 \times 10^6}$$

$$= 8.8\ \text{dB} + 0\ \text{dB} = 8.8\ \text{db}$$

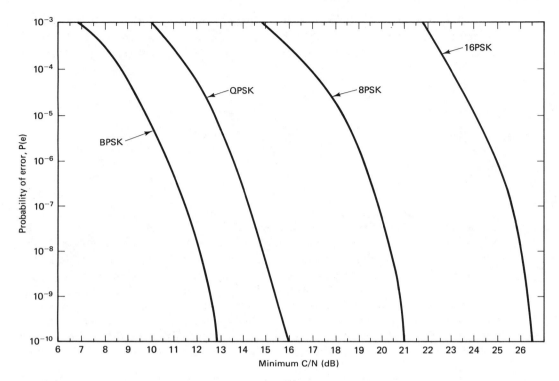

Figure 8-16 Probability of error $P(e)$ versus C/N for various digital modulation schemes. (Bandwidth equals minimum double-sided Nyquist bandwidth.)

Note: The minimum E_b/N_o equals the minimum C/N when the receiver noise bandwidth equals the minimum Nyquist bandwidth. The minimum E_b/N_o of 8.8 can be verified from Figure 8-17.

What effect does increasing the noise bandwidth have on the minimum C/N and E_b/N_o ratios? The wideband carrier power is totally independent of the noise bandwidth. Similarly, an increase in the bandwidth causes a corresponding increase in the noise power. Consequently, a decrease in C/N is realized that is directly proportional to the increase in the noise bandwidth. E_b is dependent on the wideband carrier power and the bit rate only. Therefore, E_b is unaffected by an increase in the noise bandwidth. N_o is the noise power normalized to a 1-Hz bandwidth and, consequently, is also unaffected by an increase in the noise bandwidth.

(b) Since E_b/N_o is independent of bandwidth, measuring the C/N at a point in the receiver where the bandwidth is equal to twice the minimum Nyquist bandwidth has absolutely no effect on E_b/N_o. Therefore, E_b/N_o becomes the constant in Equation 8-6 and is used to solve for the new value of C/N. Rearranging Equation 8-6 and using the calculated E_b/N_o ratio, we have

$$\frac{C}{N}\,(\text{dB}) = \frac{E_b}{N_o}\,(\text{dB}) - \frac{B}{F_b}\,(\text{dB})$$

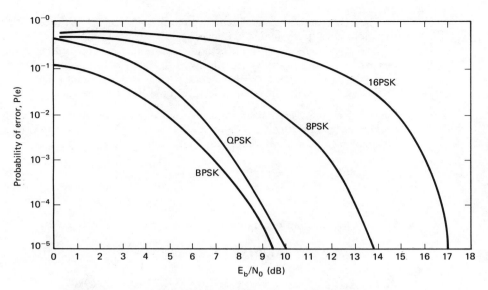

Figure 8-17 Probability of error $P(e)$ versus E_b/N_0 ratio for various digital modulation schemes.

$$= 8.8 \text{ dB} - 10 \log \frac{40 \times 10^6}{20 \times 10^6}$$

$$= 8.8 \text{ dB} - 10 \log 2$$

$$= 8.8 \text{ dB} - 3 \text{ dB} = 5.8 \text{ dB}$$

(c) Measuring the C/N ratio at a point in the receiver where the bandwidth equals three times the minimum bandwidth yields the following results for C/N.

$$\frac{C}{N} = \frac{E_b}{N_0} - 10 \log \frac{60 \times 10^6}{20 \times 10^6}$$

$$= 8.8 \text{ dB} - 10 \log 3$$

$$= 4.03 \text{ dB}$$

The C/N ratios of 8.8, 5.8, and 4.03 dB indicate the C/N ratios that would be measured at the three specified points in the receiver to achieve the desired minimum E_b/N_0 and $P(e)$.

Because E_b/N_0 cannot be directly measured to determine the E_b/N_0 ratio, the wideband carrier-to-noise ratio is measured and then substituted into Equation 8-6. Consequently, to accurately determine the E_b/N_0 ratio, the noise bandwidth of the receiver must be known.

EXAMPLE 8-7

A coherent 8PSK transmitter operates at a bit rate of 90 Mbps. For a probability of error of 10^{-5}:

(a) Determine the minimum theoretical C/N and E_b/N_o ratios for a receiver bandwidth equal to the minimum double-sided Nyquist bandwidth.

(b) Determine the C/N if the noise is measured at a point prior to the bandpass filter where the bandwidth is equal to twice the Nyquist bandwidth.

(c) Determine the C/N if the noise is measured at a point prior to the bandpass filter where the bandwidth is equal to three times the Nyquist bandwidth.

Solution (a) 8PSK has a bandwidth efficiency of 3 bps/Hz and, consequently, requires a minimum bandwidth of one-third the bit rate or 30 MHz. From Figure 8-16, the minimum C/N is 18.5 dB. Substituting into Equation 8-6, we obtain

$$\frac{E_b}{N_o} \text{ (dB)} = 18.5 \text{ dB} + 10 \log \frac{30 \text{ MHz}}{90 \text{ Mbps}}$$

$$= 18.5 \text{ dB} + (-4.8 \text{ dB}) = 13.7 \text{ db}$$

(b) Rearranging Equation 8-6 and substituting in E_b/N_o yields

$$\frac{C}{N} \text{ (dB)} = 13.7 \text{ dB} - 10 \log \frac{60 \text{ MHz}}{90 \text{ Mbps}}$$

$$= 13.7 \text{ dB} - (-1.77 \text{ dB}) = 15.47 \text{ dB}$$

(c) Again, rearranging Equation 8-6 and substituting in E_b/N_o gives us

$$\frac{C}{N} \text{ (dB)} = 13.7 \text{ (dB)} - 10 \log \frac{90 \text{ MHz}}{90 \text{ Mbps}}$$

$$= 13.7 \text{ dB} - 0 \text{ dB} = 13.7 \text{ dB}$$

It should be evident from Examples 8-6 and 8-7 that the E_b/N_o and C/N ratios are equal only when the noise bandwidth is equal to the bit rate. Also, as the bandwidth at the point of measurement increases, the C/N decreases.

When the modulation scheme, bit rate, bandwidth, and C/N ratios of two digital radio systems are different; it is often difficult to determine which system has the lower probability of error. Because E_b/N_o is independent of bit rate, bandwidth, and modulation scheme; it is a convenient common denominator to use for comparing the probability of error performance of two digital radio systems.

EXAMPLE 8-8

Compare the performance characteristics of the two digital systems listed below, and determine which system has the lower probability of error.

	QPSK	8PSK
Bit rate	40 Mbps	60 Mbps
Bandwidth	1.5 × minimum	2 × minimum
C/N	10.75 dB	13.76 dB

Solution Substituting into Equation 8-6 for the QPSK system gives us

$$\frac{E_b}{N_o}(\text{dB}) = \frac{C}{N}(\text{dB}) + 10\log\frac{B}{F_b}$$

$$= 10.75\text{ dB} + 10\log\frac{1.5 \times 20\text{ MHz}}{40\text{ Mbps}}$$

$$= 10.75\text{ dB} + (-1.25\text{ dB})$$

$$= 9.5\text{ dB}$$

From Figure 8-17, the $P(e)$ is 10^{-4}.

Substituting into Equation 8-6 for the 8PSK system gives us

$$\frac{E_b}{N_o}(\text{dB}) = 13.76\text{ dB} + 10\log\frac{2 \times 20\text{ MHz}}{60\text{ Mbps}}$$

$$= 13.76\text{ dB} + (-1.76\text{ dB})$$

$$= 12\text{ dB}$$

From Figure 8-17, the $P(e)$ is 10^{-3}.

Although the QPSK system has a lower C/N and E_b/N_o ratio, the $P(e)$ of the QPSK system is 10 times lower (better) than the 8PSK system.

Gain-to-Equivalent Noise Temperature Ratio

Essentially, *gain-to-equivalent noise temperature ratio* (G/T_e) is a figure of merit used to represent the quality of a satellite or an earth station receiver. The G/T_e of a receiver is the ratio of the receive antenna gain to the equivalent noise temperature (T_e) of the receiver. Because of the extremely small receive carrier powers typically experienced with satellite systems, very often an LNA is physically located at the feedpoint of the antenna. When this is the case, G/T_e is a ratio of the gain of the receiving antenna plus the gain of the LNA to the equivalent noise temperature. Mathematically, gain-to-equivalent noise temperature ratio is

$$\frac{G}{T_e} = \frac{A_r + A(\text{LNA})}{T_e} \tag{8-7}$$

Expressed in logs, we have

$$\frac{G}{T_e}(\text{dBK}^{-1}) = A_r(\text{dB}) + A(\text{LNA})(\text{dB}) - T_e(\text{dBK}^{-1}) \tag{8-8}$$

G/T_e is a very useful parameter for determining the E_b/N_0 and C/N ratios at the satellite transponder and earth station receivers. G/T_e is essentially the only parameter required at a satellite or an earth station receiver when completing a link budget.

EXAMPLE 8-9

For a satellite transponder with a receiver antenna gain of 22 dB, an LNA gain of 10 dB, and an equivalent noise temperature of 22 dBK^{-1}; determine the G/T_e figure of merit.

Solution   Substituting into Equation 8-8 yields

$$\frac{G}{T_e} \text{ (dBK}^{-1}) = 22 \text{ dB} + 10 \text{ dB} - 22 \text{ dBK}^{-1}$$

$$= 10 \text{ dBK}^{-1}$$

SATELLITE SYSTEM LINK EQUATIONS

The error performance of a digital satellite system is quite predictable. Figure 8-18 shows a simplified block diagram of a digital satellite system and identifies the various gains and losses that may affect the system performance. When evaluating the performance of a digital satellite system, the uplink and downlink parameters are first considered separately, then the overall performance is determined by combining them in the appropriate manner. Keep in mind, a digital microwave or satellite radio simply means the original and demodulated baseband signals are digital in nature. The RF portion of the radio is analog; that is, FSK, PSK, QAM, or some other higher-level modulation riding on an analog microwave carrier.

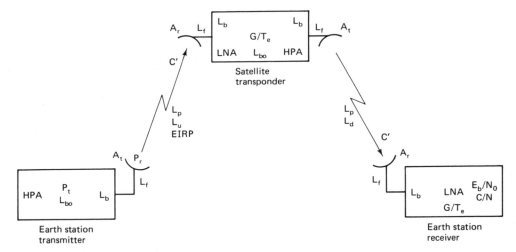

Figure 8-18   Overall satellite system showing the gains and losses incurred in both the uplink and downlink sections. HPA, High-power amplifier; P_t, HPA output power; L_{bo}, back-off loss; L_f, feeder loss; L_b, branching loss; A_t, transmit antenna gain; P_r, total radiated power = $P_t - L_{bo} - L_b - L_f$; EIRP, effective isotropic radiated power = $P_r A_t$; L_u, additional uplink losses due to atmosphere; L_p, path loss; A_r, receive antenna gain; G/T_e, gain-to-equivalent noise ratio; L_d, additional downlink losses due to atmosphere; LNA, low-noise amplifier; C/T_e, carrier-to-equivalent noise ratio; C/N_o, carrier-to-noise density ratio; E_b/N_o, energy of bit-to-noise density ratio; C/N, carrier-to-noise ratio.

LINK EQUATIONS

The following *link equations* are used to separately analyze the uplink and the downlink sections of a single radio-frequency carrier satellite system. These equations consider only the ideal gains and losses and effects of thermal noise associated with the earth station transmitter, earth station receiver, and the satellite transponder. The nonideal aspects of the system are discussed later in this chapter.

Uplink Equation

$$\frac{C}{N_0} = \frac{A_t P_r (L_p L_u) A_r}{K T_e} = \frac{A_t P_r (L_p L_u)}{K} \times \frac{G}{T_e}$$

where L_d and L_u are the additional uplink and downlink atmospheric losses, respectively. The uplink and downlink signals must pass through the earth's atmosphere, where they are partially absorbed by the moisture, oxygen, and particulates in the air. Depending on the elevation angle, the distance the RF signal travels through the atmosphere varies from one earth station to another. Because L_u and L_d represent losses, they are decimal values less than 1. G/T_e is the receiving antenna gain divided by the equivalent input noise temperature.

Expressed as a log,

$$\frac{C}{N_0} = \underbrace{10 \log A_t P_r}_{\substack{\text{EIRP} \\ \text{earth} \\ \text{station}}} - \underbrace{20 \log \left(\frac{4\pi D}{\lambda}\right)}_{\substack{\text{free-space} \\ \text{path loss}}} + \underbrace{10 \log \left(\frac{G}{T_e}\right)}_{\substack{\text{satellite} \\ G/T_e}} - \underbrace{10 \log L_u}_{\substack{\text{additional} \\ \text{atmospheric} \\ \text{losses}}} - \underbrace{10 \log K}_{\substack{\text{Boltzmann's} \\ \text{constant}}}$$

$$= \text{EIRP (dBW)} - L_p \text{ (dB)} + \frac{G}{T_e} \text{ (dBK}^{-1}) - L_u \text{ (dB)} - K \text{ (dBWK)}$$

Downlink Equation

$$\frac{C}{N_0} = \frac{A_t P_r (L_p L_d) A_r}{K T_e} = \frac{A_t A_r (L_p L_d)}{K} \times \frac{G}{T_e}$$

Expressed as a log

$$\frac{C}{N_0} = \underbrace{10 \log A_t P_r}_{\substack{\text{EIRP} \\ \text{satellite}}} - \underbrace{20 \log \left(\frac{4\pi D}{\lambda}\right)}_{\substack{\text{free-space} \\ \text{path loss}}} + \underbrace{10 \log \frac{G}{T_e}}_{\substack{\text{earth station} \\ G/T_e}} - \underbrace{10 \log L_d}_{\substack{\text{additional} \\ \text{atmospheric} \\ \text{losses}}} - \underbrace{10 \log K}_{\substack{\text{Boltzmann's} \\ \text{constant}}}$$

$$= \text{EIRP (dBW)} - L_p \text{ (dB)} + \frac{G}{T_e} \text{ (dBK}^{-1}) - L_d \text{ (dB)} - K \text{ (dBWK)}$$

LINK BUDGET

Table 8-4 lists the system parameters for three typical satellite communication systems. The systems and their parameters are not necessarily for an existing or future system; they are hypothetical examples only. The system parameters are used to construct a *link budget*. A link budget identifies the system parameters and is used to determine the projected C/N and E_b/N_0 ratios at both the satellite and earth station receivers for a given modulation scheme and desired $P(e)$.

TABLE 8-4　SYSTEM PARAMETERS FOR THREE HYPOTHETICAL SATELLITE SYSTEMS

	System A: 6/4 GHz, earth coverage QPSK modulation, 60 Mbps	System B: 14/12 GHz, earth coverage 8PSK modulation, 90 Mbps	System C: 14/12 GHz, earth coverage 8PSK modulation, 120 Mbps
Uplink			
Transmitter output power (saturation, dBW)	35	25	33
Earth station back-off loss (dB)	2	2	3
Earth station branching and feeder loss (dB)	3	3	4
Additional atmospheric (dB)	0.6	0.4	0.6
Earth station antenna gain (dB)	55	46	64
Free-space path loss (dB)	200	208	206.5
Satellite receive antenna gain (dB)	20	46	23.7
Satellite branching and feeder loss (dB)	1	1	0
Satellite equivalent noise temperature (K)	1000	800	800
Satellite G/T_e (dBK^{-1})	−10	16	−5.3
Downlink			
Transmitter output power (saturation, dBW)	18	20	30.8
Satellite back-off loss (dB)	0.5	0.2	0.1
Satellite branching and feeder loss (dB)	1	1	0.5
Additional atmospheric loss (dB)	0.8	1.4	0.4
Satellite antenna gain (dB)	16	44	10
Free-space path loss (dB)	197	206	205.6
Earth station receive antenna gain (dB)	51	44	62
Earth station branching and feeder loss (dB)	3	3	0
Earth station equivalent noise temperature (K)	250	1000	270
Earth station G/T_e (dBK^{-1})	27	14	37.7

EXAMPLE 8-10

Complete the link budget for a satellite system with the following parameters.

Uplink

1. Earth station transmitter output power at saturation, 2000 W   33 dBW

2. Earth station back-off loss   3 dB

3. Earth station branching and feeder losses   4 dB

4. Earth station transmit antenna gain (from Figure 8-19, 15 m at 14 GHz)   64 dB

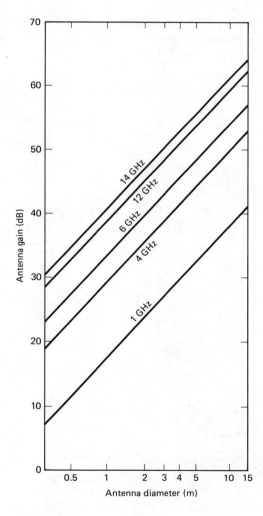

Figure 8-19 Antenna gain based on the gain equation for a parabolic antenna:

$$A \ (\text{dB}) = 10 \log \eta \ (\pi D/\lambda)^2$$

where D is the antenna diameter, $\lambda =$ the wavelength, and $\eta =$ the antenna efficiency. Here $\eta = 0.55$. To correct for a 100% efficient antenna, add 2.66 dB to the value.

5. Additional uplink atmospheric losses   0.6 dB
6. Free-space path loss (from Figure 8-20,   206.5 db
   at 14 Ghz)
7. Satellite receiver G/T_e ratio -5.3 dBK^{-1}
8. Satellite branching and feeder losses   0 dB
9. Bit rate   120 Mbps
10. Modulation scheme   8 PSK

Downlink

1. Satellite transmitter output power at satu-   10 dBW
   ration 10 W
2. Satellite back-off loss   0.1 dB
3. Satellite branching and feeder losses   0.5 dB
4. Satellite transmit antenna gain (from Fig-   30.8 dB
   ure 8-19, 0.37 m at 12 GHz)

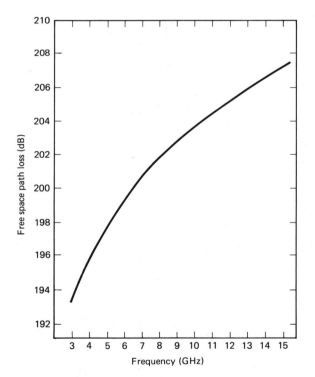

Angle	+dB
90°	0
45°	0.44
0°	1.33

Elevation angle correction:

Figure 8-20 Free-space path loss (L_p) determined from

$$L_p = 183.5 + 20 \log F \text{ (GHz)}$$

Elevation angle = 90°, distance = 35,930 km.

5. Additional downlink atmospheric losses   0.4 dB

6. Free-space path loss (from Figure 8-20,   205.6 dB
   at 12 GHz)

7. Earth station receive antenna gain (15 m,   62 dB
   12 GHz)

8. Earth station branching and feeder losses   0 dB

9. Earth station equivalent noise temperature   270 K

10. Earth station G/T_e ratio   37.7 dBK^{-1}

11. Bit rate   120 Mbps

12. Modulation scheme   8 PSK

Solution   *Uplink budget*: Expressed as a log,

$$\text{EIRP (earth station)} = P_t + A_t - L_{bo} - L_{bf}$$

$$= 33 \text{ dBW} + 64 \text{ dB} - 3 \text{ dB} - 4 \text{ dB} = 90 \text{ dBW}$$

Carrier power density at the satellite antenna:

$$C' = \text{EIRP (earth station)} - L_p - L_u$$

$$= 90 \text{ dBw} - 206.5 \text{ dB} - 0.6 \text{ dB} = -117.1 \text{ dBW}$$

C/N_0 at the satellite:

$$\frac{C}{N_0} = \frac{C}{KT_e} = \frac{C}{T_e} \times \frac{1}{K} \qquad \text{where } \frac{C}{T_e} = C' \times \frac{G}{T_e}$$

Thus

$$\frac{C}{N_0} = C' \times \frac{G}{T_e} \times \frac{1}{K}$$

Expressed as a log,

$$\frac{C}{N_0} \text{ (dB)} = C' \text{ (dBW)} + \frac{G}{T_e} \text{ (dBK}^{-1}) - 10 \log (1.38 \times 10^{-23})$$

$$\frac{C}{N_0} = -117.1 \text{ dBW} + (-5.3 \text{ dBK}^{-1}) - (-228.6 \text{ dBWK}) = 106.2 \text{ dB}$$

Thus

$$\frac{E_b}{N_0} \text{ (dB)} = \frac{C/F_b}{N_0} \text{ (dB)} = \frac{C}{N_0} \text{ (dB)} - 10 \log F_b$$

$$\frac{E_b}{N_0} = 106.2 \text{ dB} - 10 (\log 120 \times 10^6) = 25.4 \text{ dB}$$

and for a minimum bandwidth system,

$$\frac{C}{N} = \frac{E_b}{N_0} - \frac{B}{F_b} = 25.4 - 10 \log \frac{40 \times 10^6}{120 \times 10^6} = 30.2 \text{ dB}$$

Downlink budget: Expressed as a log,

$$\text{EIRP (satellite transponder)} = P_t + A_t - L_{bo} - L_{bf}$$

$$= 10 \text{ dBW} + 30.8 \text{ dB} - 0.1 \text{ dB} - 0.5 \text{ dB}$$

$$= 40.2 \text{ dBW}$$

Carrier power density at earth station antenna:

$$C' = \text{EIRP (dBW)} - L_p \text{ (dB)} - L_d \text{ (dB)}$$

$$= 40.2 \text{ dBW} - 205.6 \text{ dB} - 0.4 \text{ dB} = -165.8 \text{ dBW}$$

C/N_0 at the earth station receiver:

$$\frac{C}{N_0} = \frac{C}{KT_e} = \frac{C}{T_e} \times \frac{1}{K} \qquad \text{where } \frac{C}{T_e} = C' \times \frac{G}{T_e}$$

Thus

$$\frac{C}{N_0} = C' \times \frac{G}{T_e} \times \frac{1}{K}$$

Expressed as a log,

$$\frac{C}{N_0} \text{ (dB)} = C' \text{ (dBW)} + \frac{G}{T_e} \text{ (dBK}^{-1}) - 10 \log(1.38 \times 10^{-23})$$

$$= -165.8 \text{ dBW} + (37.7 \text{ dBK}^{-1}) - (-228.6 \text{ dBWK}) = 100.5 \text{ dB}$$

An alternative method of solving for C/N_0 is

$$\frac{C}{N_0} \text{ (dB)} = C' \text{ (dBW)} + A_r \text{ (dB)} - T_e \text{ (dBk}^{-1}) - K \text{ (dBWK)}$$

$$= -165.8 \text{ dBW} + 62 \text{ dB} - 10 \log 270 - (-228.6 \text{ dBWK})$$

$$\frac{C}{N_0} = -165.8 \text{ dBW} + 62 \text{ dB} - 24.3 \text{ dBK}^{-1} + 228.6 \text{ dBWK} = 100.5 \text{ dB}$$

$$\frac{E_b}{N_0} \text{ (dB)} = \frac{C}{N_0} \text{ (dB)} - 10 \log F_b$$

$$= 100.5 \text{ dB} - 10 \log (120 \times 10^6)$$

$$= 100.5 \text{ dB} - 80.8 \text{ dB} = 19.7 \text{ dB}$$

and for a minimum bandwidth system,

$$\frac{C}{N} = \frac{E_b}{N_0} - \frac{B}{F_b} = 19.7 - 10 \log \frac{40 \times 10^6}{120 \times 10^6} = 24.5 \text{ dB}$$

With careful analysis and a little algebra, it can be shown that the overall energy of bit-to-noise density ratio (E_b/N_0), which includes the combined effects of the uplink ratio $(E_b/N_0)_u$ and the downlink ratio $(E_b/N_0)_d$, is a standard product over the sum relationship and is expressed mathematically as

$$\frac{E_b}{N_o} \text{(overall)} = \frac{(E_b/N_o)_u \, (E_b/N_o)_d}{(E_b/N_o)_u + (E_b/N_o)_d} \tag{8-9}$$

where all E_b/N_o ratios are in absolute values. For Example 8-10, the overall E_b/N_o ratio is

$$\frac{E_b}{N_o} \text{(overall)} = \frac{(346.7)(93.3)}{346.7 + 93.3} = 73.5$$

$$= 10 \log 73.5 = 18.7 \text{ dB}$$

TABLE 8-5   LINK BUDGET FOR EXAMPLE 8-10

Uplink

1. Earth station transmitter output power at saturation, 2000 W	33 dBW
2. Earth station back-off loss	3 dB
3. Earth station branching and feeder losses	4 dB
4. Earth station transmit antenna gain	64 dB
5. Earth station EIRP	90 dBW
6. Additional uplink atmospheric losses	0.6 dB
7. Free-space path loss	206.5 dB
8. Carrier power density at satellite	-117.1 dBW
9. Satellite branching and feeder losses	0 dB
10. Satellite G/T_e ratio	-5.3 dBK^{-1}
11. Satellite C/T_e ratio	-122.4 dBWK^{-1}
12. Satellite C/N_0 ratio	106.2 dB
13. Satellite C/N ratio	30.2 dB
14. Satellite E_b/N_0 ratio	25.4 dB
15. Bit rate	120 Mbps
16. Modulation scheme	8 PSK

Downlink

1. Satellite transmitter output power at saturation, 10 W	10 dBW
2. Satellite back-off loss	0.1 dB
3. Satellite branching and feeder losses	0.5 dB
4. Satellite transmit antenna gain	30.8 dB
5. Satellite EIRP	40.2 dBW
6. Additional downlink atmospheric losses	0.4 dB
7. Free-space path loss	205.6 dB
8. Earth station receive antenna gain	62 dB
9. Earth station equivalent noise temperature	270 K
10. Earth station branching and feeder losses	0 dB
11. Earth station G/T_e ratio	37.7 dBK^{-1}
12. Carrier power density at earth station	-165.8 dBW
13. Earth station C/T_e ratio	-128.1 dBWK^{-1}
14. Earth station C/N_0 ratio	100.5 dB
15. Earth station C/N ratio	24.5 dB
16. Earth station E_b/N_0 ratio	19.7 dB
17. Bit rate	120 Mbps
18. Modulation scheme	8 PSK

As with all product-over-sum relationships, the smaller of the two numbers dominates. If one number is substantially smaller than the other, the overall result is approximately equal to the smaller of the two numbers.

The system parameters used for Example 8-10 were taken from system C in Table 8-4. A complete link budget for the system is shown in Table 8-5.

NONIDEAL SYSTEM PARAMETERS

Additional *nonideal parameters* include the following impairments: AM/AM conversion and AM/PM conversion, which result from nonlinear amplification in HPAs and limiters; *pointing error*, which occurs when the earth station and satellite antennas are not exactly aligned; *phase jitter*, which results from imperfect carrier recovery in receivers; *nonideal filtering*, due to the imperfections introduced in bandpass filters; *timing error*, due to imperfect clock recovery in receivers; and *frequency translation errors* introduced in the satellite transponders. The degradation caused by the preceding impairments effectively reduces the E_b/N_o ratios determined in the link budget calculations. Consequently, they have to be included in the link budget as equivalent losses. An in-depth coverage of the nonideal parameters is beyond the intent of this text.

QUESTIONS

8-1. Briefly describe a satellite.

8-2. What is a passive satellite? An active satellite?

8-3. Contrast nonsynchronous and synchronous satellites.

8-4. Define *prograde* and *retrograde*.

8-5. Define *apogee* and *perigee*.

8-6. Briefly explain the characteristics of low-, medium-, and high-altitude satellite orbits.

8-7. Explain equatorial, polar, and inclined orbits.

8-8. Contrast the advantages and disadvantages of geosynchronous satellites.

8-9. Define *look angles*, *angle of elevation*, and *azimuth*.

8-10. Define *satellite spatial separation* and list its restrictions.

8-11. Describe a "footprint."

8-12. Describe spot, zonal, and earth coverage radiation patterns.

8-13. Explain *reuse*.

8-14. Briefly describe the functional characteristics of an uplink, a transponder, and a downlink model for a satellite system.

8-15. Define *back-off loss* and its relationship to saturated and transmit power.

8-16. Define *bit energy*.

8-17. Define *effective isotropic radiated power*.

8-18. Define *equivalent noise temperature*.

8-19. Define *noise density*.

8-20. Define *carrier-to-noise density ratio* and *energy of bit-to-noise density ratio*.

8-21. Define *gain-to-equivalent noise temperature ratio*.

8-22. Describe what a satellite link budget is and how it is used.

PROBLEMS

8-1. An earth station is located at Houston, Texas, which has a longitude of 99.5° and a latitude of 29.5° north. The satellite of interest is Satcom 2. Determine the look angles for the earth station antenna.

8-2. A satellite system operates at 14-GHz uplink and 11-GHz downlink and has a projected $P(e)$ of 10^{-7}. The modulation scheme is 8PSK, and the system will carry 120 Mbps. The equivalent noise temperature of the receiver is 400 K, and the receiver noise bandwidth is equal to the minimum Nyquist frequency. Determine the following parameters: minimum theoretical C/N ratio, minimum theoretical E_b/N_o ratio, noise density, total receiver input noise, minimum receive carrier power, and the minimum energy per bit at the receiver input.

8-3. A satellite system operates at 6-GHz uplink and 4-GHz downlink and has a projected $P(e)$ of 10^{-6}. The modulation scheme is QPSK and the system will carry 100 Mbps. The equivalent receiver noise temperature is 290 K, and the receiver noise bandwidth is equal to the minimum Nyquist frequency. Determine the following:
  (a) The C/N ratio that would be measured at a point in the receiver prior to the BPF where the bandwidth is equal to $1\frac{1}{2}$ times the minimum Nyquist frequency.
  (b) The C/N ratio that would be measured at a point in the receiver prior to the BPF where the bandwidth is equal to 3 times the minimum Nyquist frequency.

8-4. Which system has the best projected BER?
  (a) 8QAM, $C/N = 15$ dB, $B = 2F_N$, $F_b = 60$ Mbps.
  (b) QPSK, $C/N = 16$ dB, $B = F_N$, $F_b = 40$ Mbps.

8-5. An earth station satellite transmitter has an HPA with a rated saturated output power of 10,000 W. The back-off ratio is 6 dB, the branching loss is 2 dB, the feeder loss is 4 dB, and the antenna gain is 40 dB. Determine the actual radiated power and the EIRP.

8-6. Determine the total noise power for a receiver with an input bandwidth of 20 MHz and an equivalent noise temperature of 600 K.

8-7. Determine the noise density for Problem 8-6.

8-8. Determine the minimum C/N ratio required to achieve a $P(e)$ of 10^{-5} for an 8PSK receiver with a bandwidth equal to F_N.

8-9. Determine the energy of bit-to-noise density ratio when the receiver input carrier power is -100 dBW, the receiver input noise temperature is 290 K, and a 60-Mbps transmission rate is used.

8-10. Determine the carrier-to-noise density ratio for a receiver with a -70-dBW input carrier power, an equivalent noise temperature of 180 K, and a bandwidth of 20 MHz.

8-11. Determine the minimum C/N ratio for an 8PSK system when the transmission rate is 60 Mbps, the minimum energy of bit-to-noise density ratio is 15 dB, and the receiver bandwidth is equal to the minimum Nyquist frequency.

8-12. For an earth station receiver with an equivalent input temperature of 200 K, a noise bandwidth of 20 MHz, a receive antenna gain of 50 dB, and a carrier frequency of 12 GHz, determine the following: G/T_e, N_o, and N.

8-13. For a satellite with an uplink E_b/N_o of 14 dB and a downlink E_b/N_o of 18 dB, determine the overall E_b/N_o ratio.

8-14. Complete the following link budget:

Uplink Parameters

1. Earth station transmitter output power at saturation, 1 kW
2. Earth station back-off loss, 3 dB
3. Earth station total branching and feeder losses, 3 dB
4. Earth station transmit antenna gain for a 10-m parabolic dish at 14 GHz
5. Free-space path loss for 14 GHz
6. Additional uplink losses due to the earth's atmosphere, 0.8 dB
7. Satellite transponder G/Te, −4.6 dBk
8. Transmission bit rate, 90 Mbps, 8 PSK

Downlink Parameters

1. Satellite transmitter output power at saturation, 10 W
2. Satellite station transmit antenna gain for a 0.5-m parabolic dish at 12 GHz
3. Satellite modulation back-off loss, 0.8 dB
4. Free-space path loss for 12 GHz
5. Additional downlink losses due to earth's atmosphere, 0.6 dB
6. Earth station receive antenna gain for a 10-m parabolic dish at 12 GHz
7. Earth station equivalent noise temperature, 200 K
8. Earth station branching and feeder losses, 0 dB
9. Transmission bit rate, 90 Mbps, 8 PSK

Chapter 9

SATELLITE MULTIPLE-ACCESS ARRANGEMENTS

FDM/FM SATELLITE SYSTEMS

Figure 9-1a shows a single-link (two earth stations) *fixed-frequency* FDM/FM system using a single satellite transponder. With earth coverage antennas and for full-duplex operation, each link requires two RF satellite channels (i.e., four RF carrier frequencies, two uplink and two downlink). In Figure 9-1a, earth station 1 transmits on a high-band carrier (F11, F12, F13, etc.) and receives on a low-band carrier (F1, F2, F3, etc.). To avoid interfering with earth station 1, earth station 2 must transmit and receive on different RF carrier frequencies. The RF carrier frequencies are fixed and the satellite transponder is simply an RF-to-RF repeater that provides the uplink/downlink frequency translation. This arrangement is economically impractical and extremely inefficient as well. Additional earth stations can communicate through different transponders within the same satellite structure (Figure 9-1b). Each additional link requires four more RF carrier frequencies. It is unlikely that any two-point link would require the capacity available in an entire RF satellite channel. Consequently, most of the available bandwidth is wasted. Also, with this arrangement, each earth station can communicate with only one other earth station. The RF satellite channels are fixed between any two earth stations; thus the voice band channels from each earth station are dedicated to a single destination.

In a system where three or more earth stations wish to communicate with each other, fixed-frequency or *dedicated channel* systems such as those shown in Figure 8-1 are inadequate; a method of *multiple accessing* is required. That is, each earth station using the satellite system has a means of communicating with each of the other earth stations in the system through a common satellite transponder. Multiple

305

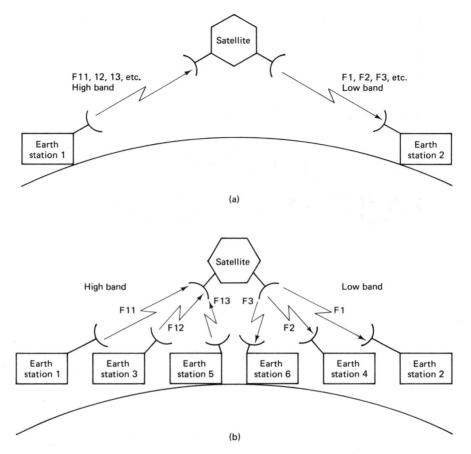

Figure 9-1   Fixed-frequency earth station satellite system: (a) single link; (b) multiple link.

accessing is sometimes called *multiple destination* because the transmissions from each earth station are received by all the other earth stations in the system. The voice band channels between any two earth stations may be *preassigned* (*dedicated*) or *demand-assigned* (*switched*). When preassignment is used, a given number of the available voice band channels from each earth station are assigned a dedicated destination. With demand assignment, voice band channels are assigned on an as-needed basis. Demand assignment provides more versatility and more efficient use of the available frequency spectrum. On the other hand, demand assignment requires a control mechanism that is common to all the earth stations to keep track of channel routing and the availability of each voice band channel.

Remember, in an FDM/FM satellite system, each RF channel requires a separate transponder. Also, with FDM/FM transmissions, it is impossible to differentiate (separate) multiple transmissions that occupy the same bandwidth. Fixed-frequency systems may be used in a multiple-access configuration by switching the RF carriers at the satellite, reconfiguring the baseband signals with multiplexing/demultiplexing equipment on board the satellite, or by using multiple spot beam antennas (reuse). All three of these methods require relatively complicated, expensive, and heavy hardware on the spacecraft.

MULTIPLE ACCESSING

Figure 9-2 shows the three most commonly used multiple accessing arrangements: frequency-division multiple accessing (FDMA), time-division multiple accessing (TDMA), and code-division multiple accessing (CDMA). With FDMA, each earth

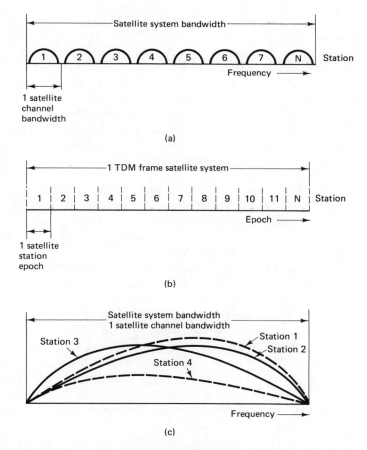

Figure 9-2 Multiple-accessing arrangements: (a) FDMA; (b) TDMA; (c) CDMA.

station's transmissions are assigned specific uplink and downlink frequency bands within an allotted satellite channel bandwidth; they may be preassigned or demand assigned. Consequently, transmissions from different earth stations are separated in the frequency domain. With TDMA, each earth station transmits a short burst of information during a specific time slot (*epoch*) within a TDMA frame. The bursts must be synchronized so that each station's *burst* arrives at the satellite at a different time. Consequently, transmissions from different earth stations are separated in the time domain. With CDMA, all earth stations transmit within the same frequency band and, for all practical purposes, have no limitation on when they may transmit or on which carrier frequency. Carrier separation is accomplished with *envelope encryption/decryption* techniques.

Frequency-Division Multiple Access

Frequency-division multiple access (FDMA) is a method of multiple accessing where a given RF channel bandwidth is divided into smaller frequency bands called *subdivisions*. Each subdivision is used to carry one voice band channel. A control mechanism is used to ensure that no two earth stations transmit on the same subdivision at the same time. Essentially, the control mechanism designates a receive station for each of the subdivisions. In demand-assignment systems, the control mechanism is also used to establish or terminate the voice band links between the source and destination earth stations. Consequently, any of the subdivisions may be used by any of the participating earth stations at any given time. Typically, each subdivision is used to carry a single 4-kHz voice band channel, but occasionally, groups, supergroups, or even mastergroups are assigned a larger subdivision.

       SPADE system.   The first FDMA demand-assignment system for satellites was developed by COMSAT for use on the INTELSAT IV satellite. This system was called *SPADE* (single-channel-per-carrier PCM multiple-access demand assignment equipment). Figures 9-3 and 9-4 show the block diagram and IF frequency assignments for SPADE, respectively.

       With SPADE, 800 PCM-encoded voice band channels separately QPSK modulate an IF carrier frequency (hence the name *single carrier per channel*, SCPC). Each 4-kHz voice band channel is sampled at an 8-kHz rate and converted to an 8-bit PCM code. This produces a 64-kbps PCM code for each voice band channel. The PCM code from each voice band channel QPSK modulates a different IF carrier frequency. With QPSK, the minimum required bandwidth is equal to one-half the input bit rate. Consequently, the output of each QPSK modulator requires a minimum bandwidth of 32 kHz. Each channel is allocated a 45-kHz bandwidth, which allows for a 13-kHz guard band between each frequency-division-multiplexed channel. The IF carrier frequencies begin at 52.0225 MHz (low-band channel 1) and increase in 45-kHz steps to 87.9775 MHz (high-band channel 400). The entire 36-MHz band (52 to 88 MHz) is divided in half, producing two 400-channel bands (a low-band and a high-band). For full-duplex operation, four hundred 45-kHz channels are used

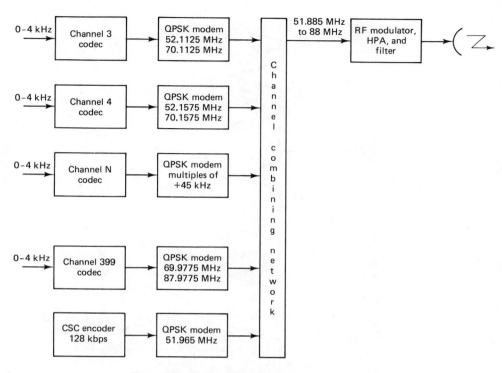

Figure 9-3 FDMA, SPADE earth station transmitter.

for one direction of transmission and 400 are used for the opposite direction. Also, channels 1, 2, and 400 from each band are left permanently vacant. This reduces the number of usable full-duplex voice band channels to 397. The 6-GHz C-band extends from 5.725 to 6.425 GHz (700 MHz). This allows for approximately nineteen 36-MHz RF channels per system. Each RF channel has a capacity of 397 full-duplex voice band channels.

    Each IF channel (Figure 9-4) has a 160-kHz *common signaling channel* (CSC). The CSC is a time-division-multiplexed transmission that is frequency-division multiplexed onto the IF spectrum below the QPSK-encoded voice band channels. Figure 9-5 shows the TDM frame structure for the CSC. The total frame time is 50 ms, which is subdivided into fifty 1-ms epochs. Each earth station transmits on the CSC channel only during its preassigned 1-ms time slot. The CSC signal is a 128-bit binary code. To transmit a 128-bit code in 1 ms, a transmission rate of 128 kbps is required. The CSC code is used for establishing and disconnecting voice band links between two earth station users when demand-assignment channel allocation is used.

EXAMPLE 9-1

    For the system shown in Figure 9-6, a user earth station in New York wishes to establish a voice band link between itself and London. New York randomly selects an idle voice

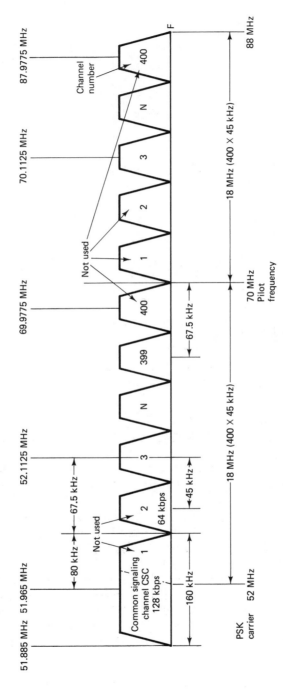

Figure 9-4 Carrier frequency assignments for the Intelsat single channel-per-carrier PCM multiple access demand assignment equipment (SPADE).

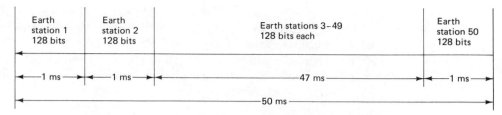

128 bits/1 ms × 1000 ms/1 s = 128 kbps or 6400 bits/frame × 1 frame/50 ms = 128 kbps

Figure 9-5 FDMA, SPADE common signaling channel (CSC).

band channel. It then transmits a binary-coded message to London on the CSC channel during its respective time slot, requesting that a link be established on the randomly selected channel. London responds on the CSC channel during its time slot with a binary code, either confirming or denying the establishment of the voice band link. The link is disconnected in a similar manner when the users are finished.

The CSC channel occupies a 160-kHz bandwidth, which includes the 45 kHz for low-band channel 1. Consequently, the CSC channel extends from 51.885 MHz to 52.045 MHz. The 128-kbps CSC binary code QPSK modulates a 51.965-MHz carrier. The minimum bandwidth required for the CSC channel is 64 kHz; this results in a 48-kHz guard band on either side of the CSC signal.

With FDMA, each earth station may transmit simultaneously within the same 36-MHz RF spectrum, but on different voice band channels. Consequently, simultaneous transmissions of voice band channels from all earth stations within the satellite network are interleaved in the frequency domain in the satellite transponder. Transmissions of CSC signals are interleaved in the time domain.

An obvious disadvantage of FDMA is that carriers from multiple earth stations

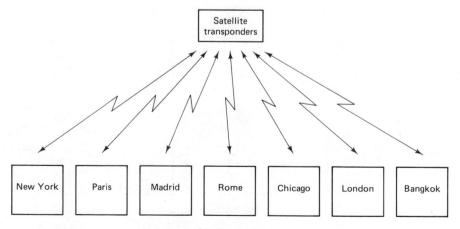

Figure 9-6 Diagram of the system for Example 9-11.

may be present in a satellite transponder at the same time. This results in cross-modulation distortion between the various earth station transmissions. This is alleviated somewhat by shutting off the IF subcarriers on all unused 45-kHz voice band channels. Because balanced modulators are used in the generation of QPSK, carrier suppression is inherent. This also reduces the power load on a system and increases its capacity by reducing the idle channel power.

Time-Division Multiple Access

Time-division multiple access (TDMA) is the predominant multiple-access method used today. It provides the most efficient method of transmitting digitally modulated carriers (PSK). TDMA is a method of time-division multiplexing digitally modulated carriers between participating earth stations within a satellite network through a common satellite transponder. With TDMA, each earth station transmits a short *burst* of a digitally modulated carrier during a precise time slot (epoch) within a TDMA frame. Each station's burst is synchronized so that it arrives at the satellite transponder at a different time. Consequently, only one earth station's carrier is present in the transponder at any given time, thus avoiding a collision with another station's carrier. The transponder is an RF-to-RF repeater that simply receives the earth station transmissions, amplifies them, and then retransmits them in a down-link beam which is received by all the participating earth stations. Each earth station receives the bursts from all other earth stations and must select from them the traffic destined only for itself.

Figure 9-7 shows a basic TDMA frame. Transmissions from all earth stations are synchronized to a *reference burst*. Figure 9-7 shows the reference burst as a

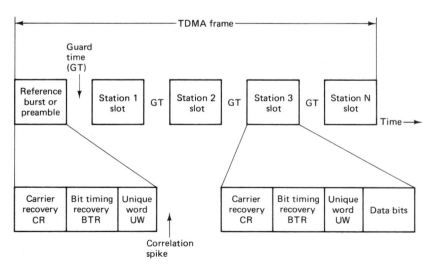

Figure 9-7   Basic time-division-multiple accessing (TDMA) frame.

separate transmission, but it may be the *preamble* which precedes a reference station's transmission of data. Also, there may be more than one synchronizing reference burst.

The reference burst contains a *carrier recovery sequence* (CRS) from which all receiving stations recover a frequency and phase coherent carrier for PSK demodulation. Also included in the reference burst is a binary sequence for *bit timing recovery* (BTR, i.e., clock recovery). At the end of each reference burst, a *unique word* (UW) is transmitted. The UW sequence is used to establish a precise time reference that each of the earth stations use to synchronize the transmission of its burst. The UW is typically a string of successive binary 1's terminated with a binary 0. Each earth station receiver demodulates and integrates the UW sequence. Figure 9-8 shows the result of the integration process. The integrator and threshold detector are designed so that the threshold voltage is reached precisely when the last bit of the UW sequence is integrated. This generates a *correlation spike* at the output of the threshold detector at the exact time the UW sequence ends.

Each earth station synchronizes the transmission of its carrier to the occurrence of the UW correlation spike. Each station waits a different length of time before it begins transmitting. Consequently, no two stations will transmit carrier at the same time. Note the *guard time* (GT) between transmissions from successive stations. This is analogous to a guard band in a frequency-division-multiplexed system. Each station precedes the transmission of data with a *preamble*. The preamble is logically equivalent

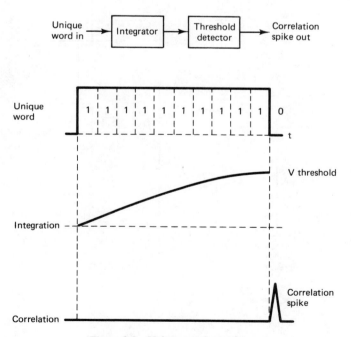

Figure 9-8 Unique word correlator.

to the reference burst. Because each station's transmissions must be received by all other earth stations, all stations must recover carrier and clocking information prior to demodulating the data. If demand assignment is used, a common signaling channel must also be included in the preamble.

CEPT primary multiplex frame.   Figures 9-9 and 9-10 show the block diagram and timing sequence for the CEPT primary multiplex frame respectively (CEPT— Conference of European Postal and Telecommunications Administrations; the CEPT sets many of the European telecommunications standards). This is a commonly used TDMA frame format for digital satellite systems.

Essentially, TDMA is a *store-and-forward* system. Earth stations can transmit only during their specified time slot, although the incoming voice band signals are continuous. Consequently, it is necessary to sample and store the voice band signals prior to transmission. The CEPT frame is made up of 8-bit PCM encoded samples from 16 independent voice band channels. Each channel has a separate codec that samples the incoming voice signals at a 16-kHz rate and converts those samples to

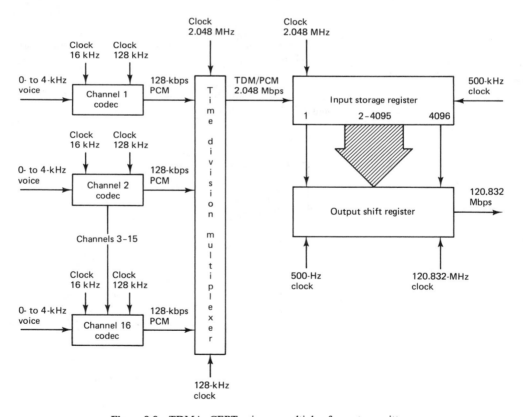

Figure 9-9   TDMA, CEPT primary multiplex frame transmitter.

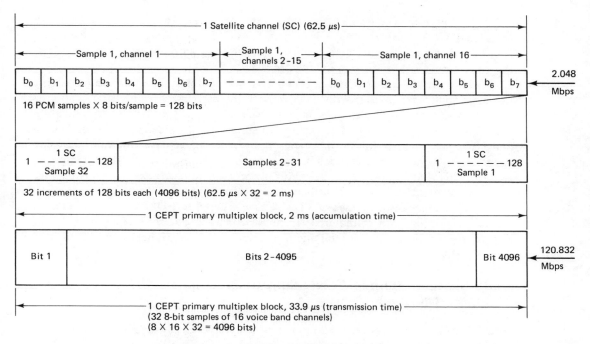

Figure 9-10   TDMA, CEPT primary multiplex frame.

an 8-bit binary code. This results in a 128-kbps transmission rate at the output of each voice channel codec. The sixteen 128-kbps transmissions are time-division multiplexed into a subframe that contains one 8-bit sample from each of the 16 channels (128 bits). It requires only 62.5 μs to accumulate the 128 bits (2.048-Mbps transmission rate). The CEPT multiplex format specifies a 2-ms frame time. Consequently, each earth station can transmit only once every 2 ms and therefore must store the PCM-encoded samples. The 128 bits accumulated during the first sample of each voice band channel are stored in a holding register while a second sample is taken from each channel and converted into another 128-bit *subframe*. This 128-bit sequence is stored in the holding register behind the first 128 bits. The process continues for 32 subframes (32 $\times$ 62.5 μs = 2 ms). After 2 ms, thirty-two 8-bit samples have been taken from each of 16 voice band channels for a total of 4096 bits (32 $\times$ 8 $\times$ 16 = 4096). At this time, the 4096 bits are transferred to an output shift register for transmission. Because the total TDMA frame is 2 ms long and during this 2-ms period each of the participating earth stations must transmit at different times, the individual transmissions from each station must occur in a significantly shorter time period. In the CEPT frame, a transmission rate of 120.832 Mbps is used. This rate is the fifty-ninth multiple of 2.048 Mbps. Consequently, the actual transmission of the 4096 accumulated bits takes approximately 33.9 μs. At the earth station receivers, the 4096 bits are stored in a holding register and shifted out to their PCM decoders at a 2.048-Mbps rate. Because all the clock rates (500 Hz, 16 kHz, 128 kHz, 2.048 MHz, and 120.832 MHz) are synchronized, the PCM codes are accumulated, stored, transmitted, received, and then decoded in perfect synchronization. To the users, the voice transmission is a continuous process.

There are several advantages of TDMA over FDMA. The first, and probably the most significant, is that with TDMA only the carrier from one earth station is present in the satellite transponder at any given time, thus reducing intermodulation distortion. Second, with FDMA, each earth station must be capable of transmitting and receiving on a multitude of carrier frequencies to achieve multiple accessing capabilities. Third, TDMA is much better suited to the transmission of digital information than FDMA. Digital signals are naturally acclimated to storage, rate conversions, and time-domain processing than their analog counterparts.

The primary disadvantage of TDMA as compared to FDMA is that in TDMA precise synchronization is required. Each earth station's transmissions must occur during an exact time slot. Also, bit and frame timing must be achieved and maintained with TDMA.

Code-Division Multiple Access (Spread-Spectrum Multiple Accessing)

With FDMA, earth stations are limited to a specific bandwidth within a satellite channel or system but have no restriction on when they can transmit. With TDMA, earth station's transmissions are restricted to a precise time slot but have no restriction on what frequency or bandwidth they may use within a specified satellite system or channel allocation. With *code-division multiple access* (CDMA), there are no restrictions on time or bandwidth. Each earth station transmitter may transmit whenever it wishes and can use any or all of the bandwidth allocated a particular satellite system or channel. Because there is no limitation on the bandwidth, CDMA is sometimes referred to as *spread-spectrum multiple access*; transmissions can spread throughout the entire allocated bandwidth spectrum. Transmissions are separated through envelope encryption/decryption techniques. That is, each earth station's transmissions are encoded with a unique binary word called a *chip code*. Each station has a unique chip code. To receive a particular earth station's transmission, a receive station must know the chip code for that station.

Figure 9-11 shows the block diagram of a CDMA encoder and decoder. In the encoder (Figure 9-11a), the input data (which may be PCM-encoded voice band signals of raw digital data) is multiplied by a unique chip code. The product code PSK modulates an IF carrier which is up-converted to RF for transmission. At the receiver (Figure 9-11b), the RF is down-converted to IF. From the IF, a coherent PSK carrier is recovered. Also, the chip code is acquired and used to synchronize the receive station's code generator. Keep in mind, the receiving station knows the chip code but must generate a chip code that is synchronous in time with the receive code. The recovered synchronous chip code multiplies the recovered PSK carrier and generates a PSK modulated signal that contains the PSK carrier plus the chip

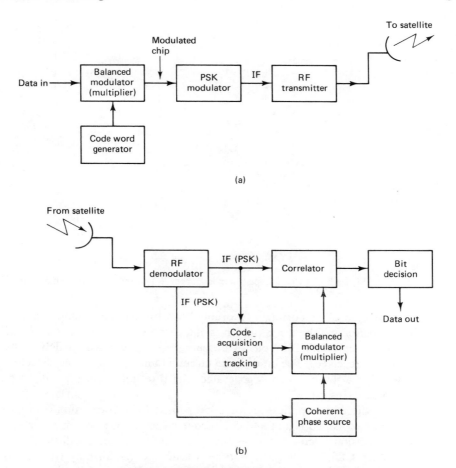

Figure 9-11 Code-division multiple access (CDMA): (a) encoder; (b) decoder.

code. The received IF signal that contains the chip code, the PSK carrier, and the data information is compared to the received IF signal in the *correlator*. The function of the correlator is to compare the two signals and recover the original data. Essentially, the correlator subtracts the recovered PSK carrier + chip code from the received PSK carrier + chip code + data. The resultant is the data.

The correlation is accomplished on the analog signals. Figure 9-12 shows how the encoding and decoding is accomplished. Figure 9-12a shows the correlation of the correctly received chip code. A +1 indicates an in-phase carrier and a −1 indicates an out-of-phase carrier. The chip code is multiplied by the data (either +1 or −1). The product is either an in-phase code or one that is 180° out of phase with the chip code. In the receiver, the recovered synchronous chip code is compared in the correlator to the received signaling elements. If the phases are the same, a +1 is produced; if they are 180° out of phase, a −1 is produced. It can be seen that if all

	← t_c →						← t_c →				
Data in			1						−1		
Chip code	1	1	−1	−1	1	1	1	1	−1	−1	1	1
Product	1	1	−1	−1	1	1	−1	−1	1	1	−1	−1
Recovered chip code	1	1	−1	−1	1	1	1	1	−1	−1	1	1
Correlation	1	1	1	1	1	1	−1	−1	−1	−1	−1	−1
			+6 V						−6 V			

(a)

Orthogonal code	1	1	−1	1	−1	−1	−1	−1	1	−1	1	1
Recovered chip code	1	1	−1	−1	1	1	1	1	−1	−1	1	1
Correlation	1	1	1	−1	−1	−1	−1	−1	−1	1	1	1
			0 V						0 V			

(b)

Figure 9-12 CDMA code/data alignment: (a) correct code; (b) orthogonal code.

the recovered chips correlate favorably with the incoming chip code, the output of the correlator will be a +6 (which is the case when a logic 1 is received). If all the code chips correlate 180° out of phase, a −6 is generated (which is the case when a logic 0 is received). The bit decision circuit is simply a threshold detector. Depending on whether a +6 or a −6 is generated, the threshold detector will output a logic 1 or a logic 0, respectively.

As the name implies, the correlator looks for a correlation (similarity) between the incoming coded signal and the recovered chip code. When a correlation occurs, the bit decision circuit generates the corresponding logic condition.

With CDMA, all earth stations within the system may transmit on the same frequency at the same time. Consequently, an earth station receiver may be receiving coded PSK signals simultaneously from more than one transmitter. When this is the case, the job of the correlator becomes considerably more difficult. The correlator must compare the recovered chip code with the entire received spectrum and separate from it only the chip code from the desired earth station transmitter. Consequently, the chip code from one earth station must not correlate with the chip codes from any of the other earth stations.

Figure 9-12b shows how such a coding scheme is achieved. If half of the chips within a code were made the same and half were made exactly the opposite, the resultant would be zero cross correlation between chip codes. Such a code is called an *orthogonal code*. In Figure 9-12b it can be seen that when the orthogonal code is compared with the original chip code, there is no correlation (i.e., the sum of the comparison is zero). Consequently, the orthogonal code, although received simultaneously with the desired chip code, had absolutely no effect on the correlation process. For this example, the orthogonal code is received in exact time synchronization with the desired chip code; this is not always the case. For systems that do not have time synchronous transmissions, codes must be developed where there is no correlation between one station's code and any phase of another station's code. For more than two participating earth stations, this is impossible to do. A code set has been developed called the *Gold code*. With the Gold code, there is a minimum correlation between different chips' codes. For a reasonable number of users, it is impossible to achieve perfect orthogonal codes. You can design only for a minimum *cross correlation* between chips.

One of the advantages of CDMA was that the entire bandwith of a satellite channel or system may be used for each transmission from every earth station. For our example, the chip rate was six times the original bit rate. Consequently, the actual transmission rate of information was one-sixth of the PSK modulation rate, and the bandwidth required is six times that required to simply transmit the original data as binary. Because of the coding inefficiency resulting from transmitting chips for bits, the advantage of more bandwidth is partially offset and is thus less of an advantage. Also, if the transmission of chips from the various earth stations must be synchronized, precise timing is required for the system to work. Therefore, the disadvantage of requiring time synchronization in TDMA systems is also present with CDMA. In short, CDMA is not all that it is cracked up to be. The only significant advantage of CDMA is immunity to interference (jamming), which makes CDMA ideally suited for military applications.

FREQUENCY HOPPING

Frequency hopping is a form of CDMA where a digital code is used to continually change the frequency of the carrier. With frequency hopping, the total available bandwidth is partitioned into smaller frequency bands and the total transmission time is subdivided into smaller time slots. The idea is to transmit within a limited frequency band for only a short period of time, then switch to another frequency band, and so on. This process continues indefinitely. The frequency hopping pattern is determined by a binary code. Each station uses a different code sequence. A typical *hopping pattern* (*frequency-time matrix*) is shown in Figure 9-13.

With frequency hopping, each earth station within a CDMA network is assigned a different frequency hopping pattern. Each transmitter switches (hops) from one frequency band to the next according to their assigned pattern. With frequency hopping, each station uses the entire RF spectrum but never occupies more than a small portion of that spectrum at any one time.

FSK is the modulation scheme most commonly used with frequency hopping. When it is a given station's turn to transmit, it sends one of the two frequencies (either mark or space) for the particular band in which it is transmitting. The number of stations in a given frequency hopping system is limited by the number of unique hopping patterns that can be generated.

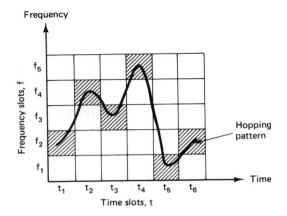

(a)

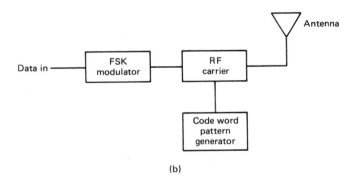

(b)

Figure 9-13   Frequency hopping: (a) frequency time-hopping matrix; (b) frequency hopping transmitter.

CHANNEL CAPACITY

Essentially, there are two methods used to interface terrestrial voice band channels with satellite channels: digital noninterpolated interfaces (DNI) and digital speech interpolated interfaces (DSI).

Digital Noninterpolated Interfaces

A *digital noninterpolated interface* assigns an individual terrestrial channel (TC) to a particular satellite channel (SC) for the duration of the call. A DNI system can carry no more traffic than the number of satellite channels it has. Once a TC has been assigned an SC, the SC is unavailable to the other TCs for the duration of the call. DNI is a form of preassignment; each TC has a permanent dedicated SC.

Digital Speech Interpolated Interfaces

A *digital speech interpolated interface* assigns a terrestrial channel to a satellite channel only when speech energy is present on the TC. DSI interfaces have *speech detectors* that are similar to *echo suppressors*; they sense speech energy, then seize an SC. Whenever a speech detector senses energy on a TC, the TC is assigned to an SC. The SC assigned is randomly selected from the idle SCs. On a given TC, each time speech energy is detected, the TC could be assigned to a different SC. Therefore, a single TC can use several SCs for a single call. For demultiplexing purposes, the TC/SC assignment information must be conveyed to the receive terminal. This is done on a common signaling channel similar to the one used on the SPADE system. DSI is a form of demand assignment; SCs are randomly assigned on an as-needed basis.

With DSI it is apparent that there is a *channel compression*; there can be more TCs assigned than there are SCs. Generally, a TC:SC ratio of 2:1 is used. For a full-duplex (two-way simultaneous) communication circuit, there is speech in each direction 40% of the time, and for 20% of the time the circuit is idle in both directions. Therefore, a DSI gain slightly more than 2 is realized. The DSI gain is affected by a phenomenon called *competitive clipping*. Competitive clipping is when speech energy is detected on a TC and there is no SC to assign it to. During the *wait* time, speech information is lost. Competitive clipping is not noticed by a subscriber if its duration is less than 50 ms.

To further enhance the channel capacity, a technique called *bit stealing* is used. With bit stealing, channels can be added to fully loaded systems by stealing bits from the in-use channels. Generally, an overload channel is generated by stealing the least significant bit from seven other satellite channels. Bit stealing results in eight channels with 7-bit resolution for the time that the *overload channel* is in use. Consequently, bit stealing results in a lower SQR than normal.

Time-Assignment Speech Interpolation

Time-assignment speech interpolation (TASI) is a form of analog channel compression that has been used for suboceanic cables for many years. TASI is very similar to DSI except that the signals interpolated are analog rather than digital. TASI also uses a 2:1 compression ratio. TASI was also the first means used to scramble voice for military security. TASI is similar to a packet data network; the voice message is chopped up into smaller segments comprised of sounds or portions of sounds. The sounds are sent through the network as separate bundles of energy, then put back together at the receive end to reform the original voice message.

QUESTIONS

9-1. Discuss the drawbacks of using FDM/FM modulation for satellite multiple-accessing systems.

9-2. Contrast *preassignment* and *demand assignment*.

9-3. What are the three most common multiple-accessing arrangements used with satellite systems?

9-4. Briefly describe the multiple-accessing arrangements listed in Question 9.3.

9-5. Briefly describe the operation of Comsat's *Spade* system.

9-6. What is meant by *single carrier per channel*?

9-7. What is a common signaling channel, and how is it used?

9-8. Describe what a reference burst is for TDMA and explain the following terms: preamble, carrier recovery sequence, bit timing recovery, unique word, and correlation spike.

9-9. Describe guard time.

9-10. Briefly describe the operation of the CEPT primary multiplex frame.

9-11. What is a store-and-forward system?

9-12. What is the primary advantage of TDMA as compared to FDMA?

9-13. What is the primary advantage of FDMA as compared to TDMA?

9-14. Briefly describe the operation of a CDMA multiple-accessing system.

9-15. Describe a chip code.

9-16. Describe what is meant by an orthogonal code.

9-17. Describe cross correlation.

9-18. What are the advantages of CDMA as compared to TDMA and FDMA?

9-19. What are the disadvantages of CDMA?

9-20. What is a Gold code?

9-21. Describe frequency hopping.

9-22. What is a frequency-time matrix?

9-23. Describe digital noninterpolated interfaces.

9-24. Describe digital speech interpolated interfaces.

9-25. What is channel compression, and how is it accomplished with a DSI system?

9-26. Describe competitive clipping.

9-27. What is meant by *bit stealing*?

9-28. Describe time-assignment speech interpolation.

PROBLEMS

9-1. How many satellite transponders are required to interlink six earth stations with FDM/FM modulation?

9-2. For the *Spade* system, what are the carrier frequencies for channel 7? What are the allocated passbands for channel 7? What are the actual passband frequencies (excluding guard bands) required?

9-3. If a 512-bit preamble precedes each CEPT station's transmission, what is the maximum number of earth stations that can be linked together with a single satellite transponder?

9-4. Determine an orthogonal code for the following chip code (101010). Prove that your selection will not produce any cross correlation for an in-phase comparison. Determine the cross correlation for each out-of-phase condition that is possible.

Chapter 10

FIBER OPTIC COMMUNICATIONS

INTRODUCTION

During the past 10 years, the electronic communications industry has experienced many remarkable and dramatic changes. A phenomenal increase in voice, data, and video communications has caused a corresponding increase in the demand for more economical and larger capacity communications systems. This has caused a technical revolution in the electronic communications industry. Terrestrial microwave systems have long since reached their capacity, and satellite systems can provide, at best, only a temporary relief to the ever-increasing demand. It is obvious that economical communications systems that can handle large capacities and provide high-quality service are needed.

Communications systems that use light as the carrier of information have recently received a great deal of attention. As we shall see later in this chapter, propagating light waves through the earth's atmosphere is difficult and impractical. Consequently, systems that use glass or plastic fiber cables to "contain" a light wave and guide it from a source to a destination are presently being investigated at several prominent research and development laboratories. Communications systems that carry information through a *guided fiber cable* are called *fiber optic* systems.

The *information-carrying capacity* of a communications system is directly proportional to its bandwidth; the wider the bandwidth, the greater its information-carrying capacity. For comparison purposes, it is common to express the bandwidth of a system as a percentage of its carrier frequency. For instance, a VHF radio system operating at 100 MHz has a bandwidth equal to 10 MHz (i.e., 10% of the carrier frequency). A microwave radio system operating at 6 GHz with a bandwidth equal to 10% of its carrier frequency would have a bandwidth equal to 600 MHz. Thus the higher the carrier frequency, the wider the bandwidth possible and consequently, the greater the information-carrying capacity. Light frequencies used in fiber optic systems are between 10^{14} and 10^{15} Hz (100,000 to 1,000,000 GHz). Ten percent of 1,000,000 GHz is 100,000 GHz. To meet today's communications needs or the needs of the foreseeable future, 100,000 GHz is an excessive bandwidth. However, it does illustrate the capabilities of fiber optic systems.

HISTORY OF FIBER OPTICS

In 1880, Alexander Graham Bell experimented with an apparatus he called a *photophone*. The photophone was a device constructed from mirrors and selenium detectors that transmitted sound waves over a beam of light. The photophone was awkward, unreliable, and had no real practical application. Actually, visual light was a primary means of communicating long before electronic communications came about. Smoke signals and mirrors were used ages ago to convey short, simple messages. Bell's contraption, however, was the first attempt at using a beam of light for carrying information.

Transmission of light waves for any useful distance through the earth's atmosphere is impractical because water vapor, oxygen, and particulates in the air absorb and attenuate the ultrahigh light frequencies. Consequently, the only practical type of optical communications system is one that uses a fiber guide. In 1930, J. L. Baird, an English scientist, and C. W. Hansell, a scientist from the United States, were granted patents for scanning and transmitting television images through uncoated fiber cables. A few years later a German scientist named H. Lamm successfully transmitted images through a single glass fiber. At that time, most people considered fiber optics more of a toy or a laboratory stunt and consequently, it was not until the early 1950s that any substantial breakthrough was made in the field of fiber optics.

In 1951, A. C. S. van Heel of Holland and H. H. Hopkins and N. S. Kapany of England experimented with light transmission through *bundles* of fibers. Their studies led to the development of the *flexible fiberscope*, which is used extensively in the medical field. It was Kapany who coined the term "fiber optics" in 1956.

The *laser* (*l*ight *a*mplification by *s*timulated *e*mission of *r*adiation) was invented in 1960. The laser's relatively high output power, high frequency of operation, and capability of carrying an extremely wide bandwidth signal make it ideally suited for high-capacity communications systems. The invention of the laser greatly accelerated research efforts in fiber optic communications, although it was not until 1967 that K. C. Kao and G. A. Bockham of the Standard Telecommunications Laboratory in England proposed a new communications medium using *cladded* fiber cables.

The fiber cables available in the 1960s were extremely *lossy* (more than 1000 dB/km), which limited optical transmissions to short distances. In 1970, Kapron, Keck, and Maurer of Corning Glass Works in Corning, New York, developed an optical fiber with losses less than 20 dB/km. That was the "big" breakthrough needed to permit practical fiber optics communications systems. Since 1970, fiber optics technology has grown exponentially. Recently, Bell Laboratories successfully transmitted 1 billion bps through a fiber cable for 75 miles without a regenerator. AT&T has projected they will have a transatlantic fiber cable installed and operational by 1988.

In the late 1970s and early 1980s, the refinement of optical cables and the development of high-quality, affordable light sources and detectors have opened the door to the development of high-quality, high-capacity, and efficient fiber optics communications systems.

FIBER OPTIC VERSUS METALLIC CABLE FACILITIES

Communications through glass or plastic fiber cables has several overwhelming advantages over communications over conventional *metallic* or *coaxial* cable facilities.

Advantages of Fiber Systems

1. Fiber systems have a greater capacity due to the inherently larger bandwidths available with optical frequencies. Metallic cables exhibit capacitance between and inductance along their conductors. These properties cause them to act like low-pass filters which limit their bandwidths.

2. Fiber systems are immune to crosstalk between cables caused by *magnetic induction*. Glass or plastic fibers are nonconductors of electricity and therefore do not have a magnetic field associated with them. In metallic cables, the primary cause of crosstalk is magnetic induction between conductors located near each other.

3. Fiber cables are immune to *static* interference caused by lightning, electric motors, fluorescent lights, and other electrical noise sources. This immunity is also attributable to the fact that optical fibers are nonconductors of electricity. Also, fiber cables do not radiate energy and therefore cannot cause interference with other communications systems. This characteristic makes fiber systems ideally suited to military applications, where the effects of nuclear weapons (EMP—electromagnetic pulse interference) has a devastating effect on conventional communications systems.

4. Fiber cables are more resistive to environmental extremes. They operate over a larger temperature variation than their metallic counterparts, and fiber cables are affected less by corrosive liquids and gases.

5. Fiber cables are safer and easier to install and maintain. Because glass and plastic fibers are nonconductors, there are no electrical currents or voltages associated with them. Fibers can be used around volatile liquids and gases without worrying about their causing explosions or fires. Fibers are smaller and much more lightweight than their metallic counterparts. Consequently, they are easier to work with. Also, fiber cables require less storage space and are cheaper to transport.

6. Fiber cables are more secure than their copper counterparts. It is virtually impossible to tap into a fiber cable without the user knowing about it. This is another quality attractive for military applications.

7. Although it has not yet been proven, it is projected that fiber systems will last longer than metallic facilities. This assumption is based on the higher tolerances that fiber cables have to changes in the environment.

8. The long-term cost of a fiber optic system is projected to be less than that of its metallic counterpart.

Disadvantages of Fiber Systems

At the present time, there are few disadvantages of fiber systems. The only significant disadvantage is the higher initial cost of installing a fiber system, although in the future it is believed that the cost of installing a fiber system will be reduced dramatically. Another disadvantage of fiber systems is the fact that they are unproven; there are no systems that have been in operation for an extended period of time.

ELECTROMAGNETIC SPECTRUM

The total electromagnetic frequency spectrum is shown in Figure 10-1. It can be seen that the frequency spectrum extends from the *subsonic* frequencies (a few hertz) to *cosmic rays* (10^{22} Hz). The frequencies used for fiber optic systems extend from approximately 10^{14} to 10^{15} Hz (infrared to ultraviolet). This frequency subspectrum is called *visible light*, although it extends above and below the actual sensitivity of the human eye.

When dealing with ultrahigh-frequency electromagnetic waves, such as light, it is common to use units of *wavelength* rather than frequency. The wavelengths

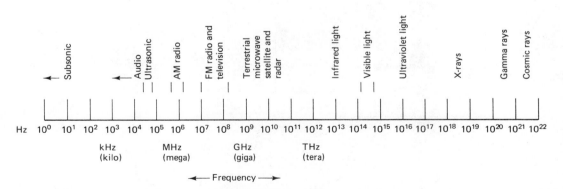

Figure 10-1 Electromagnetic frequency spectrum.

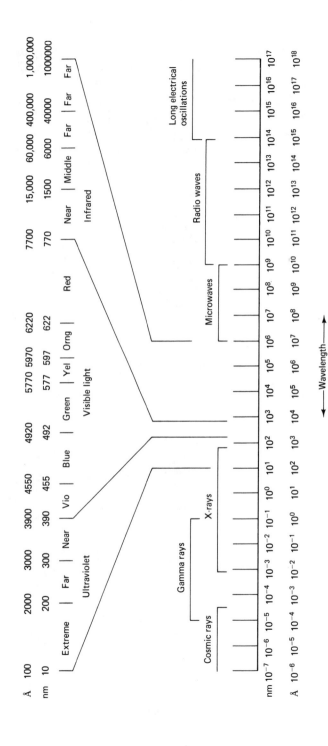

Figure 10-2 Electromagnetic wavelength spectrum.

associated with light frequencies are shown in Figure 10-2. There are two common units for wavelength: *nanometer* (nm) and *angstrom* (Å). One nanometer is 10^{-9} m, and 1 angstrom is 10^{-10} m. Therefore, 1 nanometer is equal to 10 angstroms.

FIBER OPTIC COMMUNICATIONS SYSTEM

Figure 10-3 shows a simplified block diagram of a fiber optic communications link. The three primary building blocks of the link are the *transmitter*, the *receiver*, and the *fiber guide*. The transmitter consists of an analog or digital interface, a voltage-to-current converter, a light source, and a source-to-fiber light coupler. The fiber guide is either an ultra-pure glass or plastic cable. The receiver includes a fiber-to-light detector coupling device, a photo detector, a current-to-voltage converter, an amplifier, and an analog or digital interface.

In a fiber optic transmitter, the light source can be modulated by a digital or an analog signal. For analog modulation, the input interface matches impedances and limits the input signal amplitude. For digital modulation, the original source may already be in digital form or, if in analog form, it must be converted to a digital pulse stream. For the latter case, an analog-to-digital converter must be included in the interface.

The voltage-to-current converter serves as an electrical interface between the input circuitry and the light source. The light source is either a light-emitting diode (LED) or an injection laser diode (ILD). The amount of light emitted by either an LED or an ILD is proportional to the amount of drive current. Thus the voltage-to-current converter converts an input signal voltage to a current which is used to drive the light source.

The source-to-fiber coupler is a mechanical interface. Its function is to couple the light emitted by the source into the optical fiber cable. The optical fiber consists of a glass or plastic fiber core, a cladding, and a protective jacket. The fiber-to-light detector coupling device is also a mechanical coupler. Its function is to couple as much light as possible from the fiber cable into the light detector.

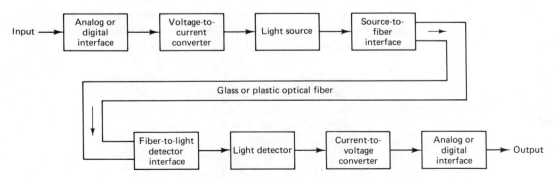

Figure 10-3 Fiber optic communications link.

330 Fiber Optic Communications   Chap. 10

The light detector is very often either a PIN (*p*ositive-*i*ntrinsic-*n*egative) diode or an APD (*a*valanche *p*hoto*d*iode). Both the APD and the PIN diode convert light energy to current. Consequently, a current-to-voltage converter is required. The current-to-voltage converter transforms changes in detector current to changes in output signal voltage.

The analog or digital interface at the receiver output is also an electrical interface. If analog modulation is used, the interface matches impedances and signal levels to the output circuitry. If digital modulation is used, the interface must include a digital-to-analog converter.

OPTICAL FIBERS

Fiber Types

Essentially, there are three varieties of optical fibers available today. All three varieties are constructed of either glass, plastic, or a combination of glass and plastic. The three varieties are:

1. Plastic core and cladding
2. Glass core with plastic cladding (often called PCS fiber, plastic-clad silica)
3. Glass core and glass cladding (often called SCS, silica-clad silica)

Presently, Bell Laboratories is investigating the possibility of using a fourth variety that uses a *nonsilicate* substance, *zinc chloride*. Preliminary experiments have indicated that fibers made of this substance will be as much as 1000 times as efficient as glass, their silica-based counterpart.

Plastic fibers have several advantages over glass fibers. First, plastic fibers are more flexible and, consequently, more rugged than glass. They are easy to install, can better withstand stress, are less expensive, and weigh approximately 60% less than glass. The disadvantage of plastic fibers is their high attenuation characteristic; they do not propagate light as efficiently as glass. Consequently, plastic fibers are limited to relatively short runs, such as within a single building or a building complex.

Fibers with glass cores exhibit low attenuation characteristics. However, PCS fibers are slightly better than SCS fibers. Also, PCS fibers are less affected by radiation and are therefore more attractive to military applications. SCS fibers have the best propagation characteristics and they are easier to terminate than PCS fibers. Unfortunately, SCS cables are the least rugged, and they are more susceptible to increases in attenuation when exposed to radiation.

The selection of a fiber for a given application is a function of specific system requirements. There are always trade-offs based on the economics and logistics of a particular application.

Fiber Construction

There are many different cable designs available today. Figure 10-4 shows examples of several fiber optic cable configurations. Depending on the configuration, the cable may include a *core*, a *cladding*, a *protective tube*, *buffers*, *strength members*, and one or more *protective jackets*.

With the *loose* tube construction (shown in Figure 10-4a) each fiber is contained in a protective tube. Inside the protective tube, a polyurethane compound encapsules the fiber and prevents the intrusion of water.

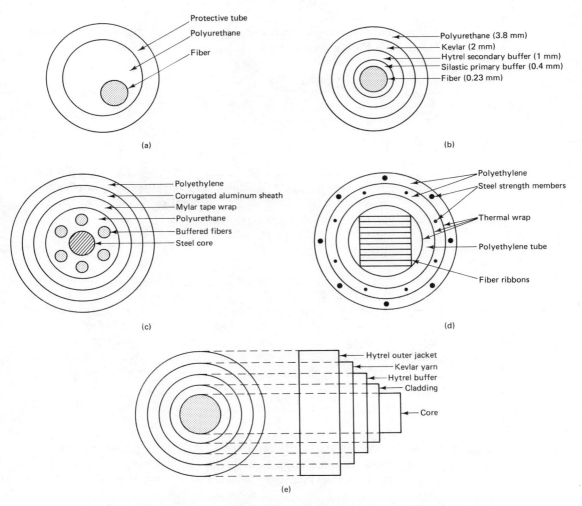

Figure 10-4 Fiber optic cable configurations: (A) loose tube construction; (b) constrained fiber; (c) multiple strands; (d) telephone cable; (e) plastic-clad silica cable.

Figure 10-4b shows the construction of a *constrained* fiber cable. Surrounding the fiber cable are a primary and a secondary buffer. The buffer jackets provide protection for the fiber from external mechanical influences which could cause fiber breakage or excessive optical attenuation. Kelvar is a yarn-type material that increases the tensile strength of the cable. Again, an outer protective tube is filled with polyurethane, which prevents moisture from coming into contact with the fiber core.

Figure 10-4c shows a *multiple-strand* configuration. To increase the tensile strength, a steel central member and a layer of Mylar tape wrap are included in the package. Figure 10-4d shows a *ribbon* configuration, which is frequently seen in telephone systems using fiber optics. Figure 10-4e shows both the end and side views of a plastic-clad silica cable.

The type of cable construction used depends on the performance requirements of the system and both the economic and environmental constraints.

LIGHT PROPAGATION

The Physics of Light

Although the performance of optical fibers can be analyzed completely by application of Maxwell's equations, this is necessarily complex. For most practical applications, Maxwell's equations may be substituted by the application of *geometric ray tracing*, which will yield a sufficiently detailed analysis.

Velocity of Propagation

Electromagnetic energy, such as light, travels at approximately 300,000,000 m/s (186,000 miles per second) in free space. Also, the velocity of propagation is the same for all light frequencies in free space. However, it has been demonstrated that in materials more dense than free space, the velocity is reduced. When the velocity of an electromagnetic wave is reduced as it passes from one medium to another medium of a denser material, the light ray is *refracted* (bent) toward the normal. Also, in materials more dense than free space, all light frequencies do not propagate at the same velocity.

Refraction

Figure 10-5a shows how a light ray is refracted as it passes from a material of a given density into a less dense material. (Actually, the light ray is not bent, but rather, it changes direction at the interface.) Figure 10-5b shows how sunlight, which contains all light frequencies, is affected as it passes through a material more dense than free space. Refraction occurs at both air/glass interfaces. The violet wavelengths are refracted the most, and the red wavelengths are refracted the least. The spectral

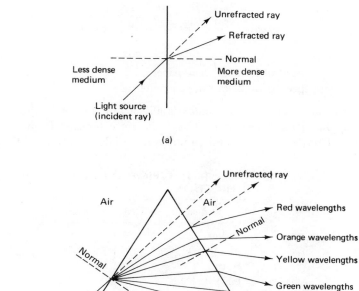

Figure 10-5   Refraction of light: (A) light refraction; (b) prismatic refraction.

separation of white light in this manner is called *prismatic refraction*. It is this phenomenon that causes rainbows; water droplets in the atmosphere act like small prisms that split the white sunlight into the various wavelengths, creating a visible spectrum of color.

Refractive Index

The amount of bending or refraction that occurs at the interface of two materials of different densities is quite predictable and depends on the *refractive index* (also called *index of refraction*) of the two materials. The refractive index is simply the ratio of the velocity of propagation of a light ray in free space to the velocity of propagation of a light ray in a given material. Mathematically, the refractive index is

$$n = \frac{c}{v}$$

where

  c = speed of light in free space
  v = speed of light in a given material

Although the refractive index is also a function of frequency, the variation in most applications is insignificant and therefore omitted from this discussion. The indexes of refraction of several common materials are given in Table 10-1.

TABLE 10-1   TYPICAL INDEXES OF REFRACTION

Medium	Index of refraction[a]
Vacuum	1.0
Air	1.0003 (1.0)
Water	1.33
Ethyl alcohol	1.36
Fused quartz	1.46
Glass fiber	1.5–1.9
Diamond	2.0–2.42
Silicon	3.4
Gallium-arsenide	3.6

[a] Index of refraction is based on a wavelength of light emitted from a sodium flame (5890 Å).

How a light ray reacts when it meets the interface of two transmissive materials that have different indexes of refraction can be explained with *Snell's law*. Snell's law simply states:

$$n_1 \sin \theta_1 = n_2 \sin \theta_2 \tag{10-1}$$

where

  n_1 = refractive index of material 1
  n_2 = refractive index of material 2
  θ_1 = angle of incidence
  θ_2 = angle of refraction

A refractive index model for Snell's law is shown in Figure 10-6. At the interface, the incident ray may be refracted toward the normal or away from it, depending on whether n_1 is less than or greater than n_2.

  Figure 10-7 shows how a light ray is refracted as it travels from a more dense (higher refractive index) material into a less dense (lower refractive index) material.

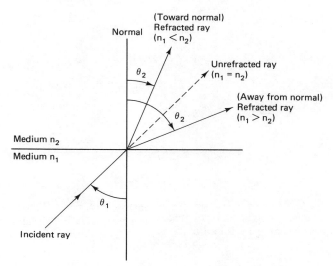

Figure 10-6   Refractive model for Snell's law.

It can be seen that the light ray changes direction at the interface, and the angle of refraction is greater than the angle of incidence. Consequently, when a light ray enters a less dense material, the ray bends away from the normal. The normal is simply a line drawn perpendicular to the interface at the point where the incident ray strikes the interface. Similarly, when a light ray enters a more dense material, the ray bends toward the normal.

EXAMPLE 10-1

In Figure 10-7, let medium 1 be glass and medium 2 be ethyl alcohol. For an angle of incidence of 30°, determine the angle of refraction.

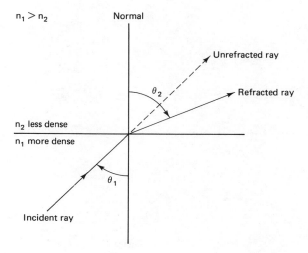

Figure 10-7   Light ray refracted away from the normal.

Solution   From Table 10-1,

$$n_1 \text{ (glass)} = 1.5$$

$$n_2 \text{ (ethyl alcohol)} = 1.36$$

Rearranging Equation 10-1 and substituting for n_1, n_2, and θ_1 gives us

$$\frac{n_1}{n_2} \sin \theta_1 = \sin \theta_2$$

$$\frac{1.5}{1.36} \sin 30 = 0.5514 = \sin \theta_2$$

$$\theta_2 = \sin^{-1} 0.5514 = 33.47°$$

The result indicates that the light ray refracted (bent) or changed direction by 3.47° at the interface. Because the light was traveling from a more dense material into a less dense material, the ray bent away from the normal.

Critical Angle

Figure 10-8 shows a condition in which an *incident ray* is at an angle such that the angle of refraction is 90° and the refracted ray is along the interface. (It is important to note that the light ray is traveling from a medium of higher refractive index to a medium with a lower refractive index.) Again, using Snell's law,

$$\sin \theta_1 = \frac{n_2}{n_1} \sin \theta_2$$

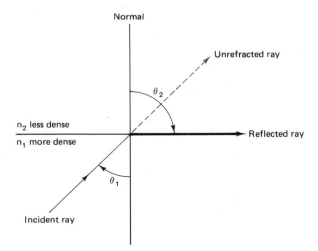

Figure 10-8   Critical angle reflection.

With $\theta_2 = 90°$,

$$\sin \theta_1 = \frac{n_2}{n_1} \,(1) \qquad \text{or} \qquad \sin \theta_1 = \frac{n_2}{n_1}$$

and

$$\sin^{-1} \frac{n_2}{n_1} = \theta_1 = \theta_c \qquad\qquad (10\text{-}2)$$

where θ_c is the critical angle.

The *critical angle* is defined as the minimum angle of incidence at which a light ray may strike the interface of two media and result in an angle of refraction of 90° or greater. (This definition pertains only when the light ray is traveling from a more dense medium into a less dense medium.) If the angle of refraction is 90° or greater, the light ray is not allowed to penetrate the less dense material. Consequently, total reflection takes place at the interface, and the angle of reflection is equal to the angle of incidence. Figure 10-9 shows a comparison of the angle of refraction and the angle of reflection when the angle of incidence is less than or more than the critical angle.

PROPAGATION OF LIGHT THROUGH AN OPTICAL FIBER

Light can be propagated down an optical fiber cable by either reflection or refraction. How the light is propagated depends on the *mode of propagation* and the *index profile* of the fiber.

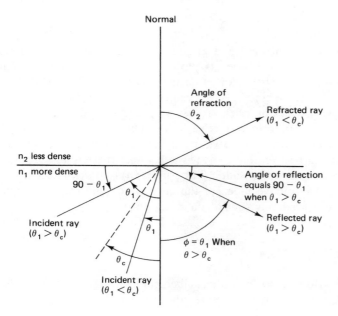

Figure 10-9 Angle of reflection and refraction.

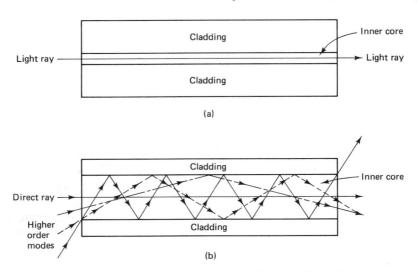

Figure 10-10   Modes of propagation: (a) single mode; (b) multimode.

Mode of Propagation

In fiber optics terminology, the word *mode* simply means path. If there is only one path for light to take down the cable, it is called *single mode*. If there is more than one path, it is called *multimode*. Figure 10-10 shows single and multimode propagation of light down an optical fiber.

Index Profile

The index profile of an optical fiber is a graphical representation of the refractive index of the core. The refractive index is plotted on the horizontal axis and the radial distance from the core axis is plotted on the vertical axis. Figure 10-11 shows the core index profiles of three types of fiber cables.

There are two basic types of index profiles: step and graded. A *step-index fiber* has a central core with a uniform refractive index. The core is surrounded by an outside cladding with a uniform refractive index less than that of the central core. From Figure 10-11 it can be seen that in a step-index fiber there is an abrupt change in the refractive index at the core/cladding interface. In a *graded-index fiber* there is no cladding, and the refractive index of the core is nonuniform; it is highest at the center and decreases gradually toward the outer edge.

OPTICAL FIBER CONFIGURATIONS

Essentially, there are three types of optical fiber configurations: single-mode step-index, multimode step-index, and multimode graded-index.

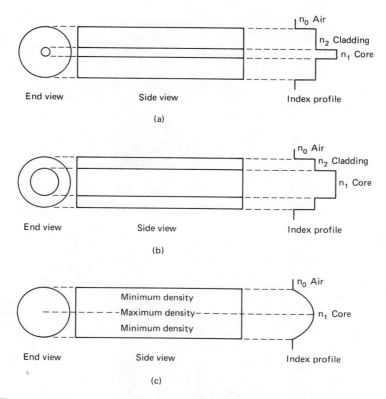

Figure 10-11 Core index profiles: (a) single-mode step index; (b) multimode step index; (c) multimode graded index.

Single-Mode Step-Index Fiber

A *single-mode step-index fiber* has a central core that is sufficiently small so that there is essentially only one path that light may take as it propagates down the cable. This type of fiber is shown in Figure 10-12. In the simplest form of single-mode step-index fiber, the outside cladding is simply air (Figure 10-12a). The refractive index of the glass core (n_1) is approximately 1.5, and the refractive index of the air cladding (n_0) is 1. The large difference in the refractive indexes results in a small critical angle (approximately 42°) at the glass/air interface. Consequently, the fiber will accept light from a wide aperture. This makes it relatively easy to couple light from a source into the cable. However, this type of fiber is typically very weak and of limited practical use.

A more practical type of single-mode step-index fiber is one that has a cladding other than air (Figure 10-12b). The refractive index of the cladding (n_2) is slightly less than that of the central core (n_1) and is uniform throughout the cladding. This type of cable is physically stronger than the air-clad fiber, but the critical angle is

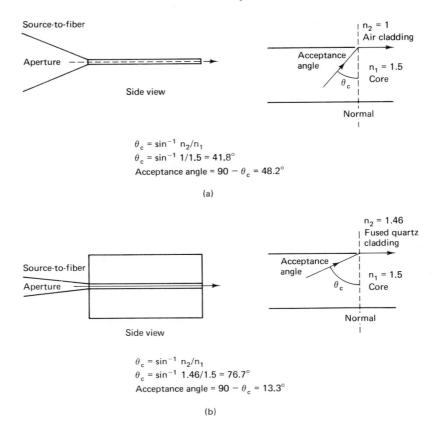

Figure 10-12   Single-mode step-index fibers: (a) air cladding; (b) glass cladding.

also much higher (approximately 77°). This results in a small acceptance angle and a narrow source-to-fiber aperture, making it much more difficult to couple light into the fiber from a light source.

With both types of single-mode step-index fibers, light is propagated down the fiber through reflection. Light rays that enter the fiber propagate straight down the core or, perhaps, are reflected once. Consequently, all light rays follow approximately the same path down the cable and take approximately the same amount of time to travel the length of the cable. This is one overwhelming advantage of single-mode step-index fibers and will be explained in more detail later.

Multimode Step-Index Fiber

A *multimode step-index fiber* is shown in Figure 10-13. It is similar to the single-mode configuration except that the center core is much larger. This type of fiber has a larger light-to-fiber aperture and, consequently, allows more light to enter the

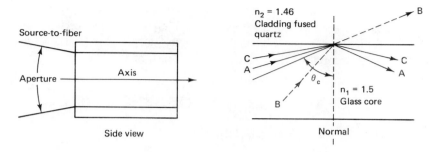

Figure 10-13 Multimode step-index fiber.

cable. The light rays that strike the core/cladding interface at an angle greater than the critical angle (ray A) are propagated down the core in a zigzag fashion, continuously reflecting off the interface boundary. Light rays that strike the core/cladding interface at an angle less than the critical angle (ray B) enter the cladding and are lost. It can be seen that there are many paths that a light ray may follow as it propagates down the fiber. As a result, all light rays do not follow the same path and, consequently, do not take the same amount of time to travel the length of the fiber.

Multimode Graded-Index Fiber

A *multimode graded-index fiber* is shown in Figure 10-14. A multimode graded-index fiber is characterized by a central core that has a refractive index that is nonuniform; it is maximum at the center and decreases gradually toward the outer edge. Light is propagated down this type of fiber through refraction. As a light ray propagates diagonally across the core, it is continually intersecting a less-dense-to-more dense interface. Consequently, the light rays are constantly being refracted, which results in a continuous bending of the light rays. Light enters the fiber at many different angles. As they propagate down the fiber, the light rays that travel in the outermost area of the fiber travel a greater distance than the rays traveling near the center.

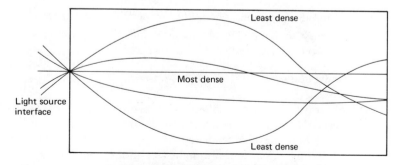

Figure 10-14 Multimode graded-index fiber.

Because the refractive index decreases with distance from the center and the velocity is inversely proportional to the refractive index, the light rays traveling farthest from the center propagate at a higher velocity. Consequently, they take approximately the same amount of time to travel the length of the fiber.

COMPARISON OF THE THREE TYPES OF OPTICAL FIBERS

Single-Mode Step-Index Fiber

Advantages

1. There is minimum dispersion. Because all rays propagating down the fiber take approximately the same path, they take approximately the same amount of time to travel down the cable. Consequently, a pulse of light entering the cable can be reproduced at the receiving end very accurately.
2. Because of the high accuracy in reproducing transmitted pulses at the receive end, larger bandwidths and higher information transmission rates are possible with single-mode step-index fibers than with the other types of fibers.

Disadvantages

1. Because the central core is very small, it is difficult to couple light into and out of this type of fiber. The source-to-fiber aperture is the smallest of all the fiber types.
2. Again, because of the small central core, a highly directive light source such as a laser is required to couple light into a single-mode step-index fiber.
3. Single-mode step-index fibers are expensive and difficult to manufacture.

Multimode Step-Index Fiber

Advantages

1. Multimode step-index fibers are inexpensive and simple to manufacture.
2. It is easy to couple light into and out of multimode step-index fibers; they have a relatively large source-to-fiber aperture.

Disadvantages

1. Light rays take many different paths down the fiber, which results in large differences in their propagation times. Because of this, rays traveling down this type of fiber have a tendency to spread out. Consequently, a pulse of light propagating down a multimode step-index fiber is distorted more than with the other types of fibers.

2. The bandwidth and rate of information transfer possible with this type of cable are less than the other types.

Multimode Graded-Index Fiber

Essentially, there are no outstanding advantages or disadvantages of this type of fiber. Multimode graded-index fibers are easier to couple light into and out of than single-mode step-index fibers but more difficult than multimode step-index fibers. Distortion due to multiple propagation paths is greater than in single-mode step-index fibers but less than in multimode step-index fibers. Graded-index fibers are easier to manufacture than single-mode step-index fibers but more difficult than multimode step-index fibers. The multimode graded-index fiber is considered an intermediate fiber compared to the other types.

ACCEPTANCE ANGLE AND ACCEPTANCE CONE

In previous discussions, the *source-to-fiber aperture* was mentioned several times, and the *critical* and *acceptance* angles at the point where a light ray strikes the core/cladding interface were explained. The following discussion deals with the light-gathering ability of the fiber, the ability to couple light from the source into the fiber cable.

Figure 10-15 shows the source end of a fiber cable. When light rays enter the fiber, they strike the air/glass interface at normal A. The refractive index of air is 1 and the refractive index of the glass core is 1.5. Consequently, the light entering at the air/glass interface propagates from a less dense medium into a more dense

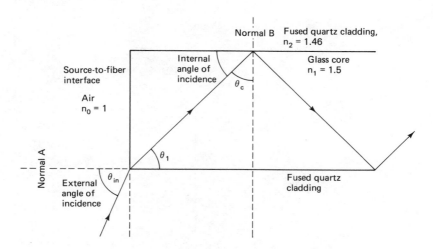

Figure 10-15 Ray propagation into and down an optical fiber cable.

medium. Under these conditions and according to Snell's law, the light rays will refract toward the normal. This causes the light rays to change direction and propagate diagonally down the core at an angle (θ_c) which is different than the external angle of incidence at the air/glass interface (θ_{in}). In order for a ray of light to propagate down the cable, it must strike the internal core/cladding interface at an angle that is greater than the critical angle (θ_c).

Applying Snell's law to the external angle of incidence yields the following expression:

$$n_0 \sin \theta_{in} = n_1 \sin \theta_1 \qquad (10\text{-}3)$$

and

$$\theta_1 = 90 - \theta_c$$

Thus

$$\sin \theta_1 = \sin (90 - \theta_c) = \cos \theta_c \qquad (10\text{-}4)$$

Substituting Equation 10-4 into Equation 10-3 yields the following expression:

$$n_0 \sin \theta_{in} = n_1 \cos \theta_c$$

Rearranging and solving for $\sin \theta_{in}$ gives us

$$\sin \theta_{in} = \frac{n_1}{n_0} \cos \theta_c \qquad (10\text{-}5)$$

Figure 10-16 shows the geometric relationship of Equation 10-5.

From Figure 10-16 and using the Pythagorean theorem, we obtain

$$\cos \theta_c = \frac{\sqrt{n_1^2 - n_2^2}}{n_1} \qquad (10\text{-}6)$$

Substituting Equation 10-6 into Equation 10-5 yields

$$\sin \theta_{in} = \frac{n_1}{n_0} \frac{\sqrt{n_1^2 - n_2^2}}{n_1}$$

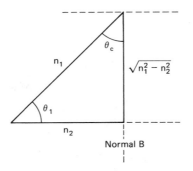

Normal B

Figure 10-16   Geometric relationship of Equation 10-5.

Reducing the equation gives

$$\sin \theta_{in} = \frac{\sqrt{n_1^2 - n_2^2}}{n_0} \tag{10-7}$$

and

$$\theta_{in} = \sin^{-1} \frac{\sqrt{n_1^2 - n_2^2}}{n_0} \tag{10-8}$$

Because light rays generally enter the fiber from an air medium, n_0 equals 1. This simplifies Equation 10-8 to

$$\theta_{in(max)} = \sin^{-1} \sqrt{n_1^2 - n_2^2} \tag{10-9}$$

θ_{in} is called the *acceptance angle* or *acceptance cone* half-angle. It defines the maximum angle in which external light rays may strike the air/fiber interface and still propagate down the fiber with a response that is no greater than 10 dB down from the peak value. Rotating the acceptance angle around the fiber axis describes the acceptance cone of the fiber. This is shown in Figure 10-17.

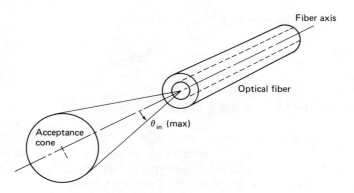

Figure 10-17 Acceptance cone of a fiber cable.

Numerical Aperture

Numerical aperture (NA) is a figure of merit that is used to measure the light-gathering or light-collecting ability of an optical fiber. The larger the magnitude of NA, the greater the amount of light accepted by the fiber from the external light source. For a step-index fiber, numerical aperture is mathematically defined as the sin of the acceptance half-angle. Thus

$$NA = \sin \theta_{in}$$

and

$$NA = \sqrt{n_1^2 - n_2^2} \tag{10-10}$$

Also,

$$\sin^{-1} NA = \theta_{in}$$

For a graded index, NA is simply the sin of the critical angle:

$$NA = \sin \theta_c$$

EXAMPLE 10-2

For this example refer to Figure 10-15. For a multimode step-index fiber with a glass core ($n_1 = 1.5$) and a fused quartz cladding ($n_2 = 1.46$), determine the critical angle (θ_c), acceptance angle (θ_{in}), and numerical aperture. The source-to-fiber media is air.

Solution   Substituting into Equation 10-2, we have

$$\theta_c = \sin^{-1} \frac{n_2}{n_1} = \sin^{-1} \frac{1.46}{1.5} = 76.7°$$

Substituting into Equation 10-9 yields

$$\theta_{in} = \sin^{-1} \sqrt{n_1^2 - n_2^2} = \sin^{-1} \sqrt{1.5^2 - 1.46^2}$$

$$= 20.2°$$

Substituting into Equation 10-10 gives us

$$NA = \sin \theta_{in} = \sin 20.2$$

$$= 0.344$$

LOSSES IN OPTICAL FIBER CABLES

Transmission losses in optical fiber cables are one of the most important characteristics of the fiber. Losses in the fiber result in a reduction in the light power and thus reduce the system bandwidth, information transmission rate, efficiency, and overall system capacity. The predominant fiber losses are as follows:

1. Absorption losses
2. Material or Rayleigh scattering losses
3. Chromatic or wavelength dispersion
4. Radiation losses
5. Modal dispersion
6. Coupling losses

Absorption Losses

Absorption loss in optical fibers is analogous to power dissipation in copper cables; impurities in the fiber absorb the light and convert it to heat. The ultrapure glass used to manufacture optical fibers is approximately 99.9999% pure. Still, absorption losses between 1 and 1000 dB/km are typical. Essentially, there are three factors that contribute to the absorption losses in optical fibers: ultraviolet absorption, infrared absorption, and ion resonance absorption.

Ultraviolet absorption. Ultraviolet absorption is caused by valence electrons in the silica material from which fibers are manufactured. Light *ionizes* the valence electrons into conduction. The ionization is equivalent to a loss in the total light field and, consequently, contributes to the transmission losses of the fiber.

Infrared absorption. Infrared absorption is a result of *photons* of light that are absorbed by the atoms of the glass core molecules. The absorbed photons are converted to random mechanical vibrations typical of heating.

Ion resonance absorption. Ion resonance absorption is caused by OH^- ions in the material. The source of the OH^- ions is water molecules that have been trapped in the glass during the manufacturing process. Ion absorption is also caused by iron, copper, and chromium molecules.

Figure 10-18 shows typical losses in optical fiber cables due to ultraviolet, infrared, and ion resonance absorption.

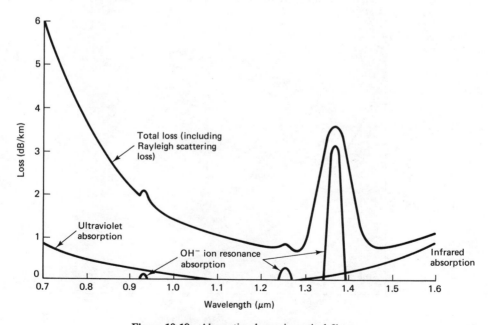

Figure 10-18 Absorption losses in optical fibers.

Material or Rayleigh Scattering Losses

During the manufacturing process, glass is extruded (drawn into long fibers of very small diameter). During this process, the glass is in a plastic state (not liquid and not solid). The tension applied to the glass during this process causes the cooling glass to develop submicroscopic irregularities that are permanently formed in the fiber. When light rays that are propagating down a fiber strike one of these impurities, they are *diffracted*. Diffraction causes the light to disperse or spread out in many directions. Some of the diffracted light continues down the fiber and some of it escapes through the cladding. The light rays that escape represent a loss in light power. This is called *Rayleigh scattering loss*. Figure 10-19 graphically shows the relationship between wavelength and Rayleigh scattering loss.

Chromatic or Wavelength Dispersion

As stated previously, the refractive index of a material is wavelength dependent. Light-emitting diodes (LEDs) emit light that contains a combination of wavelengths. Each wavelength within the composite light signal travels at a different velocity. Consequently, light rays that are simultaneously emitted from an LED and propagated down an optical fiber do not arrive at the far end of the fiber at the same time. This results in a distorted receive signal and is called *chromatic distortion*. Chromatic distortion can be eliminated by using a monochromatic source such as an injection laser diode (ILD).

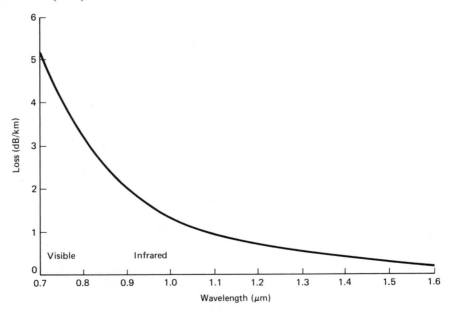

Figure 10-19   Rayleigh scattering loss as a function of wavelength.

Radiation Losses

Radiation losses are caused by small bends and kinks in the fiber. Essentially, there are two types of bends: microbends and constant-radius bends. *Microbending* occurs as a result of differences in the thermal contraction rates between the core and cladding material. A microbend represents a discontinuity in the fiber where Rayleigh scattering can occur. *Constant-radius bends* occur when fibers are bent during handling or installation.

Modal Dispersion

Modal dispersion, or *pulse spreading*, is caused by the difference in the propagation times of light rays that take different paths down a fiber. Obviously, modal dispersion can occur only in multimode fibers. It can be reduced considerably by using graded-index fibers and almost entirely eliminated by using single-mode step-index fibers.

Modal dispersion can cause a pulse of light energy to spread out as it propagates down a fiber. If the pulse spreading is sufficiently severe, one pulse may fall back on top of the next pulse (this is an example of intersymbol interference). In a multimode step-index fiber, a light ray that propagates straight down the axis of the fiber takes the least amount of time to travel the length of the fiber. A light ray that strikes the core/cladding interface at the critical angle will undergo the largest number of internal reflections and, consequently, take the longest time to travel the length of the fiber.

Figure 10-20 shows three rays of light propagating down a multimode step-index fiber. The lowest-order mode (ray 1) travels in a path parallel to the axis of the fiber. The middle-order mode (ray 2) bounces several times at the interface before traveling the length of the fiber. The highest-order mode (ray 3) makes many trips back and forth across the fiber as it propagates the entire length. It can be seen that ray 3 travels a considerably longer distance than ray 1 as it propagates down the fiber. Consequently, if the three rays of light were emitted into the fiber at the same time and represented a pulse of light energy, the three rays would reach the

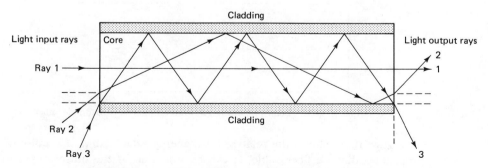

Figure 10-20   Light propagation down a multimode step-index fiber.

far end of the fiber at different times and result in a spreading out of the light energy in respect to time. This is called modal dispersion and results in a stretched pulse which is also reduced in amplitude at the output of the fiber. All three rays of light propagate through the same material at the same velocity, but ray 3 must travel a longer distance and, consequently, takes a longer period of time to propagate down the fiber.

Figure 10-21 shows light rays propagating down a single-mode step-index fiber. Because the radial dimension of the fiber is sufficiently small, there is only a single path for each of the rays to follow as they propagate down the length of the fiber. Consequently, each ray of light travels the same distance in a given period of time and the light rays have exactly the same time relationship at the far end of the fiber as they had when they entered the cable. The result is no *modal dispersion* or *pulse stretching*.

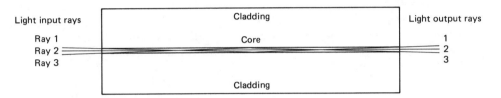

Figure 10-21   Light propagation down a single-mode step-index fiber.

Figure 10-22 shows light propagating down a multimode graded-index fiber. Three rays are shown traveling in three different modes. Each ray travels a different path but they all take approximately the same amount of time to propagate the length of fiber. This is because the refractive index of the fiber decreases with distance from the center, and the velocity at which a ray travels is inversely proportional to the refractive index. Consequently, the farther rays 2 and 3 travel from the center of the fiber, the faster they propagate.

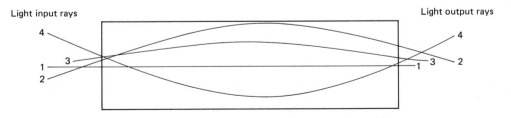

Figure 10-22   Light propagation down a multimode graded-index fiber.

Figure 10-23 shows the relative time/energy relationship of a pulse of light as it propagates down a fiber cable. It can be seen that as the pulse propagates down the fiber, the light rays that make up the pulse spread out in time, which causes a

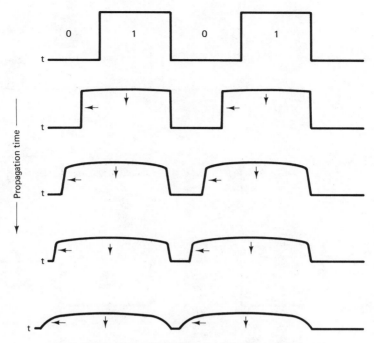

Figure 10-23   Pulse-width dispersion in an optical fiber cable.

corresponding reduction in the pulse amplitude and stretching of the pulse width. It can also be seen that as light energy from one pulse falls back in time, it will interfere with the next pulse. This is called *pulse spreading* or *pulse-width dispersion* and causes errors in digital transmission.

Figure 10-24a shows a unipolar return-to-zero (UPRZ) digital transmission. With UPRZ transmission (assuming a very narrow pulse) if light energy from pulse A were to fall back (*spread*) one bit time (T_b), it would interfere with pulse B and change what was a logic 0 to a logic 1. Figure 10-24b shows a unipolar nonreturn-to-zero (UPNRZ) digital transmission where each pulse is equal to the bit time. With UPNRZ transmission, if energy from pulse A were to fall back one-half of a bit time, it would interfere with pulse B. Consequently, UPRZ transmissions can tolerate twice as much delay or spread as UPNRZ transmissions.

The difference between the absolute delay times of the fastest and slowest rays of light propagating down a fiber is called the *pulse-spreading constant* (Δt) and is generally expressed in nanoseconds per kilometer (ns/km). The total pulse spread (ΔT) is then equal to the pulse spreading constant (Δt) times the total fiber length (L). Mathematically, ΔT is

$$\Delta T \text{ (ns)} = \Delta t \left(\frac{\text{ns}}{\text{km}}\right) \times L \text{ (km)} \qquad (10\text{-}11)$$

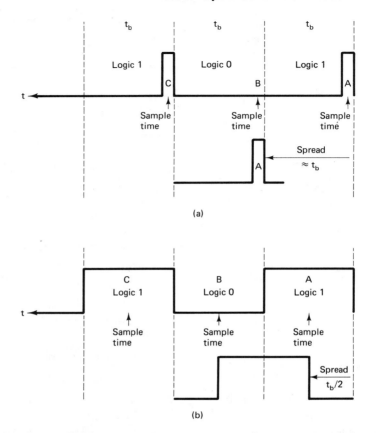

Figure 10-24 Pulse spreading of digital transmissions: (a) UPRZ; (b) UPNRZ.

For **UPRZ** transmissions, the maximum data transmission rate in bits per second (bps) is expressed as

$$F_b \text{ (bps)} = \frac{1}{\Delta t \times L} \tag{10-12}$$

and for **UPNRZ** transmissions, the maximum transmission rate is

$$F_b \text{ (bps)} = \frac{1}{2 \, \Delta t \times L} \tag{10-13}$$

EXAMPLE 10-3

For an optical fiber 10 km long with a pulse-spreading constant of 5 ns/km, determine the maximum digital transmission rates for (a) return-to-zero and (b) nonreturn-to-zero transmissions.

Solution (a) Substituting into Equation 10-12 yields

$$F_b = \frac{1}{5 \text{ ns/km} \times 10 \text{ km}} = 20 \text{ Mbps}$$

(b) Substituting into Equation 10-13 yields

$$F_b = \frac{1}{(2 \times 5 \text{ ns/km}) \times 10 \text{ km}} = 10 \text{ Mbps}$$

The results indicate that the digital transmission rate possible for this optical fiber is twice as high (20 Mbps versus 10 Mbps) for UPRZ as for UPNRZ transmission.

Coupling Losses

In fiber cables coupling losses can occur at any of the following three types of optical junctions: light source-to-fiber connections, fiber-to-fiber connections, and fiber-to-photodetector connections. Junction losses are most often caused by one of the following alignment problems: lateral misalignment, gap misalignment, angular misalignment, and imperfect surface finishes. These impairments are shown in Figure 10-25.

Lateral misalignment. This is shown in Figure 10-25a and is the lateral or axial displacement between two pieces of adjoining fiber cables. The amount of loss can be from a couple of tenths of a decibel to several decibels. This loss is generally negligible if the fiber axes are aligned to within 5% of the smaller fiber's diameter.

Gap misalignment. This is shown in Figure 10-25b and is sometimes called *end separation*. When *splices* are made in optical fibers, the fibers should actually touch. The farther apart the fibers are, the greater the loss of light. If two fibers are joined with a connector, the ends should not touch. This is because the two ends rubbing against each other in the connector could cause damage to either or both fibers.

Angular misalignment. This is shown in Figure 10-25c and is sometimes called *angular displacement*. If the angular displacement is less than 2°, the loss will be less than 0.5 dB.

Imperfect surface finish. This is shown in Figure 10-25d. The ends of the two adjoining fibers should be highly polished and fit together squarely. If the fiber ends are less than 3° off from perpendicular, the losses will be less than 0.5 dB.

LIGHT SOURCES

Essentially, there are two devices commonly used to generate light for fiber optic communications systems: light-emitting diodes (LEDs) and injection laser diodes (ILDs). Both devices have advantages and disadvantages and selection of one device over the other is determined by system requirements.

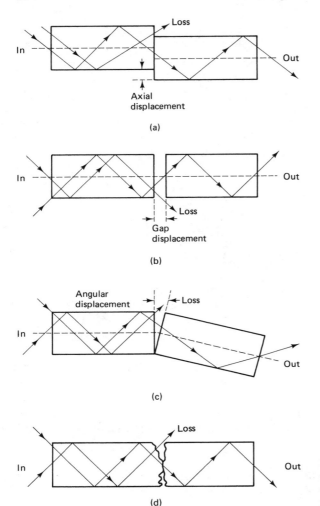

Figure 10-25 Fiber alignment impairments: (a) lateral misalignment; (b) gap displacement; (c) angular misalignment; (d) surface finish.

Light-Emitting Diodes

Essentially, a *light-emitting diode* (LED) is simply a P-N junction diode. It is usually made from a semiconductor material such as aluminum-gallium-arsenide (AlGaAs) or gallium-arsenide-phosphide (GaAsP). LEDs emit light by spontaneous emission; light is emitted as a result of the recombination of electrons and holes. When forward biased, minority carriers are injected across the *p-n* junction. Once across the junction, these minority carriers recombine with majority carriers and give up energy in the form of light. This process is essentially the same as in a conventional diode except that in LEDs certain semiconductor materials and dopants are chosen such that

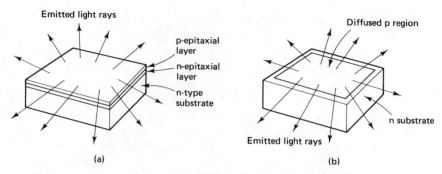

Figure 10-26 Homojunction LED structures: (a) silicon-doped gallium arsenide; (b) planar diffused.

the process is radiative; a photon is produced. A photon is a quantum of electromagnetic wave energy. Photons are particles that travel at the speed of light but at rest have no mass. In conventional semiconductor diodes (germanium and silicon, for example), the process is primarily nonradiative and no photons are generated. The energy gap of the material used to construct an LED determines whether the light emitted by it is invisible or visible and of what color.

The simplest LED structures are homojunction, epitaxially grown, or single-diffused devices and are shown in Figure 10-26. *Epitaxially grown LEDs* are generally constructed of silicon-doped gallium-arsenide (Figure 10-26a). A typical wavelength of light emitted from this construction is 940 nm, and a typical output power is approximately 3 mW at 100 mA of forward current. *Planar diffused (homojunction) LEDs* (Figure 10-26b) output approximately 500 μW at a wavelength of 900 nm. The primary disadvantage of homojunction LEDs is the nondirectionality of their light emission, which makes them a poor choice as a light source for fiber optic systems.

The *planar heterojunction LED* (Figure 10-27) is quite similar to the epitaxially grown LED except that the geometry is designed such that the forward current is concentrated to a very small area of the active layer. Because of this the planar heterojunction LED has several advantages over the homojunction type. They are:

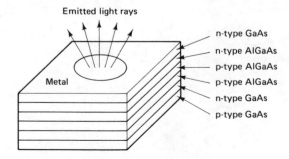

Figure 10-27 Planar heterojunction LED.

1. The increase in current density generates a more brilliant light spot.
2. The smaller emitting area makes it easier to couple its emitted light into a fiber.
3. The small effective area has a smaller capacitance, which allows the planar heterojunction LED to be used at higher speeds.

Burrus etched-well surface-emitting LED.   For the more practical applications, such as telecommunications, data rates in excess of 100 Mbps are required. For these applications, the etched-well LED was developed. Burrus and Dawson of Bell laboratories developed the etched-well LED. It is a surface-emitting LED and is shown in Figure 10-28. The Burrus etched-well LED emits light in many directions. The etched well helps concentrate the emitted light to a very small area. Also, domed lenses can be placed over the emitting surface to direct the light into a smaller area. These devices are more efficient than the standard surface emitters and they allow more power to be coupled into the optical fiber, but they are also more difficult and expensive to manufacture.

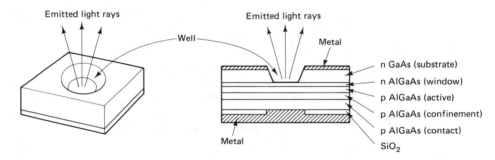

Figure 10-28   Burrus etched-well surface-emitting LED.

Edge-emitting LED.   The edge-emitting LED, which was developed by RCA, is shown in Figure 10-29. These LEDs emit a more directional light pattern than

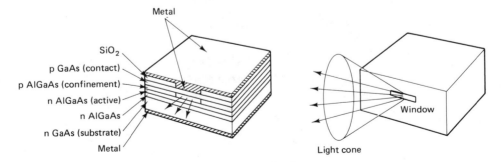

Figure 10-29   Edge-emitting LED.

do the surface-emitting LEDs. The construction is similar to the planar and Burrus diodes except that the emitting surface is a stripe rather than a confined circular area. The light is emitted from an active stripe and forms an elliptical beam. Surface-emitting LEDs are more commonly used than edge emitters because they emit more light. However, the coupling losses with surface emitters are greater and they have narrower bandwidths.

The *radiant* light power emitted from an LED is a linear function of the forward current passing through the device (Figure 10-30). It can also be seen that the optical output power of an LED is, in part, a function of the operating temperature.

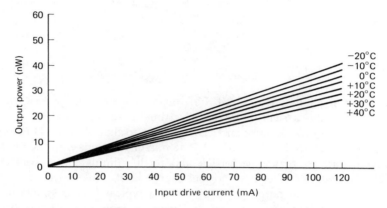

Figure 10-30 Output power versus forward current and operating temperature for an LED.

Injection Laser Diode

The word *laser* is an acronym for *l*ight *a*mplification by *s*timulated *e*mission of *r*adiation. Lasers are constructed from many different materials, including gases, liquids, and solids, although the type of laser used most often for fiber optic communications is the semiconductor laser.

The *injection laser diode* (ILD) is similar to the LED. In fact, below a certain threshold current, an ILD acts like an LED. Above the threshold current, an ILD oscillates; lasing occurs. As current passes through a forward-biased *p-n* junction diode, light is emitted by spontaneous emission at a frequency determined by the energy gap of the semiconductor material. When a particular current level is reached, the number of minority carriers and photons produced on either side of the *p-n* junction reaches a level where they begin to collide with already excited minority carriers. This causes an increase in the ionization energy level and makes the carriers unstable. When this happens, a typical carrier recombines with an opposite type of carrier at an energy level that is above its normal before-collision value. In the process, two photons are created; one is stimulated by another. Essentially, a gain in the

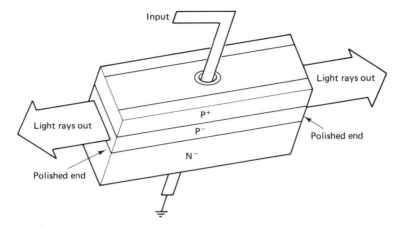

Figure 10-31   Injection laser diode construction.

number of photons is realized. For this to happen, a large forward current that can provide many carriers (holes and electrons) is required.

The construction of an ILD is similar to that of an LED (Figure 10-31) except that the ends are highly polished. The mirror-like ends trap the photons in the active region and, as they reflect back and forth, stimulate free electrons to recombine with holes at a higher-than-normal energy level. This process is called *lasing*.

The radiant output light power of a typical ILD is shown in Figure 10-32. It can be seen that very little output power is realized until the threshold current is reached; then lasing occurs. After lasing begins, the optical output power increases dramatically, with small increases in drive current. It can also be seen that the magnitude of the optical output power of the ILD is more dependent on operating temperature than is the LED.

Figure 10-33 shows the light radiation patterns typical of an LED and an ILD.

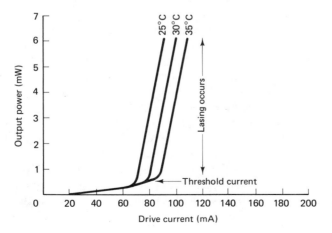

Figure 10-32   Output power versus forward current and temperature for an ILD.

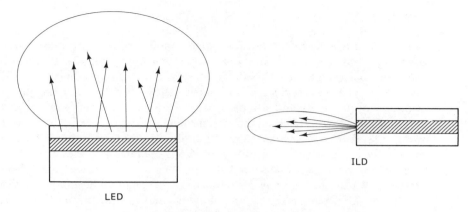

Figure 10-33 LED and ILD radiation patterns.

Because light is radiated out the end of an ILD in a narrow concentrated beam, it has a more direct radiation pattern.

Advantages of ILDs

1. Because ILDs have a more direct radiation pattern, it is easier to couple their light into an optical fiber. This reduces the coupling losses and allows smaller fibers to be used.
2. The radiant output power from an ILD is greater than that for an LED. A typical output power for an ILD is 5 mW (7 dBm) and 0.5 mW (−3 dBm) for LEDs. This allows ILDs to provide a higher drive power and to be used for systems that operate over longer distances.
3. ILDs can be used at higher bit rates than can LEDs.
4. ILDs generate monochromatic light, which reduces chromatic or wavelength dispersion.

Disadvantages of ILDs

1. ILDs are typically on the order of 10 times more expensive than LEDs.
2. Because ILDs operate at higher powers, they typically have a much shorter lifetime than LEDs.
3. ILDs are more temperature dependent than LEDs.

LIGHT DETECTORS

There are two devices that are commonly used to detect light energy in fiber optic communications receivers; PIN (positive-intrinsic-negative) diodes and APD (avalanche photodiodes).

PIN Diodes

A *PIN diode* is a *depletion-layer photodiode* and is probably the most common device used as a light detector in fiber optic communications systems. Figure 10-34 shows the basic construction of a PIN diode. A very lightly doped (almost pure or intrinsic) layer of *n*-type semiconductor material is sandwiched between the junction of the two heavily doped *n*- and *p*-type contact areas. Light enters the device through a very small window and falls on the carrier-void intrinsic material. The intrinsic material is made thick enough so that most of the photons that enter the device are absorbed by this layer. Essentially, the PIN photodiode operates just the opposite of an LED. Most of the photons are absorbed by electrons in the valence band of the intrinsic material. When the photons are absorbed, they add sufficient energy to generate carriers in the depletion region and allow current to flow through the device.

Photoelectric effect.   Light entering through the window of a PIN diode is absorbed by the intrinsic material and adds enough energy to cause electrons to move from the valence band into the conduction band. The increase in the number of electrons that move into the conduction band is matched by an increase in the number of holes in the valence band. To cause current to flow in a photodiode, sufficient light must be absorbed to give valence electrons enough energy to jump the energy gap. The energy gap for silicon is 1.12 eV (electrode volts). Mathematically, the operation is as follows.

For silicon, the energy gap (E_g) equals 1.12 eV:

$$1 \text{ eV} = 1.6 \times 10^{-19} \text{ J}$$

Thus the energy gap for silicon is

$$E_g = (1.12 \text{ eV}) \left(1.6 \times 10^{-19} \, \frac{\text{J}}{\text{eV}} \right) = 1.792 \times 10^{-19} \text{ J}$$

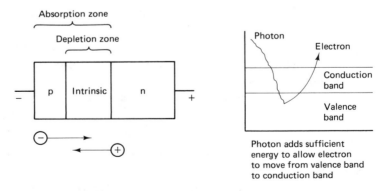

Figure 10-34   PIN photodiode construction.

and

$$\text{energy } (E) = hf$$

where

$h = $ Planck's constant $= 6.6256 \times 10^{-34}$ J/Hz
$f = $ frequency (Hz)

Rearranging and solving for f yields

$$f = \frac{E}{h}$$

For a silicon photodiode,

$$f = \frac{1.792 \times 10^{-19} \text{ J}}{6.6256 \times 10^{-34} \text{ J/Hz}}$$

$$= 2.705 \times 10^{14} \text{ Hz}$$

Converting to wavelength yields

$$\lambda = \frac{c}{f} = \frac{3 \times 10^8 \text{ m/s}}{2.705 \times 10^{14} \text{ Hz}} = 1109 \text{ nm/cycle}$$

Consequently, light wavelengths of 1109 nm or shorter, or light frequencies of 2.705 $\times 10^{14}$ Hz or higher, are required to generate enough electrons to jump the energy gap of a silicon photodiode.

Avalanche Photodiodes

Figure 10-35 shows the basic construction of an *avalanche photodiode* (APD). An APD is a *pipn* structure. Light enters the diode and is absorbed by the thin, heavily doped *n*-layer. This causes a high electric field intensity to be developed across the *i-p-n* junction. The high reverse-biased field intensity causes impact ionization to occur near the breakdown voltage of the junction. During impact ionization, a carrier can gain sufficient energy to ionize other bound electrons. These ionized carriers, in turn, cause more ionizations to occur. The process continues like an avalanche and is, effectively, equivalent to an internal gain or carrier multiplication. Consequently, APDs are more sensitive than PIN diodes and require less additional amplification. The disadvantages of APDs are relatively long transit times and additional internally generated noise due to the avalanche multiplication factor.

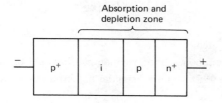

Figure 10-35 Avalanche photodiode construction.

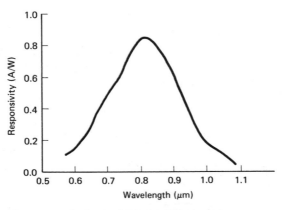

Figure 10-36 Spectral response curve.

Characteristics of Light Detectors

The most important characteristics of light detectors are:

Responsitivity.   This is a measure of the conversion efficiency of a photodetector. It is the ratio of the output current of a photodiode to the input optical power and has the unit of amperes/watt. Responsivity is generally given for a particular wavelength or frequency.

Dark current.   This is the leakage current that flows through a photodiode with no light input. Dark current is caused by thermally generated carriers in the diode.

Transit time.   This is the time it takes a light-induced carrier to travel across the depletion area. This parameter determines the maximum bit rate possible with a particular photodiode.

Spectral response.   This parameter determines the range or system length that can be achieved for a given wavelength. Generally, relative spectral response is graphed as a function of wavelength or frequency. Figure 10-36 is an illustrative example of a spectral response curve. It can be seen that this particular photodiode more efficiently absorbs energy in the range 800 to 820 nm.

QUESTIONS

10-1. Define a fiber optic system.

10-2. What is the relationship between information capacity and bandwidth.

10-3. What development in 1951 was a substantial breakthrough in the field of fiber optics? In 1960? In 1970?

10-4. Contrast the advantages and disadvantages of fiber optic cables and metallic cables.

10-5. Outline the primary building blocks of a fiber optic system.

10-6. Contrast glass and plastic fiber cables.

10-7. Briefly describe the construction of a fiber optic cable.

10-8. Define the following terms: velocity of propagation, refraction, and refractive index.

10-9. State Snell's law for refraction and outline its significance in fiber optic cables.

10-10. Define *critical angle*.

10-11. Describe what is meant by mode of operation; by index profile.

10-12. Describe a step-index fiber cable; a graded-index cable.

10-13. Contrast the advantages and disadvantages of step-index, graded-index, single-mode propagation, and multimode propagation.

10-14. Why is single-mode propagation impossible with graded-index fibers?

10-15. Describe the source-to-fiber aperture.

10-16. What are the acceptance angle and the acceptance cone for a fiber cable?

10-17. Define *numerical aperture*.

10-18. List and briefly describe the losses associated with fiber cables.

10-19. What is *pulse spreading*?

10-20. Define *pulse spreading constant*.

10-21. List and briefly describe the various coupling losses.

10-22. Briefly describe the operation of a light-emitting diode.

10-23. What are the two primary types of LEDs?

10-24. Briefly describe the operation of an injection laser diode.

10-25. What is lasing?

10-26. Contrast the advantages and disadvantages of ILDs and LEDs.

10-27. Briefly describe the function of a photodiode.

10-28. Describe the photoelectric effect.

10-29. Explain the difference between a PIN diode and an APD.

10-30. List and describe the primary characteristics of light detectors.

PROBLEMS

10-1. Determine the wavelengths in nanometers and angstroms for the following light frequencies.
   (a) 3.45×10^{14} Hz
   (b) 3.62×10^{14} Hz
   (c) 3.21×10^{14} Hz

10-2. Determine the light frequency for the following wavelengths.
   (a) 670 nm
   (b) 7800 Å
   (c) 710 nm

10-3. For a glass ($n = 1.5$)/quartz ($n = 1.38$) interface and an angle of incidence of 35°, determine the angle of refraction.

10-4. Determine the critical angle for the fiber described in Problem 10-3.

10-5. Determine the acceptance angle for the cable described in Problem 10-3.

10-6. Determine the numerical aperture for the cable described in Problem 10-3.

10-7. Determine the maximum bit rate for RZ and NRZ encoding for the following pulse-spreading constants and cable lengths.

(a) $\Delta t = 10$ ns/m, $L = 100$ m

(b) $\Delta t = 20$ ns/m, $L - 1000$ m

(c) $\Delta t = 2000$ ns/km, $L = 2$ km

10-8. Determine the lowest light frequency that can be detected by a photodiode with an energy gap $= 1.2$ eV.

SOLUTIONS TO ODD-NUMBERED PROBLEMS

CHAPTER 1

1-1. 10 megabaud; minimum bandwidth = 40 MHz

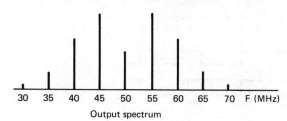

Output spectrum

1-3.

I	Q	Output expression
0	0	$-\sin \omega_c t + \cos \omega_c t$
0	1	$-\sin \omega_c t - \cos \omega_c t$
1	0	$+\sin \omega_c t + \cos \omega_c t$
1	1	$+\sin \omega_c t - \cos \omega_c t$

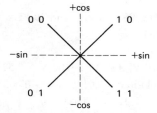

1-5. 6.67 megabaud; minimum bandwidth = 6.67 MHz

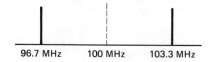

1-7. 5 megabaud; minimum bandwidth = 5 MHz

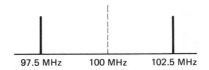

97.5 MHz   100 MHz   102.5 MHz

1-9. (a) 2 bps/Hz   (b) 3 bps/Hz   (c) 4 bps/Hz

CHAPTER 2

2-1

	D A T A Sp	C O M M U N I C A T I O N S	EXT	BCS
b_0	0 1 0 1 0	1 1 1 1 1 0 1 1 1 0 1 1 0 1	1	0
b_1	0 0 0 0 0	1 1 0 0 0 1 0 1 0 0 0 1 1 1	1	0
b_2	1 0 1 0 0	0 1 1 1 1 1 0 0 0 1 0 1 1 0	0	0
b_3	0 0 0 0 0	0 1 1 1 0 1 1 0 0 0 1 1 1 0	0	0
b_4	0 0 1 0 0	0 0 0 0 1 0 0 0 0 1 0 0 0 1	0	0
b_5	0 0 0 0 1	0 0 0 0 0 0 0 0 0 0 0 0 0 0	0	1
b_6	1 1 1 1 0	1 1 1 1 1 1 1 1 1 1 1 1 1 1	0	0
VRC	1 1 0 1 0	0 0 1 1 1 1 0 0 1 0 0 0 1 1	1	1

2-3. Four Hamming bits

2-5. 10111001 binary or B9H

2-7. 1110001000011111 11110011111 0100111101011111 11111 1001011

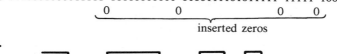

inserted zeros

2-9.

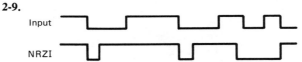

Input

NRZI

CHAPTER 4

4-1. 4 kHz, FS = 8 kHz and 10 kHz, FS = 20 kHz

4-3. 10 kHz

4-5. 511 or 54 dB

4-7. −1.16 V, 0.12 V, +0.04 V, −2.52 V, +0.24 V

4-9. Resolution = 0.01 V, Qe = 0.005 V

4-11. **(a)** +0.01 to +0.03V

      (b) −0.01 to 0 V

      (c) +10.23 to +10.25 V

      (d) −10.23 to −10.25 V

      (e) +5.13 to +5.15 V

      (f) +6.81 to +6.83 V

CHAPTER 5

5-1. (a) 1.521 Mbps **5-3.**
   (b) 760.5 kHz

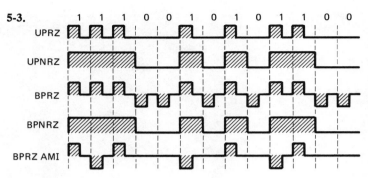

CHAPTER 6

6-1. Ch 1, 108 kHz; Ch 2, 104 kHz; Ch 3, 100 kHz; Ch 4, 96 kHz; Ch 5, 92 kHz; Ch 6, 88 kHz; Ch 7, 84 kHz; Ch 8, 80 kHz; Ch 9, 76 kHz; Ch 10, 72 kHz; Ch 11, 68 kHz; Ch 12, 64 kHz

6-3. Ch = 96 to 100 kHz, GP = 464 to 468 kHz, SG = 1144 to 1148 kHz, MG = 5100 to 5104 kHz

6-5. 5495.92 kHz   **6-7.** 4948 to 5188 kHz

CHAPTER 7

7-1. -99.23 dBm   **7-7.** -1.25 dB

7-3. 28.9 dB   **7-9.** 6.58 dB

7-5. 35 dBm

CHAPTER 8

8-1. Elevation angle = 40°, azimuth = 29° west of south

8-3. (a) 11.74 dB   (b) 19.2 dB

8-5. Radiated power = 28 dBW, EIRP = 68 dBW

8-7. -200.8 dBW   **8-11.** 19.77 dB

8-9. 26.18 dB   **8-13.** 12.5 dB

CHAPTER 9

9-1 15 transponders   **9-3.** 45 stations

CHAPTER 10

10-1. (a) 869 nm, 8690 Å   **10-7.** (a) RZ = 1 Mbps, NRZ = 500 kbps
   (b) 828 nm, 8280 Å   (b) RZ = 50 kbps, NRZ = 25 kbps
   (c) 935 nm, 9350 Å   (c) RZ = 250 kbp, NRZ = 125 kbps

10-3. 38.57°

10-5. 36°

INDEX